Anthony,

ENERGISE!

JAMES WOUDHUYSEN
& JOE KAPLINSKY

Enjoy the jokes!

James @
Woudhuysen . con

SES 14

A BEAUTIFUL SPECIAL

First published 2009.

A Beautiful Special.

www.beautiful-books.co.uk

Beautiful Books Limited
36-38 Glasshouse Street
London W1B 5DL

ISBN 9781905636273

9 8 7 6 5 4 3 2 1

Cover and book design by Studio Dempsey
Set in Franklin Gothic and Schnellfetter
Diagrams by Michael Robinson
Printed in Great Britain by Quadracolor of London.

To Dido and Nazila

Acknowledgements

Many people helped the two authors write this book.

We would particularly like to thank Alice Woudhuysen for her editorial skills, our publisher Simon Petherick for his patience, and Mike Dempsey and Stephanie Jerey for their design work.

We are also grateful to Ian Abley, Daniel Ben-Ami, Bernhard Blauel, Robert Clowes, Sean Collins, Bill Durodié, Claire Fox, Frank Furedi, Tony Gilland, John Gillott, Alex Gourevich, Mark Harrop, Philip Hammond, Rob Lyons, Kevin McCullagh, Phil Mullan, Peter Sammonds, Paul Seaman, Antti Silvast and Phil Slade.

CONTENTS

Before the financial crisis

of autumn 2008, soaring Chinese demand for oil led some commentators to predict a rosy future for renewable energy. Then, after the Crash of 2008, others suspected that new renewables firms would falter through lack of finance, and that prospects for renewable energy in general would recede. Yet the $700bn bailout of the US financial system, agreed in October 2008, was accompanied by important tax credits for renewables, and for plug-in hybrid vehicles. [1]

It's a difficult moment to forecast the future of energy. Completed in the weeks that saw the collapse of Lehman Brothers and the climax of the US Presidential campaign, this book tries to take the longer-term view.

In the past 100 years, energy forecasters have pretty much failed to get their predictions right. [2] But as Alan Kay, architect of the Graphical User Interface, so memorably said:

'The best way to predict the future is to invent it.'

This book, therefore, has a pragmatic intent. We want to help invent a future of rational energy supply. Our emphasis is on the *politics* of energy innovation. That's also why we've put some of the more technical matters around energy and climate change 'into grey-tinted panels'.

About this book

This book is a riposte to the endless doctrine that you are personally responsible for climate change and must curb your consumption of energy. *Energise!* argues that consuming more energy isn't a problem if the right kind of supply can be arranged. With the right supply, climate *won't* run out of control. But so long as the state's ineffective, moralistic policy on energy is left unchallenged, it's the state's interventions in our everyday lives that look set to run out of control.

For the lay reader, climate science appears to be a

discipline so vast that it's impenetrable. So, to summarise the state of climate science in a handy manner, the end of Chapter 1 presents tables that give a bird's eye view on some of the main forecasts and recommendations that have been made about global warming. In these tables, we also present our own ideas.

In Chapter 2 we establish *why people see energy as a problem of individual consumption more than one of supply*. This is a concept that must be understood if a rational politics of supply is ever to win through.

Chapter 3 is about climate change, and presents a new interpretation of it.

In Chapters 4, 5 and 6 – on nuclear, carbon-based and renewables technologies – we make suggestions about which energy technologies will make the most sense, both generally and in terms of emissions of greenhouse gases (GHGs). [3] We also follow Alan Kay's activist spirit and suggest roughly how, and by how much, different technologies could triumph, *if people mobilise political backing for them.*

At the end of each of Chapters 4, 5 and 6, then, we present tables that give an overview of the advantages and disadvantages of key technologies. We then provide simple ratings, out of 10, for each technology considered, both today, and *in a better future*.

Finally, in Chapter 7, we move somewhat beyond the energy sector. Both inside and outside it, in fact, we compare our proposals for *transforming* the planet with those of the United Nations Intergovernmental Panel on Climate Change (IPCC), the consultants McKinsey, and Bjørn Lomborg, the world's most prominent critic of Green thinking.

How we approach climate change

Necessarily, this book deals with the science of climate change. It also deals with something rather different – the *politics* of climate change. We very much favour science, but very much oppose the manipulation of science in the cause of political point scoring.

Ironically, it was free-market ideologues and members of the energy establishment who, when climate was first raised with them, pioneered the idea that scientific evidence could substitute for political argument and thus refute the idea of man-made global warming. Thereafter, the Left and many environmentalists adopted the same tactic to advance their solutions to global warming. More recently, in a rearguard action, an old Conservative – Baron Nigel Lawson of Blaby, Britain's former Chancellor of the Exchequer – has indulged in a little deification of science. To back up his main argument that government policy on global warming is a denial of personal liberty, he has used some very partial data about average world temperatures to bolster those who are sceptical about man-made climate change: *climate sceptics.* [4]

This book differs from both environmentalism and climate sceptics. It offers a radically new perspective on energy and climate change. It covers not just the technology, economics, science and politics of these two issues, but also their *sociology*: how people perceive energy and how they organise it. Our main focus is on humanity's need for a lot more energy, and a lot more innovation in energy supply.

With this focus, *Energise!* is unlike mainstream books on climate change, in which the pattern is: first, identify what level of climate change is dangerous; second, identify the maximum level of GHGs compatible with that; and third, propose measures to ensure that this limit is not exceeded.

The standard book on climate change tends to build its conclusions into its premises. Beginning with somewhat arbitrary definitions of what is dangerous, it typically uses science to calculate the 'right' emission levels, and then feeds those levels into dubious and opaque economic models to calculate how costly CO_2 taxes or CO_2 permits should be to keep emissions below those levels.

If such an approach sounds boring and technocratic, that's because it is. Ours is different. We concentrate on climate change, but we put it within a social context. That social context begins with the world's growing energy requirements.

Austerity and the sociology of energy

In the wake of the Crash of 2008, climate change is destined to become *more*, not less important, to political and economic decisions. However much European industry would like emissions regulation to be delayed, and however much consumers will need to focus on tightening their belts, the weather will not stop; and neither will the contemporary impulse to connect absolutely everything with climate change. Indeed, there's already evidence that the authorities will paint the austerity of 2009 onward in feelgood shades of Green.

Tightening belts, it is now said, is a good thing – because all individuals have a responsibility to conserve energy, and, in that cause, improve their behaviour. EdF, a French energy company, offers to engage Britons in what it insists is a 'coaching programme' on how to save energy. [5] Others want more radical steps to be taken. One 'radical fantasy' suggests not just that people will 'earn less and consume less,' but that the Crash of 2008 has given them 'a chance to start again.' [6]

The compatibility between capitalism and Green thinking is something that Green thinkers themselves have long been keen to promote. [7] In the next few years, people can expect to hear a lot more about how:

- going Green saves money, which is something everyone must do
- slower growth is wiser growth
- the world must not exhaust finite supplies of energy too fast.

Whether the general public finds these arguments for austerity credible, though, remains a very open question.

Like climate change, energy is set to become an increasingly important factor in people's lives. In looking at energy, however, we're not overly concerned with burying the reader in statistics on oil reserves, for example. These statistics are readily available.

Nor do we mull over whether it was speculation that really drove up oil prices in much of 2008. But we *are* interested when Barack Obama tells Fox TV that, had he been President during 9/11, he would have asked Americans not to shop, as George W Bush did, but to 'tap into the feeling that everybody has been caught up in'. In other words, to tap into America's need for a bold energy policy. [8]

Obama said that, after 9/11, he would have proposed that all Americans 'make commitments' to increase fuel efficiency in their cars and in their homes, somehow. The government would have worked 'in partnership' with them in the cause of decreasing America's dependence on foreign oil by 20 or 40 per cent over a decade or two. [9]

This book suggests that *the call to arms to cut energy use is not going to go away*. Most likely, it will grow more urgent.

One example of a call to arms in energy is 'fuel poverty' in the UK – officially, circumstances in which a household spends 10 or more per cent of its income on energy. In 2001, New Labour promised to end this newly defined condition by 2018. [10] Following that, not much was heard about it. But by October 2008, Friends of the Earth (FoE), along with the charity Help the Aged, was ready to sue the government in the High Court for its failure to meet its own targets. Claiming that more than five million households now suffered from fuel fuel poverty, an FoE spokesman announced:

> 'A massive energy efficiency programme is needed. This will keep people warm, cut bills and help meet our targets for tackling climate change.' [11]

Yet Prime Minister Gordon Brown is already planning a massive programme to promote energy efficiency in British homes. All that can be surmised is that, for government and critics alike, rallying the nation around energy conservation is what now passes for a political cause.

That fact, too, is part of our sociology of energy.

The meaning of energy

Given that in the UK, old-fashioned poverty has been transfigured into fuel poverty and is supposed to afflict nearly a fifth of households there, it ought to be clear that the precise meaning of energy among men and women is pretty malleable. But with every shift in its social significance, energy looks poised to count for more than it ever did in the past.

The meaning of energy is what conventional treatments of energy, just like conventional books on climate change, tend to avoid. This book doesn't make that mistake.

In the downturn that has followed the Crash, the meaning of energy has changed again. Politicians have rediscovered the Depression economics of John Maynard Keynes, the merits of state spending, and the merits of state spending on energy in particular. In his election campaign, Barack Obama said he wanted to spend $150bn on renewables over the next decade, so that this source of energy produces a quarter of US electricity by 2025. British Chancellor Alistair Darling proclaims that in switching his spending priorities, energy is one of the 'areas that make a difference'. It's an area, indeed, 'where people are feeling squeezed at the moment,' and spending on it would create jobs. [12]

Here energy acquires a new meaning. It's now about creating jobs. But before people sign up for the Keynesian management of economic demand through spending, investment and job creation around energy, consider two facts. First, the number of jobs likely to be created in the UK renewable energy sector is set to be very limited. Second, and more importantly, job creation for the few will be accompanied by renewed cries that *everyone* cut their demand for energy.

This book makes no apology for its historical dimension. That allows us to see where the future of energy is headed. In all the euphoria around applying Keynesian principles to energy, it's worth recalling what Keynes actually said about his policy. In the preface to the German edition of his most famous work, *The General Theory of Employment, Interest and Money* (1936), he expressed the hope that his book would help German economists develop a theory 'designed to meet specifically German conditions'. His book's theory was, he said, 'much more easily adapted to the conditions of a totalitarian state' than were theories premised on free competition and a large measure of *laissez-faire*. [13]

Britain and the US today don't face the advent of a totalitarian state. However, the state's intervention in personal demand for energy, and its insistence that energy use is cut back, could well turn out to be an authoritarian exercise.

We hope you enjoy *Energise!*

Human beings need lots more cheap energy

If the world could be more thoughtful about energy supply, individuals could be thoughtless about their energy use

1

Irresponsible?

No. The authors of this book acknowledge that climate change exists and is largely man-made. We accept that there's a problem with greenhouse gases (GHGs). But we believe that these concerns must be seen in perspective.

Energise! is about the science and technology of energy. Our starting point, however, is the uniqueness of human beings. To us, humans will always want to do more than simply survive. They will always want more home comforts, better-lit streets and greater mobility. But to get all of this – now and in the future – they will need more cheap energy. In energy matters, therefore, a far bigger and more urgent challenge than global warming lies in *thoughtfully supplying the world's populations and organisations with lots more cheap energy*. If people can do that right, then they will be able to overcome man-made climate change in the process.

Energy innovations can do so much more than simply slow global warming. They can help humanity thrive, not just survive.

Before the Crash of 2008, several enthusiasts for free markets breezily suggested that oil priced at $130 a barrel or more had one merit: it would force people to conserve energy. [1] After the Crash, others observed that people would worry less about climate change during a downturn, especially if it turned out to be deep and prolonged. [2] In fact, both of these views are complacent.

When oil can only be extracted, refined and piped with difficulty, producing and transporting the world's food becomes expensive. When energy in general is expensive, steel and cement cost more to make, inflating the price of buildings, roads, rail systems and even wind turbines. To put it simply, *every sector and every nation has an interest in more cheap energy*.

On the other hand, concerns about climate change will outlive the current period of financial turmoil. These concerns are deep-seated not just in large swathes of the population of the West, or with Barack Obama, but also among elites in China, India and the East. The Crash of 2008 will make the world focus more on the East's leadership – not least, around the issue of global warming. We are certain that climate change will regain its prominence in national and international politics.

People are constantly being told that they live in a consumer society. Yet for most adults under 65, the main event in life remains *work* – the realm of wealth generation, production and the different kinds of waste products that go with that.

It's the same with energy.

Too often, governments and environmentalists address us as ignorant consumers, telling us to curb our driving and flying, eat local food, switch things off and insulate our homes. But in fact, *the human input into climate change is best dealt with not in people's personal lives, but at source* – in the world's energy supply sector (see panel below). And even if climate change disappeared tomorrow, energy supply would still deserve much more investment over the next 30 years than it has had over the past 30.

Without a large new round of investment in advanced energy technologies, human beings face power cuts. There's no need to be alarmist about these, nor, as we show later on in this book, attribute them to an alleged 'peak' in oil supplies. But in 2008 alone, power cuts occurred in places as varied as South Africa, Pakistan, China and the UK.

Worse, society simply won't develop. Even the conservative World Bank estimates that, without a change in energy policy, 60 per cent of sub-Saharan Africans will lack access to electricity in 2030. [3]

It's time to get a grip on these facts and stop feeling guilty about climate change. *Thoughtful ingenuity, not changes in consumer awareness or behaviour,* is the way to exit today's energy crisis – and the way to deal with a warming planet.

Climate change is best fixed at source – in the energy supply sector

In 2004, the world's road transport created about four billion tonnes – four Gigatonnes (Gt) – of CO_2 emissions. However, electricity plants contributed more than 10 Gt, and oil refineries a further 2 Gt. In sum, the world's energy supply sector emitted three times as much CO_2 as its motor vehicles.

Direct worldwide emissions of CO_2, by sector, 1970–2004 [4]

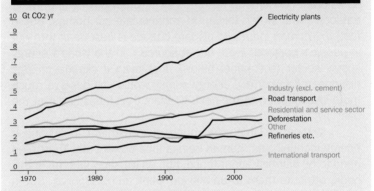

As the IPCC puts it, by 2004 CO_2 emissions from power generation represented more than 27 per cent of all man made CO_2 emissions; indeed, the power sector was 'by far' the most important source of such emissions. [5]

Widen out from CO_2 to GHGs as a whole, and agriculture appears as a significant emitter of CH_4 and N_2O. Nevertheless, in 2004, energy supply was unequivocally the main source of GHGs.

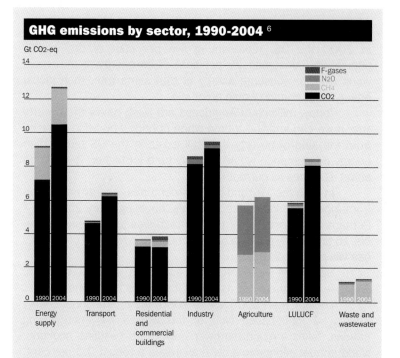

GHG emissions by sector, 1990-2004 [6]

Gt CO2-eq

Legend: F-gases, N2O, CH4, CO2

Categories: Energy supply · Transport · Residential and commercial buildings · Industry · Agriculture · LULUCF · Waste and wastewater (each showing 1990 and 2004)

Between 1970 and 2004, the IPCC reports, GHG emissions from energy supply rose by more than 145 per cent, while those from general transport rose by 120 per cent. Interestingly, GHG emissions associated with residential and commercial property experienced the slowest growth – just 26 per cent in 34 years. [7]

The basic figures show that emissions from road transport are growing fast, but those from power generation are much larger – and are growing faster. Even with transport and buildings, people need to look beyond immediate use, and back to *energy supply*. A vehicle or building supplied with clean energy would have zero emissions of CO_2.

The amount of energy the world will need

In 2000 the US, representing five per cent of the world's population, consumed 25 per cent of its energy. If the whole world were to consume energy at that same rate, then global energy consumption would quintuple.

Today, millions of Americans are living below the poverty line. If we can imagine a time when nobody worldwide suffered from the kind of poverty that still exists in the US today, then energy consumption would not only be a lot higher: it would be about *10 times its current rate.*

In both the 19th and the 20th centuries, energy production from modern sources rose 16 times, doubling every 25 years. [8] Perhaps the constancy of the increase was just a coincidence, but 16 times may not be a bad estimate for how much our energy use will grow by 2100.

Some will write off such estimates as absurdly high. *Energise!* sees them not as forecasts, but as ambitions.

Looking from 2005 just to 2050, the International Energy Agency, a Paris-based club of the world's big energy-using nations, takes a much narrower view. In its 'baseline' scenario, world annual economic growth averages a robust 3.3 per cent, quadrupling to $227 trillion by 2050. But world final energy demand doesn't even quite triple. Indeed, the IEA believes that because of changes in world economic structure, and, even more, increases in the efficiency with which energy is *used*, world energy demand will in practice just double. [9]

That seems to us improbable. In a moment, we will deal with the merits and limits of improvements in energy efficiency; but even with such measures, and certainly with the aggressive programme of thoughtful innovations put forward in this book, world energy demand could *and should* double in 25 years, not in 50.

For convenience, through good choice of technique

The industrialisation of the West brought with it man-made emissions. But it also brought new products, and, even more, innovations in the process of production. Industrialisation gave us the whole idea of *convenience* – of not having to scrape around to build a fire, but instead having hot running water, and eventually central heating. Finally, too, industrialisation brought with it a special form of convenience: *mobility*.

Convenience is still something worth fighting for – especially convenience in the use of energy. People should *not* have to spend their time watching 'smart meters' that tell them how much CO_2 they are generating every time they make a cup of coffee. Instead, they should be looking forward, as *Energise!* does, to a world where energy is:

- cheap, always on, and to hand
- available to everyone, wherever they are
- delivered so unobtrusively that nobody worries about it.

As far as possible, the means of delivering energy should be invisible, or simply part of the furniture.

In developed countries, few worry about the humble sockets that deliver electricity to their appliances. The householder does not pause to maintain, repair, or clean an electricity socket, in the same way that the family with roof-mounted solar panels must spend time up a ladder fiddling with them. [10]

People should know how energy works, but they shouldn't have to think energy all the time. Life is too much fun for that.

The idea that people should now start to sacrifice convenience in the cause of energy *conservation* is also particularly insulting to women. Even today, the women of the world do most of its cooking, washing and food shopping. In truth they need all the convenient gadgets and all the energy they can get. [11]

Prigs move in

Around the world, priggish politicians and celebrities – Bono, Bob Geldof, Sienna Miller, Leonardo DiCaprio – urge everyone to consume less and conserve more energy. Some suggest that the way to beat climate change is to stop families having more than two children. [12] Meanwhile, Sir Paul McCartney suggests that everyone stop eating meat. [13]

This is all bad news. When politicians and celebrities insist that people adopt their kind of etiquette of energy use, they bolster the state's growing interference with people's personal lives. In practice, their liberal-sounding demand that people make 'informed choices' about energy is an authoritarian affront. Why should people listen to what these dignitaries say about how we should behave? What do they know about the potential for new energy technologies to bring *convenience, mobility and fun* to billions of people?

Politicians and celebrities are not the only problem. Educationalists in particular seek to come between parents and children. As the urban critic Austin Williams has shown, since the 1992 United Nations Conference on Environment and Development in Rio de Janeiro, many educationalists have tailored school curricula to environmentalist ends. [14] Worryingly, pupils are sometimes expected to upbraid their parents for failing to live ecologically correct lives.

Yet mankind does not yet face a Greenhouse Apocalypse, from which the only way out is to cut back on energy use immediately – to tax it harder, make travel by car or air unacceptable, or introduce personal carbon allowances. Most people will not give up their energy-using habits that easily, in any case.

Instead of consumer cutbacks as a one-size-fits-all alternative to global warming, human beings in fact face a still open-ended *choice of technique* in energy supply. Here, in contrast with energy use, it makes sense to think hard.

Green misanthropes (1): Paul Ehrlich and Amory Lovins – more energy as 'mischief'

Energise! believes that the world needs cheap, abundant energy. But two of America's most prominent environmentalists reacted against that simple, humanistic idea back in 1975. Giving society such energy, Stanford University professor Paul Ehrlich wrote, would be the moral equivalent of 'giving an idiot child a machine gun'. [15] In another vivid metaphor, he lamented that mankind was likely to follow the 'pied pipers of technology' to 'destruction'.

In casting human users of energy as imbeciles and rats, Ehrlich was first in a long line of Green misanthropes.

In 1976, Amory Lovins, an American physicist representing Friends of the Earth in the UK, took a similar anti-energy, anti-human line. Attacking US electricity generation for its inefficiency and its capital cost, he proposed that it be cut by 60 per cent. Lovins also hoped that in the long term, a modest, zero, or negative growth in America's rate of energy use would be realistic. He favoured 'soft' energy: sources that were renewable, diverse, low-tech, small-scale and geared to end user needs. Like Ehrlich, he reserved particular contempt for nuclear fusion, arguing:

> 'We should prefer energy sources that give us enough for our needs while denying us the excesses of concentrated energy with which we might do mischief to the earth or to each other.' [16]

So mankind should seek softness in energy, because too much powerful and concentrated energy will only lead to mischief.

Mischief-free softness is a fascinating concept, but one that has proved rather elastic for Amory Lovins. He co-founded the influential, non-profit Rocky Mountain Institute in Snowmass, Colorado, in 1982.

Since that year, he's consulted for that enormously soft, mischief-free, low-energy force for good – the Pentagon.

Don't fear the East – celebrate it

On 11 May 2007, the US House of Representatives authorised the compilation of a National Intelligence Estimate on climate change. Diplomatic tensions on climate preceded that date; but, compounded by the subsequent US sub-prime crisis and the credit crunch, 2007 was the year when those tensions broke into the open.

Today it's clear that many of the West's general fears centre on the East. The Crash of 2008 made Wall Street vulnerable to Eastern financial institutions; and there is always the chance

Godfathers of Green: Paul Ehrlich and Amory Lovins

that these may move more decisively into the West's energy sector. When it thinks energy, the West thinks East. When the West looks East, it sees energy and climate problems. [17]

The growing part of the world's oil that today comes from the Organisation of the Petroleum Exporting Countries (OPEC) has re-focused attention on the Middle East, and on security of supply in energy. There is also concern about Europe's dependence on Russian gas. Last, the West has made much of the fact that China overtook the US in 2007 as the world's largest emitter of CO_2. [18]

It's true that NASA scientist James Hansen has blamed Britain for doing the most to boost the world's accumulated stock of man-made CO_2 emissions. For Hansen, Britain's pioneering Industrial Revolution has made it emit even more CO_2 since 1751 than the US. [19] Yet what most worries Western planners is Eastern demand for energy.

In choice of technique in energy supply, elites in North America and Europe fret about China and India's fondness for coal-fired power generation. But the dread that billions of Asians will one day drive cars and travel by plane looms still larger.

Like most fears in society, this one must be resisted.

First, the East wants, *and deserves*, all that we have in the West. Second, the East simply won't allow the West to dictate to it. Third and above all, to see the populous East just as billions of consumers is a mistake.

If the world can think through energy supply, it can be entirely sanguine about Asian energy use. Indeed, Asia promises to be an important source of energy innovation and investment in the future. It's well known that China easily leads the world in solar water-heating panels: it has 52 million square metres of them and wants four times that by 2015. [20] But what isn't so well known, for example, is that China's work in 'fourth generation' (4G) nuclear technology has already drawn significant interest in the US, and might one day figure in a revitalised nuclear program for that country. [21]

It would be idle to imagine that, on climate change, the West's diplomacy toward the East will be motivated merely by environmental concerns. Even before the July 2008 collapse of the Doha talks on trade, Bill Emmott, former editor of the *Economist*, gave a vivid sketch of how West-East economic antagonisms are likely to intertwine with diplomacy on GHG emissions. [22] Already, too, the West entertains imposing 'carbon border taxes' on Eastern exporters it conveniently deems a danger to the planet. [23]

The general prospect is for Western leaders to use climate change to try to control the pace and direction of growth in the East. But from the point of view of humanity, that would be a great shame. The thought and the engineers that the East has to offer the world are precious. They should not be jeopardised by Western highhandedness on the issue of climate change.

Given good science and technology, scarcity isn't an absolute

At the December 2007 Bali conference on climate change, where there were profound disputes between West and East, it was agreed that the rate of transfer from West to East of technological innovations in energy supply should speed up.

Yet to be thoughtful about energy supply means thinking hard not just about advances in energy, but also about the general business of technological innovation. And, as a concept, technological *innovation* is far too exciting to be reduced, in the manner of the Bali conference, to technology *transfer*.

Technological innovations aren't just moved around from one nation, sector of industry or organisation to another. They are also produced in the first place. They therefore rely on fresh thinking, and upon a whole series of prototypes, experiments and refinements. New technologies, therefore, rely on new scientific insights, together with a willingness to take practical risks, both in the laboratory and elsewhere.

Exactly the same is true of new energy technologies. Environmentalists and the media focus on the personal use of energy. But this book upholds science, technological innovation, research and development (R&D), and indeed, what is today derided as the 'technical fix'.

Environmentalists love to say how the science of climatology has reached a consensus that will tolerate no 'denial.' And to point to the limits of the world's resources, the environmental group WWF likewise insists that if the world's inhabitants shared the UK population's lifestyle, three planets would be needed to support their needs and their waste. [24] Swept away by their desire to go carbon accounting and thus moralising about consumer excess, too many environmentalists ignore new scientific insights beyond those of climatology, and ignore, too, how thoughtful supply-side technologies can overcome the alleged scarcity of the Earth's energy resources.

Scarcity isn't an absolute. The IEA and BP make generous estimates of the world's likely reserves of oil and non-conventional oil (heavy, or from tar sands, shale and the Arctic). [25] But leave aside oil reserves. Overleaf, we describe how solar power can be used both to make hydrogen from water, and to strip carbon out of atmospheric CO_2. It's also true that two types of planned nuclear reactors will be able to generate hydrogen (see the table at the end of Chapter 4). In principle, then, *zero-carbon renewable and nuclear energy can be used to separate out hydrogen and carbon from water and air, and then combine them so as to make many Earths' worth of new, compact and powerful hydrocarbon fuels.* By perhaps 2050, those artificial carbon-based fuels will start to power more and more transport vehicles. When consumed, they'll emit CO_2; but over the whole process of getting hold of carbon from the atmosphere, combining it with hydrogen, and burning the result to go places, no new CO_2 will be created. Artificial fuels will join biofuels in gradually making transport a limitless, carbon-neutral affair.

Since 1972, when the English economist Barbara Ward and the French-American microbiologist René Dubos published *Only One Earth*, environmentalism has monotonously repeated how finite the planet's riches are, compared with mankind's infinite capacity for causing havoc. [26] Yet it's really the imagination of too many environmentalists that is finite. Just two current research projects in energy hint at boundless possibilities:

1. At the Paul Scherrer Institut, Villigen, Switzerland, 170 scientists have learnt how to generate a lot of high-energy neutrons. In principle, such particles can turn long-life nuclear waste into short-life or even stable elements [27]

2. At Sandia National Laboratories, New Mexico, solar collectors irradiate giant rings that rotate at one revolution per minute and contain a metal oxide. Cooled from 1500°C to 1000°C, then exposed to superheated steam, the scorched rust generates free hydrogen. In the same labs, solar power is used to split CO_2 into oxygen and carbon monoxide. The hydrogen and carbon monoxide can then be used to synthesise hydrocarbons. [28]

For most environmentalists, the world has already reached a tipping point, so no faith can be placed in exploring these two projects. After all, in 2005 some scientists said that even the *current* stock of accumulated greenhouse gases would, in the long term, heat the planet by 2°C above pre-industrial levels. Any more greenhouse gas would make for big shifts in climate variability. [29]

In this urgent Green framework, then, even a 'nearly ready' kind of energy technology like Carbon Capture and Storage (CCS) will take too long to make a difference. To build a new round of nuclear power stations would similarly take too long.

Chapter 3 of this book shows, however, that a Second Flood is *not* just round the corner. In any case, the problem that Greens have with energy innovation isn't that it's *too slow*. Green objections are designed precisely to slow up the building of new nuclear power stations. Rather, the problem Greens have with energy innovation is that it's *too risky*.

Taking risks to leave the low-CO$_2$ cave

Imagine that someone comes to you with a thoughtful energy innovation. He says it's powerful, sometimes lethal, and that he doesn't yet know precisely how it works. It brings new problems, certainly; but it could also bring enormous benefits for civilisation. Do you say yes, even if you want to know more about the safety of the innovation? Or do you dismiss it as a 'technical fix,' and instead follow the Precautionary Principle, which gained legal and political prominence at the UN Rio Summit of 1992 and was adopted by the European Commission in 2000?

To date, the conduct of the UN and the Brussels Commission does not tell in favour of the Precautionary Principle. The Principle means that with this technological breakthrough, as with any other, fears about even the remotest possibility of irreversible harm to the environment or humans must come before turning the breakthrough into a mass-market affair.

But what if the energy breakthrough you're offered is in fact *fire*? Fire is something that mankind first tamed nearly 790,000 years ago – if we are to believe seven Israeli scientists who have dated some of the wood, bark, fruits, seeds and flints that were burned during the Lower and Middle Pleistocene, or Ice Age. [30]

Fire is also risky stuff. In Greek mythology, Prometheus was chained to a rock and had his liver pecked out each day by an eagle, as punishment for stealing fire from the gods and giving it to the human race. Fire burns children easily and can destroy whole communities. But the fact is that the domestication of fire on the shores of an ancient lake, in the middle of the Levantine Corridor from Africa to Europe, is what first may have allowed *homo sapiens* to move north of the Mediterranean. It was by not having the Precautionary Principle around that Africans first took fire, with all its risks, to help keep themselves warm when they migrated to and colonised the colder lands of Europe. [31]

From fire onward, technological innovations in energy supply have brought risks. But they have also brought human beings out of the cave.

The caves of old were low-carbon. But they were just that – caves. Society needs to remember and uphold the historic and progressive role of energy supply in colonising the natural world and making it comfortable and convenient to live in.

Campaign for energy supply and energy R&D

Energy innovation has been weak these past 30 years. Holding fast to the Precautionary Principle, the West has developed a deep cultural aversion to risk, technological innovation, and energy innovations in particular.

Jonathan Leake, the respected science correspondent of *The Sunday Times*, London, highlights the skittishness of Western culture when he notes that several different answers to climate change have had their 15 minutes of fame. [32] As solutions, planting trees and carbon trading aren't especially technological; but just like nuclear power, CCS, biofuels and wind farms, each has had its Andy Warhol moment.

Yet there is a solid reason behind this flirting with choice of technique. The West lacks the confidence to make serious investments – either in general technology, or in energy innovations.

Between 1988 and 2006 in the US, gross expenditure on R&D as a proportion of Gross Domestic Product (GDP) stagnated at below 2.7 per cent. The commitment to R&D made by members of the European Union (EU) was even worse, and now lies at a trifling 1.8 per cent of GDP. [33]

Across the 30 members of the Organisation for Economic Cooperation and Development (OECD), both public and private sector expenditure on energy R&D has declined. Indeed, between 1991 and 2002, R&D expenditure as a fraction of the energy sector's total turnover dropped by more than a half – to just 0.33 per cent. [34]

So much for the much-vaunted 'knowledge economy'. These statistics suggest that there has been a stark dumbing down of energy research. With that in mind, *Energise!* believes that everyone should:

- refuse to be stigmatised as energy wastrels
- campaign for more of society's money and brains to go into energy supply and energy R&D in both the private and public sectors.

Renewable sources of energy, it is said, can both save money and make money. Yet if they're so inherently profitable, why have they been avoided for so long?

Late in 2007, Al Gore's $1.2bn investment fund, Generation, linked up with a $200m Silicon Valley venture capital fund to bring Green innovations to market. But around the same time, the head of General Electric's energy business had this to say to the London *Financial Times* about what he called 'wind, solar and so on':

> 'I don't see a disruptive new technology that changes the game in the next 20-30 years. It is not the nature of this industry… Everything that has been developed so far… has taken decades to come to fruition. My expectation is that it will remain that way.' [35]

The pace of innovation in renewable energy is still sluggish, no matter how strident the authorities' calls for personal self-denial have grown.

Why the slow pace? Not, as Greens repeatedly allege, because the usual neo-conservative clique of business chiefs and their pawns in government have conspired to kill off Green innovation – all in a Wall Street-style quest for short-term profit.

No. In fact, renewable sources of energy have taken decades to develop because they only become economically viable when they are built on a grand scale – a scale which today's culture in the West often finds too daunting. We explore this further in Chapter 6.

But there's something else, too. The slowness to introduce Green energy innovations reflects Western fear of, and sloth around, technological innovation in general.

All parts of the energy sector need to free themselves from this sad culture of the past.

Doing better than Carter and his sweater

The capitalist spirit now favours Green energy innovations, but the capitalist flesh looks like it might take 20-30 years to properly introduce them.

Sensitivity to risk makes Western elites anxious: not just about nuclear fission and fossil fuels, but also about serious innovations in renewables. Take biofuels, for instance. As late as August 2006, some environmentalists routinely endorsed them as a remedy for global warming. [36] Yet now, despite the fact that they come in an enormous variety, the whole biofuels sector gets a bad rap. [37] According to various authorities, they can lead to:

- deforestation, and thus a net addition to CO_2 [38]
- the destruction of natural habitats [39]
- the marginalisation of women farmers [40]
- food price inflation. [41]

Similarly, many Greens object to a proposed tidal barrage for the Severn Estuary in the UK because it would kill the wildlife in the area. [42] And wind power done at scale? One scientist has already noticed a snag: it could have 'non-negligible' impacts on climate. [43]

For every low carbon solution, a major problem is found. That's because *many Greens want people to change more than the energy supply.*

On 2 February 1977, environmentalists gained an enduring inspiration for their cause. Then, a newly inaugurated US president came on television in a sweater, asking Americans to save energy by turning down their thermostats – and by

---

Fireside chat: President Jimmy Carter sports a jersey so as to try to get people to conserve energy in the home

donning sweaters.

Sometimes it can be wise to turn the heating down, and to dress warmly indoors. But to turn these individual choices into a presidential policy is to negate the whole concept of convenience in favour of labour-intensive, mindless toil. This puts us in a medieval world in which, by pulling sweaters on and off all day, humans repeatedly have to concede to their environment, instead of just getting on with things.

It's time to wave goodbye to the impoverished Jimmy Carter approach to energy.

Humans do waste energy. But personal struggles to conserve energy do little more than waste time (see panel below). Made into a habit, they represent not 'awareness', but a rather unthinking obedience to the state.

The run-round to save energy with UK consumer electronics

Leave a mobile phone charger on all day without a phone to charge, and you waste one watt. Britain's government-backed Energy Saving Trust (EST) thinks that if the UK's (roughly) 25 million households didn't make that mistake with their 25 million chargers, they'd save 25 megawatts (MW) of electricity – enough to power 66,000 homes. [44]

That sounds impressive. But suppose the average UK householder – pensioners included – takes 10 seconds a day to get to a charger and switch it on and off. A national effort to do that switching would absorb 26.37 million hours a year. The energy saved would power 66,000 homes, but that's still only 0.25 per cent of UK households.

Now, scale up the EST's fidgety philosophy across the full range of hateful, electricity-guzzling consumer electronics devices in the home. In 2004, the average UK household owned 2.4 televisions, 1.9 video recorders, 0.5 set-top boxes and 5.2 external power supply units. [45] The total today is at least 10 gadgets per home.

But if all British households spent 10 seconds fiddling with each gadget, they'd spend about 264 million hours a year switching. Electricity savings would power a majestic 2.5 per cent of UK households.

In the three months to July 2008, which we may take as a reasonably typical period, Britain's 29.54 million people in employment worked 947 million hours a week – equivalent to 135 million hours a day over seven days. [46] If those people were, now, really to take on society's supplication to the socket, they would be spending roughly the equivalent of two extra working days a year engaged in fruitless, labour-intensive overtime around electricity supply in the home.

Most British householders, however, might prefer to leave their consumer electronics on standby, and instead move on to a higher or more pleasurable purpose.

The three accusations made against the energy sector

Why, in energy, do environmentalists tend to favour personal conservation over industrial innovation? Well, unlike new technologies, consumer cutbacks cost nothing (or at least nothing financially). They're also supposed to bring immediate benefits to the Earth's climate. Finally, many Greens are more interested in being sanctimonious about other people's behaviour than in *actually doing something about energy*.

We deal with these issues in the next chapter. For the moment though, let's tackle something even more basic: *environmentalism's deep-rooted disdain for generating any energy at all.*

Environmentalists believe that energy generation:

* consumes resources
* pollutes the Earth
* is marked by what are known as 'negative externalities'.

Certainly, energy generation consumes resources – but as we have seen, scarcity isn't an absolute. Through sunlight, the Earth each day receives an almost unlimited supply of energy in a diffuse form. But that rather useless kind of energy contrasts strongly with human beings' desire, need, and ability to *concentrate and order* energy to pursue tasks that are more and more intricate. [47] Those energy intensive tasks include:

* etching silicon chips
* performing laser eye surgery
* flying long haul to see an ailing aunt just before she dies
* cutting pollution.

In the same way, mankind will most probably need to expend a lot of energy, and even generate a lot of CO_2, to build the low or zero-carbon power sources and carbon traps of tomorrow.

It's ironic, but society's main use of energy is to extract, refine, process and purify energy itself. The energy industry is mankind's largest energy hog. But that's a good thing: it has led not just to convenience, but also to civilisation.

What about the pollution caused by the energy industry? As we saw earlier in the chapter, the industry is indeed responsible for the largest share of man-made emissions of GHGs. However, environmentalists often exaggerate the energy industry's misdeeds: from the wreck of the *Torrey Canyon* supertanker in 1967, through the *Exxon Valdez* disaster of 1989, to the *Brent Spar* fiasco in 1995, Greens have made rather too much of oil spills at sea. [48]

Yes, energy corporations stand in need of a much more thoughtful pollution regime. But by itself, such a regime will not deliver more energy, which is what the world needs. The energy industry will only pollute less if it becomes *generally* more thoughtful, investment-orientated, and focused on R&D.

In 1986, after going 100 metres beneath the North Sea, one of the authors of this book exposed for *The Economist* how Shell UK's Brent Alpha offshore oil platform was at its most dangerous during periods of shutdown, when frenzied maintenance work could easily lead to mistakes. [49] Shell nearly sued – yet in the 6 July 1988 explosion of Piper Alpha, a North Sea oil platform run by Armand Hammer's Occidental Petroleum Corporation, 167 men died precisely during such a shutdown period.

Following Piper Alpha, a host of new safety regulations came in. Yet despite this, years of underinvestment have once again made regulators issue a 'stark warning' about the lack of safety on North Sea oil platforms. [50]

Safety on the North Sea contains clear lessons for energy industry pollution. Neither safety nor pollution is just about geology and chemistry: both are much more about the state of technology, science, management, priorities, and funding.

Energy industry pollution, like corporate pollution everywhere, is a social question. Rather than hatred, or new state laws, it demands thought, *and a rational programme of innovation.*

review
ON THE ECONOMICS
OF CLIMATE CHANGE

Imposing face of the New Scientism:
Nicholas Stern. In 2006, his massive and thoroughly
neutral treatise on what to do about global warming
focused on a scenario in which there was an almost
10 per cent chance of mankind being made extinct
by 2100 (*Stern Review*, page 47). Within 12 months
of this gloomy speculation, Stern was given a life
peerage. In the House of Lords, he sits as a
crossbencher, because he is not party-political

So what, thirdly, are the energy sector's 'negative externalities'? In his famous 700-page report on climate change published in 2006, London School of Economics professor Nicholas Stern mentioned 'externalities' more than 70 times. [51] But the concept in fact dates back to 1890 (see below).

That energy industry pollution carries an external cost to society seems commonsensical. But can right-minded technocrats accurately tax pollution, or successfully price and run markets for CO_2? The market is too chaotic not to pollute the world, but state bureaucracies also act chaotically in their attempts to beat pollution with paperwork or court fines.

Stern saw climate change as *the greatest market failure ever.* [52] Yet it's not just markets that have failed: energy corporations have also failed to innovate – thoughtfully, consistently, and therefore cleanly.

Climate change speaks also of state failure. As we show in Chapter 3, the possibility of man-made global warming began to emerge strongly in US research in 1956. However, Western governments have taken decades to do something about it.

When first conceived, externalities were about costs imposed on others. With the rise of environmentalism, however, externalities turned out to be costs imposed on nature, which was represented as something plundered. Today, environmentalist obsessions ensure that externalities and nature often become the very starting points of economics. Assigning a quantitative value to nature, environmentalists believe it represents a 'natural capital' that mankind consumes, or 'eco-system services' that mankind pollutes. [53]

But it is human activities, including those of the energy industry, which actually add value to capital and supply services. Nature builds no machines, high-speed trains, or the Internet. Mankind is not a negative blot sucking up energy from the landscape, but a positive force for progress. *The ultimate energy on the planet is human.*

Al Gore was right about one thing. Political will *is* a renewable resource. And the same is true of the human will to innovate.

CO_2 taxes, CO_2 markets:
the inside story on externalities

In 1890 the English economist Alfred Marshall praised not the intrinsic creation of value through production and innovation, but positive factors *external* to that: the goodwill surrounding a business, and the spin-off from transport infrastructure. [54]

After the conflict of 1914-18, opinion darkened, focusing instead on *negative* externalities. For Marshall's Cambridge successor Arthur Cecil Pigou, factory smoke inflicted 'a heavy uncharged loss on the community'. In 1920, Pigou fathered the modern idea that state taxes on CO_2 emissions can *internalise* their external costs to society. [55]

In 1960 the Chicago economics professor Ronald Coase, a Swedish social democrat enthused with American capitalism, took issue with Pigou. For Coase, governments might fix smoke more cheaply than private organisations, but governments themselves could also, 'on occasion', be 'extremely costly'. [56]

In effect, Coase rejected pollution taxes, preferring that *the state put a price on the right to pollute*. Later the US Congress established a market-based cap and trade mechanism. From 1995 onward, the Acid Rain Program capped the total sulphur dioxide emitted in American electricity generation, and, at the US Environmental Protection Agency each March, auctioned off permits to emit nearly three per cent of that total. [57] Similarly under George W Bush, the EPA issued, in 2004, the first federal rule to cap and trade emissions of mercury from coal-fired power plants. [58]

Today, CO_2 emissions are traded through the EU's Emissions Trading Scheme (ETS). Taking a leaf from that scheme, Barack Obama has also promised to implement a cap and trade programme to reduce GHG emissions in the US.

In fact, both state taxes and state caps on CO_2 reflect undue faith in the state, undue faith in the market, and not enough faith in energy innovation. We hope that Barack Obama knows what he's proposing. As Chapter 3 shows, the ETS has largely been a fiasco.

Citizens, not consumers

There's a big contrast between

- contemporary culture, which ridicules humans' ambitions as hubristic, warns that nature will take 'her' revenge, and insists that the limits imposed by nature on man can never be breached

and

- the logic of this book, which highlights how, depending on the state of civilisation, humans have a remarkable record of overcoming what are perceived as immutable limits.

For politicians, climate change means that nothing is certain but *death, energy meters, and carbon taxes*. Politicians want people to atone for their shocking selfishness: they want to *add to the sum total of guilt in the world* (though they don't seem to feel very guilty themselves). They seek legitimacy through the truly limp cry: 'Let's survive! It's in everyone's interest!'

Meanwhile celebrities set themselves up as role models, favouring the chic politics of the prominent gesture. Pompous and narcissistic about their energy selflessness, they feel no guiltier than politicians.

As for the energy industry, it's on the back foot. Nuclear interests refuse to make a bold case for their role in creating much more energy, instead pleading that their plants have only a modest pollution impact. Against this defensive argument, Jimmy Carter's pullover will always win.

Oil, gas and coal are cast as pariahs. And renewable sources of energy are dogged by delays and inconclusive debates.

Finally, people are disempowered by the doctrine that they are greedy consumers of energy.

This book refuses to look at people that way. People are

Putting their back into it: hippies erect a solar panel on a timber house in Cornwall, England. At the level of the individual household, Green technologies demand a lot of time and effort for the amount of electricity they actually produce

not just consumers; they can and should be *energetic citizens,* with lives that are convenient enough to be expansive, not spent watching energy meters. They can and should be able to vigorously debate, vote on and act upon choice of technique in energy supply.

The desire to do something about energy is fair enough. To make a better world, however, people can do more than go through the motions with energy at home, in the shops or on their travels. It's right to:

* feel that voting for a politician every few years doesn't help society much
* feel that following every twist and turn of celebrity gossip doesn't do much for society either
* want personal transport that's convenient, cheap and clean.

But human-powered bicycles won't solve the world's transport problems. People can do more for energy in Africa or Bangladesh than switch to the most ethical supplier of household gas.

To give something back to society and make much more than a difference, people need to mobilise for the proper kind of energy commitments – and mobilise on the basis of nobler and more profound feelings than consumer disgust with energy companies.

Solar panels on a roof, like a Toyota Prius outside a doorway, can look cool to neighbours and friends; but to uphold the microgeneration of energy by a panel, or energy efficiency in a car, is thinking too small. In practice, humans will always have larger ambitions. Certainly they will want to do more than just survive.

Is it irresponsible to let people be thoughtless in energy use? No. To neglect the energy innovations that the world requires – now *that* would be irresponsible.

What is a horsepower, anyway?

Energy is measured in **Joules**. A Joule is roughly what it takes to lift an apple one metre. The more usual main units are:

1 kilojoule (kJ) = 1,000 Joules
1 Megajoule (MJ) = 1,000,000 Joules
1 Gigajoule (GJ) = 1,000,000,000 Joules

In 2005, the world consumed 500,000,000,000 GJ of energy. [59]
 Per kilogram, coal contains 20-30 MJ. Oil contains 50 MJ. Oil packs the bigger punch – it has a higher **energy density**.
 Power is measured in **Watts**. One Watt is use of energy at the rate of one Joule per second. A fairy light on a Christmas tree gives out energy at a rate of about one Watt. James Watt's vivid coinage, **horsepower**, approximates to 750 W.
 One kilowatt (1 kW), or 1,000 Watts, approximates to the power consumed by a single-bar electrical heater.
 One kW-hour on your electricity bill is 3.6 MJ. The average American consumes power in all its forms at 10 kW per person, amounting to about 300 GJ per person per year.
 Given a strong wind, a 1 MW (1,000,000 Watts), industrial-scale windmill produces energy equivalent to all of the needs of about 100 Americans. For a coal or nuclear power station running at 1 GW (1,000,000,000 Watts), that figure rises to about 100,000 Americans.
 When a power station burns coal, the coal produces heat. Each kW of power generated is abbreviated as 1 kWt, to show that this takes the form of thermal power – heat. The thermal power is then converted into electrical power. Losses in heat-to-electricity **conversion** (typically about 40 per cent) and power-station-to-consumer **transmission** (typically seven per cent) mean that *less than* 1 kW of electricity – abbreviated as 1kWe – is actually delivered to the consumer. Conversely, to deliver 1 kWe of electricity to the end-user takes more than 1 kWt of power from coal.

Primary energy is the total energy used up in the production and distribution of energy to end-users; it's the energy in the coal that enters a power station, or in the oil that enters a refinery. **Secondary energy** is the amount finally delivered to end-users, such as the energy delivered by a light bulb, or a car engine. Thus secondary energy is equal to primary energy, less losses in transmission and conversion.

Most primary energy still comes from fossil fuels.

Shares of total primary energy supply, by type of energy, per cent [60]

WORLD:	1973	2005	OECD:	1973	2005
Oil	46.2	35.0		53.0	40.6
Coal	24.4	25.3		22.4	20.4
Gas	16.0	20.7		18.8	21.8
Nuclear	0.9	6.3		1.3	11.0
Hydroelectric	1.8	2.2		2.1	2.0
Other	10.7	10.5		2.4	4.2

Converting energy into different forms

Energy comes in different forms: chemical, electrical; as heat, light, motion and so on. Society makes use of energy by using different technologies to convert it from one form to another. For example, coal-fired power plants turn the chemical energy in coal into heated steam, which, through a turbine generator, is converted into electrical energy. In the home, lamps and motors convert that electrical energy into light and motion.

At each conversion, however, a little energy inevitably escapes. That escaped energy isn't destroyed – the **First Law of Thermodynamics** states that *energy can be neither created nor destroyed*. But the escaped energy does become unavailable for human use. As a result, more energy must always be put in at the beginning of a chain of conversions than will be derived at the end of that chain.

While most oil is used for transport, and most gas for heating and cooking, history suggests that electricity – a particularly concentrated and flexible form of energy – will grow in significance. The largest source of electricity is coal, but nuclear, gas and hydroelectric are also important. Renewables, meanwhile, should make themselves felt in electricity generation in years to come.

As consumed by the end-user, more and more energy takes the form of electricity.

Shares of final energy consumption, by type of energy, per cent [61]

WORLD:	1973	2005	OECD:	1973	2005
Oil	48.2	43.4		56.7	51.9
Coal	13.1	8.3		10.1	3.3
Gas	14.3	15.6		18.2	19.2
Other	15.1	16.4		3.6	5.6
Electricity	**9.3**	**16.3**		**11.4**	**20.0**

The numbers show a rise in electricity used by the consumer, but understate its significance for two reasons.

First, electricity only makes up 20 per cent of the energy directly used by consumers in the developed world; but generating that electricity in power stations itself absorbs about three times as much energy. Electricity, in other words, is a highly refined form of energy. Consumers don't get to see all the energy behind the electricity they use.

Second, electricity is the most versatile and convenient form of energy. That's the reason why using a lot of energy to generate it is worthwhile.

Like other forms of energy, electricity can provide power for heat, light and motion. In these applications, however, it typically gives superior performance and controllability. For example, an electric motor has the same power rating as a steam engine, but is more compact. It also has a turning force that can be set more precisely.

In terms of convenience, switching on a light is a whole lot simpler than lighting a match. Similarly, laying electrical cables is easier than installing pipes to carry fuels.

Electricity is also essential for information technology. Computers are nothing but miniaturised electrical circuits, and electrical energy is necessary to drive electrons around them.

The world needs a lot more energy, a lot more electricity – and a lot more of those unsung heroes, the electrical engineers. [62]

Capacities and load following

Choice of technique in energy supply isn't just a matter of how much energy each technique can *potentially* produce when working flat out, but how much each can *actually* produce over time. Also critical is whether energy is available *at the time it is needed*.

Peak potential energy generation is known as **nameplate capacity**. That's what's generally quoted when new power production is discussed. The proportion of that energy actually produced is called the **capacity factor**.

Typical capacity factors, per cent

Technique	
Nuclear	90
Coal	90
Hydroelectric	50
Wind	30
Solar	15

These figures mean that 1 GW of nameplate capacity will produce three times as much energy in a nuclear plant than in a wind farm. That doesn't mean that wind turbines only work a third of the time – just that they rarely reach full power.

Two main considerations determine capacity factors. The first is **availability:** how much a plant is actually doing its work. All techniques of energy generation require some downtime for maintenance. Nuclear and carbon-based generation also require downtime for refuelling. In the case of renewables, downtime occurs when the wind isn't blowing or the sun isn't shining.

A second consideration is **load following.** Because demand for energy fluctuates over each day, it's necessary to have spare generating power that's available, but not put to use – **idle capacity.** That inevitably means running a plant at a lower capacity factor.

A mix of technical issues and economic ones determines what a plant's idle capacity should be.

On the technical side, different power sources come on line in different amounts of time. In the case of hydroelectric power, for example, water flowing through a dam can begin generating energy in seconds.

By contrast, it isn't wise to turn large nuclear or coal-fired power stations on and off too much. They take a long time to heat up and cool down. If needed at short notice, they have to be kept hot, wastefully, in a condition known as 'spinning reserve.'

Because they're smaller than those in nuclear and coal-fired plants, turbines in gas-fired power stations can be started and stopped more quickly. They're a better source of reserve power.

Load following requires the kind of power that's available on demand, or 'dispatchable' by grid managers. A gas power station that runs 30 per cent of the time but helps cope with peak loads is more useful than a wind turbine that runs 30 per cent of the time because that's the proportion of time during which the wind blows.

On the economic side, it makes sense to run capital-intensive power at higher capacity. Most of the cost of nuclear plants is paid up front, while relatively little goes on fuel; so operators try to run them 24/7. The opposite is true of gas, where fuel makes up a significant proportion of costs. That's another reason gas is used to cover peak loads.

What this shows is that low capacity factors are not in themselves a problem. They are most problematic when power output is *intermittent* – that is, unpredictable, as with wind. Next best is power that's sometimes on but always predictable: tidal energy is an example here. Best of all is power available any time, on demand, such as gas or nuclear.

Energy efficiency and energy conservation

Energy efficiency means doing more with less. Narrowly, it's the proportion of energy put in that goes on a designated task. Broader measures of energy look not just at energy conversions, but also at *human goals.* For example:

- when an efficient home boiler meets a *poorly-insulated roof*, the overall home is rendered inefficient
- an efficient vehicle engine powering a *heavy vehicle* spends most of that power on moving it, rather than its cargo and passengers.

The difference between the two measures is clearest in the energy industry itself, where there's great scope for increasing efficiency. Here the narrow definition is appropriate. New turbine technology is raising the efficiency of gas fired electricity generation. In the 20th century, about 35 per cent of the energy in gas was converted into electricity. Today, new technology is pushing that number above 60 per cent.

For the economy as a whole, the broad measure of efficiency is appropriate. Here energy is being converted not just into other forms of energy, but into mobile phones, orange juice and everything else. The overall energy efficiency of a national economy is measured by its **energy intensity**, or the amount of energy required to produce one unit of GDP. Lower energy intensities mean higher efficiencies.

Efficiency doesn't say anything about how much energy is used overall. Conservation, however, means using less energy. Efficiency is one route to conservation: installing roof insulation, for instance, conserves gas by making use of central heating more efficiently. But cutting back on energy, and often accepting inconvenience in the process, need not involve improvements in energy efficiency at all.

Progress in energy efficiency is exactly that – progress. Significantly, the IEA observes:

> 'Energy efficiency in OECD countries has been improving at just below one per cent per year in recent times. A sharp decline from the rate achieved in the years immediately following the oil price shocks of the early 1970s.' [63]

But it's wrong to believe, as so many do, that efficiency is the main means of curbing CO_2 emissions. There are limits to efficiency. It takes a certain amount of energy to move objects, say, or to heat them up. The infamous **Second Law of Thermodynamics** sets further limits of how efficiently energy can be converted from one form to another. [64] Once efficiencies have reached the maximum allowed by the laws of physics then, to do even more, there's no choice but to generate more energy.

For the IEA, improvements in energy efficiency account for no less than 36-44 per cent of the emissions cuts it seeks to make on its 'baseline' scenario for 2050; nuclear, just six per cent; CCS, 14-19 per cent, and renewables, 21 per cent. These figures betray an extraordinary reliance on efficiency measures – what the IEA itself describes as a 'first step' – rather than on those related to energy supply. [65] In both of the IEA's preferred scenarios, energy efficiency improvements in buildings, appliances, transport, industry and power generation represent 'the largest and least costly savings'. [66]

The IEA recognises that higher efficiency means more demand. Historically, efficiency gains have led to *more* energy use rather than less: if you can do more with your energy, it becomes more expendable. Mainstream economists like to point out that higher efficiencies lower the price of energy and increase demand for it.

To progress beyond the IEA's goals will require greater investment in the clean production of energy.

The question of embodied energy

'Don't Buy That New Prius! Test-Drive a Used Car Instead', said *Wired* magazine in 2008. Its reasoning:

> 'Pound for pound, making a Prius contributes more carbon to the atmosphere than making a Hummer, largely due to the environmental cost of the 30 pounds of nickel in the hybrid's battery.' [67]

Greens tend to generalise this logic. The investment you propose to make in a Green house or car, they argue, will in fact use up more energy or generate more CO_2 than staying as you are.

It's true that energy isn't just about what's consumed when products are in use. It's also about what's consumed during production. The jargon has this as 'embodied energy'. It's not necessarily *physically* embodied in that house or car; but for Greens it adds to the awfulness of such items. Rather than make energy invisible and taken for granted, the concept of embodied energy is another way of trying to make energy weigh more heavily on your conscience.

The argument presumes a low-growth or no-growth world. But do people want to live like the Cubans under economic blockade, keeping second-hand cars running for decades?

Making production more energy-efficient is a problem for engineers – and one that they pay considerable attention to. That might help bring down prices of consumer goods. But if society invests enough in energy supply, it isn't something consumers should worry about.

Growth means building more goods – including houses and cars. That will use up energy. Efficiency will not mean using less energy overall. It will only mean using less energy than would have been the case with old-fashioned technology. The bottom line is still that *the world needs more energy.*

As the *Wired* example of the Prius battery shows, the 'embodied energy' argument is also applied to energy generation itself. It's said that your proposed investment in clever energy supply will use up more energy or put out more CO_2 than you generate.

The jargon is 'energy return on energy invested', or EROEI. It's modelled on a financial measure, 'return on investment' (ROI). ROI is used by business to decide on where to put its money.

It does indeed take energy to produce a barrel of oil, a biocrop, a windmill, and a solar panel. Moreover, energy needs to be produced efficiently. But EROEI studies are often suspect, reflecting politics as much as physical realities.

With investment, you're supposed to get more out than you put in. But things are more clear-cut in the world of money than they are in the world of energy. You know what counts as money invested. But exactly what energy goes into running an oil rig in the North Sea? Running the machinery, and providing heat and light for the oil workers, certainly. But what about the steel to build the rig? What about the food for the workers? What about food for the miners who dug the steel that built the rig?

Overall, society does need to capture net energy. But the preoccupation with EROEI reflects an obsession with quantity over quality. More energy goes into a refinery or a power station than comes out, giving an EROEI of less than one. But the energy that comes out emerges in a form *more useful to human beings.*

Electricity is more useful than the coal that precedes it. Petrol is more useful than a barrel of crude oil.

That's what needs to be remembered.

Carbon intensity and the 1997 Kyoto Protocol

In 1992 in Rio de Janeiro, the world's governments agreed that GHGs gases must be limited to a level that would 'avoid dangerous climate change'. Just what 'dangerous climate change' might be, let alone the levels of GHGs needed to avoid them, has been argued over ever since. But in 1997, the world's governments confirmed in Japan what they had really committed themselves to back in 1992.

The resulting 28 Articles and two Annexes of the 11 December 1997 Protocol to the UN Framework Convention on Climate Change – the Kyoto Protocol – dodged the question of how much GHG was ultimately acceptable, but insisted that developed countries cut their emissions.

Different countries were given different targets, but the collective average was a five per cent cut below 1990 levels for developed countries. Developing countries – including China and India – weren't given targets, but the Clean Development Mechanism instead (see Chapter 3).

After George Bush senior signed the Framework document in 1992, George Bush junior famously backed out of ratifying Kyoto.

The latter's alternative was to set targets for decreasing *carbon intensity*. As with energy, carbon intensity measures the amount of CO_2 emitted per unit of GDP. In 2002 Bush set out a voluntary target for the US to reduce its carbon intensity by 18 per cent by 2012. [68]

Greens ridiculed Bush. They made the point that if GDP is increasing, cutting carbon intensity gives no guarantee of lower CO_2 emissions. But in April 2008 Bush added a 'new national goal' that emissions should peak by 2025.

In fact carbon intensity is more important than Greens allow. Emissions can be cut simply by doing less. Carbon intensity can only be lowered by doing at least as much, but better.

Understanding concentrations of GHGs

In 2008 the level of CO_2 in the atmosphere stood at 384 parts per million (ppm). Bubbles trapped in cores of ice have revealed that, in the 10,000 years between the end of the most recent Ice Age and the industrial revolution, CO_2 concentrations have ranged from 260 to 280ppm. Over the past 650,000 years, ice cores show that CO_2 ranged between 180 and 210ppm during ice ages, and from 280 to 300ppm during warmer interglacial periods – such as the world enjoys at present. [69]

As we've explained, there are also other man-made GHGs in the atmosphere. The effect of these can be converted into a *carbon dioxide equivalent*, or CO_2eq. This is the concentration of CO_2 that would have the same effect as the combined power of these other GHGs. Note that this equivalence doesn't mean that the effect of a quantity of man-made GHGs other than CO_2 is identical to the effect of another quantity of CO_2 in every respect. For example, and rather importantly: after being emitted, CO_2 remains in the atmosphere for centuries. Methane, by contrast, stays for only decades.

The total level of GHGs in the atmosphere today is about 440ppm of CO_2eq. Of this, about 280ppm is the natural pre-industrial level of CO_2. About 100ppm is man-made CO_2, while the remaining 60 ppm is from other man-made GHGs.

It makes sense to convert to CO_2eq because only one number needs to be thought about. Note, though, that when commentators talk about a low- or zero carbon economy, the talk is loose. Only two-thirds of the rise in GHGs consists of CO_2.

Rises in Parts Per Million of CO_2eq, temperature, and sea levels: how fast?

At the moment, CO_2eq is rising at about 2-3ppm a year. Now, simply extrapolating such annual rises to 2100 would take levels beyond 700ppm; but in fact, economic growth is likely to take emissions far higher than that – even if developed and industrialising countries stick just to present-day technologies, which they won't.

Human beings need to take control of global warming. Nature could always throw something unexpected at them, but it's likely that, as emissions are got under control and then reduced, so warming will slow and temperatures, though higher, will stabilise.

We anticipate that the planet can take a doubling of GHGs above their pre-industrial level. That would raise the concentration of GHGs from 280 to 560ppm of CO_2eq. Stabilising concentrations at 560ppm by 2100 would ultimately lead, sometime before 2200, to a temperature rise of about 3° C.

Of course, to look up to 2200 is to look a long time ahead. We agree with the climate scientists Myles Allen and David Frame when they write:

> 'Uncertainties in how the available policy levers translate into global emissions, and how emissions translate into concentrations through the carbon cycle, are so large that uncertainty in the final concentration we are aiming for in 2200 is probably the least of our worries.' [70]

We agree with Allen and Frame, too, that humanity's descendants will revise their targets in the light of the climate changes they actually observe. Indeed, provided those descendants have the sense to alter course in response to what Allen and Frame term 'the emerging climate change signal', then they 'probably won't care' about the climate change uncertainties that detain their ancestors today. [71]

In the two tables below, we give an overview of

1. Estimates, by *Energise!* and by key individuals and institutions, of how fast the situation is deteriorating
2. The targets that we, and others, believe need to be adopted.

Overview of main climate estimates and targets (1): concentrations of CO_2eq [72]

Source, date	*Energise!*	Kyoto, 1997	EU, 2007	UK, 2008
PPM CO_2eq estimated, plus the date the estimate is made for. Premised on continuing 'business as usual'	About 700, 2050	Emissions cuts are not based on specific projections	Emissions cuts are not based on specific projections	Emissions cuts are not based on specific projections
Ultimate long term stabilisation target for stock of GHGs in the atmosphere	560	Ultimate targets left for further negotiations	Ultimate targets left to post-Kyoto negotiations	Based on target that emissions should be 'sustainable' [76]
Emissions targets, date	1990 GHG emissions rise by about 40 per cent, 2030; decline to 1990 levels, 2050. Stabilisation, 2100	Developed countries to cut GHG emissions 5.2 per cent below 1990 levels, 2012 [81]	Cut GHG emissions 20 per cent below 1990 levels, 2020, rising to 30 per cent if other industrial countries agree [82]	Cut GHG emissions 26-32 per cent below 1990 levels, 2020, and by at least 80 per cent, 2050 [83]

Overview of main climate estimates and targets (2): rises in temperature and sea levels

Source, date	*Energise!*	Kyoto, 1997	EU, 2007	UK, 2007
Rise in ° C estimated, plus the date the estimate is made for. Premised on continuing 'business as usual'		Not applicable		As Stern
Rise in ° C, target, date	Stabilise at 3 around 2200	Avoid 'dangerous' interference with climate	Stabilise at less than 2	Less than 2°
Rise in sea level, cm, estimated, plus the date the estimate is made for. Premised on continuing 'business as usual'				As Stern

IPCC, 2007	Obama 2008	IEA, 2008	Nicholas Stern, 2006	James Hansen, 2008
Illustrative scenarios range from 600 to 1550 by 2100 [73]	Emissions cuts are not based on specific projections	550ppm of CO_2 by 2050.[74] The IEA doesn't account for other GHGs, but rises at the same rate would equate to 700 ppm CO_2eq	630 by 2035 [75]	As IPCC
Doesn't set targets	'The amount scientists say is necessary' [77]	450 ppm of CO_2.[78] The IEA doesn't account for other GHGs, but rises at the same rate would equate to about 550ppm CO_2eq	Stabilise between 450 and 550 [79]	Cut current levels to 350 ppm CO_2. [80] Believes that a target for CO_2 alone, rather than all GHGs, is most appropriate
Doesn't set targets	Cut 'carbon emissions' 80 per cent below 1990 levels, 2050 [84]	The more ambitious of two scenarios envisions halving CO_2 emissions relative to 1990, 2050 [85]	To stabilise at 550 ppm of CO_2eq, Emissions should peak 2016-26, then fall by about 1-3 per cent per year. Emissions need to be about 25 per cent lower than 2006 levels, 2050 [86]	End all new non-CCS coal burning. End existing coal-based power stations, 2028 [87]

IPCC, 2007	IEA, 2008	Nicholas Stern, 2006	Al Gore 2006	James Hansen, 2008
Best estimate, 1.8-4.0 above 1980-99 at 2090-99. Unlikely: less than 1.1 or more than 6.4 [88]	6 [89]	2-5 or even higher, eventually, after 2050 [90]	'Off the chart'	As IPCC
Doesn't set targets	2-3 [91]			1 already dangerous [92]
18-59 above 1980-99 at 2090-99. Melting Greenland and Antarctica might add an extra 10-20 [93]	'Significant change in all aspects of life and irreversible change in the natural environment' [94]	If Greenland's ice begins melting irreversibly, world is committed to 700 eventually [95]	If West Antarctica was to go, 609.7 (20 feet) [96]	100-200 by 2100; 'more in the pipeline would be practically a dead certainty.' [97]

In reading the tables, it's important to distinguish between the total *stock* of GHGs in the atmosphere, and *emissions*, which are a *flow* of GHGs into the atmosphere. It's the former that makes temperatures rise. Even if emissions begin to fall, they will still contribute to a rising stock – until, that is, they reach levels so low, the earth will naturally remove them from the atmosphere. So unless human beings find ways of directly reducing the stock of GHGs, there will in the very long term have to be a cut of about 80 per cent in emissions.

Often targets for emissions are stated in terms of cutting them with respect to the level reached in 1990. In that year, the world's emissions amounted to 39.4 Gigatonnes of CO_2eq. Roughly three quarters of them consisted of CO_2, while the rest was made up other GHGs. [98]

Runaway, irreversible, dangerous: how likely is it that climate will change in these ways?

Climate needs serious attention now. But we are less precautionary in our targets than Allen and Frame. Why are we so relaxed about a doubling of GHGs above their pre-industrial level, to 560ppm of CO_2eq, when others would find such a target intolerable?

Governments and environmentalists fear that, above certain concentrations of GHGs, the greenhouse effect may lead to what they call *runaway, irreversible* and *dangerous* climate change. They hold that if a little warming leads to a little more, and that in turn to a lot more, then the only wise option is to stop the process before it starts. For them, the possibility of 'runaway' warming gives good grounds for dramatic action now.

This, however, is *alarmism*. Its basis lies in two key points:

1. Alarmists fear that a doubling in GHGs will produce enormous rises in temperature – up to 10° C or more. Put differently, they worry about the factor known as *climate sensitivity*
2. They fear the *consequences* of those temperature rises.

In line with mainstream scientific opinion, we concur that climate sensitivity – the temperature rise likely to accompany a doubling of GHGs above pre-industrial levels, to 560ppm – is about 3° C. Where alarmists err is on the magnitude and significance of the *uncertainty* attached to 3° C.

The IPCC gives a likely range for climate sensitivity of 1.5-4.5° C, adding that 'high values are consistently found to be less likely than values of around 2.0° C to 3.5° C.' Crucially, it adds that studies 'cannot rule out' values above 4.5° C. It says that the upper bound to its range is 'difficult to constrain', because

1.	The ultimate long-term warming has a complex relationship with the observed short-term response
2.	With climate change, observational records are limited in length
3.	Observations of ocean heat uptake and of the behaviour of aerosols are subject to particularly large uncertainties. [99]

A high value of climate sensitivity, some add, would result in *runaway* change.

There are economists who believe that the uncertain possibility of runaway change dictates that mankind adopt what they call an insurance policy. A panel in Chapter 3 deals with this argument. Here, we just want to suggest that the IPCC's interpretation of the evidence on climate sensitivity is too apprehensive.

It's hard to rule out, in principle, the possibility of runaway climate change. Yet very little, in life and in science, should ever be completely ruled out. What's more productive is to look at where the accumulation of evidence on climate change is pointing.

No single line of evidence is enough to exclude high climate sensitivity. A more fertile approach, however, is to combine a number of different lines of evidence. At the Frontier Research Center for Global Change, Yokohama, Japan, James Annan and Julia Hargreaves have done this using data on warming in the

20th century, volcanic cooling, the last ice age, and the Maunder Minimum period of low solar activity (the years 1645-1715). They draw bounds for climate sensitivity that are tighter than those of the IPCC, around a central value of 2.9° C. [100]

While the IPCC cites Annan and Hargreaves alongside other studies, we think that their approach of integrating different lines of evidence gives a qualitatively better answer, deserving more weight.

If climate sensitivity *does* turn out to be far above 3° C, then a small amount of warming would imply a great deal. Annan and Hargreaves already suggest that this is unlikely – and the longer-term climate record supports the same conclusion. If it were really balanced on a knife-edge, climate should have, over hundreds of millennia, left evidence of a number of different episodes of sudden change.

The idea that humanity faces *runaway* warming implies that if the world waits, or things turn out worse than expected, nothing can be done. This same idea is emphasised in the idea that change may be *irreversible*.

It's true that climate changes are likely to be irreversible. If, for example, warming were to melt the Earth's ice caps, then just reversing that warming would not necessarily result in their return. Similarly, if warming makes a species extinct, cooling will not bring it back to life.

But irreversible processes are common in both nature and society. The fact that the future is inevitably different from the past does not mean that it's inevitably worse. Whether and how people respond to warming, and what they make of a warming world, matters much more than irreversibility.

Even in the unlikely event that warming runs away, there will still be much that people can do to make a difference. Greens tend to dismiss the idea of adapting civilisation to large temperature rises. We think that adaptation is fair enough; indeed our own concept of *transforming* the planet goes further than that (see Chapter 7).

For the sake of the argument, suppose that climate

sensitivity really turns out much higher than 3° C. In that case, our descendants will most likely notice, and, if necessary, take action. Anyway, civilisation should be able to deal with temperature rises larger than 3° C.

This leads to our second difference with the alarmist perspective. Apart from being less bothered by uncertainty around climate sensitivity, we think that the consequences of temperature increases will be far less serious than generally portrayed.

For Barack Obama, global warming 'is a fact that is melting our glaciers and setting off dangerous weather patterns as we speak'. [101] But for us climate change is not the danger he makes out. The basis for our optimism here is not so much an alternative reading of science, nor even the important point that climate change cannot be held responsible for particular incidents of weather (see Chapter 3). We are optimistic about the dangers of climate change because we have confidence – at least as much confidence as Obama – in the talent human beings have to thrive in a very wide variety of conditions.

In 2004, Prime Minister Tony Blair announced a conference on avoiding dangerous climate change, in preparation for making climate the theme of the UK's presidency of the G8 group of nations. 'More than just another scientific conference', he said, the gathering would address the big questions 'on which we need to pool the answers available from the science'. In particular, Blair imagined that *science* could answer the question 'What level of greenhouse gases in the atmosphere is self-evidently too much?'. [102]

Blair's conference, then, was billed as an attempt to use science to pin down the slippery question of just what constitutes 'dangerous' climate change. Yet when the conference proceedings were published, the editors noted that:

> 'The conference did not attempt to identify a single level of greenhouse gas concentrations to be avoided... consideration of the question requires value judgments by

societies and international debate... It would be expecting too much of the scientific community to act as the arbiter of society's preferences as reflected in the valuation metrics actually employed and the decision processes actually implemented.' [103]

We agree. Danger isn't just a scientific issue, but a political one. Furthermore, judgments about danger are inevitably informed by wider assessments of human beings' vulnerability, resilience, and capacities for innovation.

For changes in temperature that are at all likely – even changes of 3° C or more – the pace of innovation and development, not climate, will be the most important determinant of human well-being.

The specific consequences of climate change *for humanity* are exaggerated. First, Bjørn Lomborg is right to point out that, in terms of the deaths that are today caused directly by changes in temperature, those from cold far outnumber those from heat. Second, global warming's deleterious effects on a disease like malaria, for instance, is not nearly so important an issue for humanity as the eradication of malaria itself (see Chapter 7).

There's no need for mankind to lose sleep about its fate. Nevertheless, the consequences of climate change *for the rest of the biological world*, beyond human beings, are important. How the biosphere will respond to rising temperatures is still more uncertain than the physics of heating gases and fluids. But it's clear that many ecosystems may be at risk at 3° C and above.

Nature is less adaptable than humanity. Here there is reason to be cautious about temperatures rising too far and too fast. There are both economic and aesthetic grounds for conserving nature – or, more accurately, for *managing* it.

With the non-human kingdom, however, Greens exaggerate the doom ahead. Their theory of ecosystem services certainly overstates the economic importance of nature.

Many Greens would add a moral case for conserving

nature. That's their right. While we like nature as much as they do, however, our moral universe is centred on human beings.

To capture losses in transmission

In alternating current (AC) used for high voltage, long distance transmission, electrons move back and forth along the 'skin' of cables, changing direction 50 times per second. At the moment, an average of seven per cent of electrical energy is lost in AC transmission and distribution. [104] As grids grow larger and stretch over longer distances, that figure could rise, and technologies to reduce it will become more significant.

One technology that makes particularly good sense for long distance transmission is direct current. With DC, electrons move in one direction only, but do so through the whole body of cables and not just their skins. That results in lower losses.

In undersea cables, AC interacts with salty water and is beset by bigger losses than those that occur on land. Here, therefore, DC has an added advantage: it doesn't interact with salt water. Particularly over long distances undersea, therefore, DC is superior to AC. Undersea cables to bring power from offshore wind to land, or undersea interconnectors linking up international grids, are ideal candidates for DC.

The transmission of what is called high voltage direct current (HVDC) has been made possible by the development of semiconductor power electronics. That kind of electronics scales up the chips that control the flow of electrons in computers so that they can handle the flow of electricity through grids.

About to lay down a wire: an ocean-going ship loads high voltage direct current cable made by ABB. Over long distances at sea, HVDC loses less energy than alternating current

Transformer: a Siemens HVDC power transformer at one end of the world's longest undersea cable, which runs over 290km from Australia to Tasmania. The cable handles 600 MW of power

Energy consumers: No need to feel guilty

You're not a needy, greedy, energy addict. You shouldn't worry about your personal carbon footprint. The world of work, and especially the energy industry, is where CO_2 emissions are mostly produced

People see energy as a problem of individual consumption.

It's hard to say exactly who first coined the phrase 'consumer society', but the idea of consumption as a way of life was certainly given an airing at the height of the Cold War. On 24 July 1959, on a stand at the American National Exhibition in Moscow, US Vice President Richard Nixon teased Soviet Premier Nikita Krushchev with, of all things, *a floor-sweeping home robot*. Moderating Cold War military tensions with consumerism, Nixon proclaimed:

> 'Would it not be better to compete in the relative merits of washing machines than in the strength of rockets?' [1]

During the credit- and property-led economic boom of the 1980s, Nixon's doctrine triumphed. Consumer society, personal lifestyle and the market won a popular victory over state intervention. Britain's Prime Minister Margaret Thatcher said that that there was 'no such thing' as society. Later, she clarified that the 'real sinews' of society were 'the acts of individuals and families'. That reflected her 'fundamental belief in personal responsibility and choice'. [2]

But Thatcher was wrong to think that consumer choice is what characterises modern society.

It's a waste of personal energy to change your lifestyle in the belief that 'the consumer,' when added up into millions, has clout (see panel below). Like it or not, you cannot change climate by refusing to buy blueberries air-freighted from Chile.

On standby: British Prime Minister Gordon Brown, at the UN headquarters, New York, 16 April 2008, four days before he issued his warning about the climate consequences of consumer electronics

Changing your home habits makes little difference to CO_2

Gordon Brown once warned UN ambassadors that the leaving of consumer goods on standby accounted for one per cent of global CO_2 emissions. [3] So, would a more moral approach to your everyday home habits make a difference to CO_2?

Turning lights and appliances on adds to CO_2 back at power stations of varying carbon intensity. By contrast, the other three quarters of household-based emissions are released on site – by gas-fired boilers for central heating and hot water:

UNITED STATES

Household-related CO_2 emissions, UK: Megatonnes and percentage shares, by technology, 2002 [4]

Technology	Mt CO_2	share of total CO_2, per cent
Space heating	76.1	51.6
Lights and appliances	36.3	24.6
Hot water	30.2	20.5
Cooking	5.1	3.4
Total	**147.6**	**100**

Now, say every household in the UK never switched on its lights again. Next, consider two facts:

1. Lighting accounts for perhaps 10 per cent of the CO_2 emissions associated with the category 'lights and appliances' [5]
2. Household-related CO_2 emissions take just 27 per cent of the UK total. [6]

So, if homes went dark forever, that would lower the UK's total CO_2 emissions by a maximum of 10 per cent of 24.6 per cent of 27 per cent, or 0.6 per cent.

That's a small reward. Of course, everyone could make a bigger difference to CO_2 by never using central heating or having a hot shower ever again. But again the penalty in terms of loss of convenience far outweighs the saving in CO_2.

Consumers can do little about climate because their role in the economy is relatively modest

As late as November 2007, despite the fact that the sub-prime crisis in US housing finance was already in full swing, the Bureau of Labor Statistics (BLS), Washington, proclaimed that US consumer spending:

- rose from 64.8 to about 70 per cent of GDP from 1970-2006, and would stay at about that level over 2006-16 [7]
- would account for 2.08 per cent of the 2.8 annual percentage change in real GDP over 2006-16. [8]

However, consumer spending doesn't play the weighty economic role that the BLS would have Americans believe. The commonly used 'expenditure measure' of GDP only computes the demand for final goods and services. It includes goods and services delivered to the consumer such as energy in the home, but to avoid double counting, it excludes 'intermediate inputs' *such as the huge delivery of energy to businesses.*

Making energy for business absorbs resources just as much as making energy for consumers does, and generates both jobs and consumer incomes. So once intermediate inputs like energy for business *are* taken into account, corporate spending vastly exceeds consumer spending.

Because of its long rise, not least in housing, consumer expenditure still dominates the Western economic imagination. But after 11 September 2001, George W Bush shouldn't have looked to real estate salesmen, supermarkets and car dealerships to turn the US economy around. Even Americans don't live in a consumer society – the colossal investments of business and government make sure of that. It's upon these investments, along with those made by energy firms, that thoughtful strategy in energy supply should focus.

Consumption is responsible for about a quarter of CO_2

Homes and transport emit very few GHGs – except for CO_2. Looking at the places where that gas is actually emitted, the consumer accounts for a maximum of 39 per cent of UK emissions.

UK CO_2 emissions, Mt and percentage shares, by site [9]

	1970	1990	2006
Energy supply	260	242	221
Business	204	107	92
Industrial processes (cement, etc)	21	13	14
Public sector	24	13	10
Military aviation and shipping	4.5	5.3	2.8
Road transport	60	109	120
Residential	96	80	81
Aviation	0.7	1.2	2.3
Rail	2.7	2.2	2.2
Shipping	3.6	4.1	5.5
TOTAL	**685**	**592**	**555**

Percentage shares:	1970	1990	2006
Roads	9	18	22
Homes	14	14	15
Planes	0.1	0.2	0.4
Trains	0.4	0.4	0.4
Ships	0.5	0.6	1
TOTAL TRANSPORT AND HOME	**24**	**33**	**39**

We say a *maximum* of 39 per cent, because all UK transport and home use cannot be reduced to family life, shopping and consumer leisure. Given the freight, corporate fleets, number of people who drive or take a train to work, and number of people who work from home, the CO_2 directly caused by *consumption*, rather than by *wealth creation*, probably comprises about 25 per cent of the UK national total.

Founding father, English reaction:
the Reverend Thomas Robert Malthus

A vulgar, 2D view: scarce supply, ever more greedy consumer demand

It's time to shelve the two-dimensional view of energy that sees only scarce resource supply confronted with ever more greedy consumer demand.

Economists and sociologists have long been obsessed with consumer demand. Turning now to the most popular thinkers on the subject, we begin with the historic inspiration for Green thinking – the right-wing English country parson, Thomas Malthus (1766-1834).

In his widely read 1798 *Essay on the Principle of Population*, Malthus started from what he took as two fixed laws of human nature: the need for food, and what was termed in his day 'the passion between the sexes'. Reflecting the largely agricultural conditions of an England that nevertheless stood on the brink of rapid urbanisation, Malthus then contrasted a relative scarcity of food with population numbers growing 'unchecked'. He attacked the 'carelessness, and want of frugality' of the lower classes, all the while upholding what he called 'unproductive consumption' among landlords and capitalists. [10] Malthus also accused 'some men of the highest mental powers' of being *addicted* to the pleasures of sensual love. [11]

Writing in reaction to the French Revolution, Malthus returned to sex, 'vice' and 'moral restraint' nearly as much as to his main theme: the natural limits to human consumption.

Today, University of California geography professor Jared Diamond is no Malthus. But he does contrast a world 'already running out of resources' such as oil, and 'total world consumption, the sum of all local consumptions, which is the product of local population times the local per capita consumption rate'. [12]

Unlike Malthus, 'per capita' is how today's environmentalists like to think. They personalise energy consumption – and stigmatise it, too.

It's a horrible fate – to be an energy fatty

In 2001 British ecologists, supported by the National Federation of Women's Institutes and the Faculty of Public Health Medicine of the Royal College of Physicians, published a major attack on 'food miles'. Their conclusion:

> 'Every time we eat, we are all essentially "eating oil".' [13]

Five years later, Northern Ireland Electricity and the government-backed Energy Saving Trust began to link excessive appetite for energy with... appetite. The two bodies encouraged Northern Ireland householders to stop being 'energy obese,' adding: 'Slimming down your energy use is a sure way to save pounds... of the money variety.' [14] In 2007 the Centre for Alternative Technology, Wales, made the same breakthrough. Britain, it said, was energy obese:

> 'Far more is used than is actually required to deliver wellbeing. Years of cheap, abundant petrochemicals have led to highly wasteful practices and attitudes.' [15]

Philosophies built around energy and the body were bogus when first developed in the 19th century. [16] But today, when British Greens hatefully connect energy use with eating, they show an unprecedented disgust for their fellow human beings. The philistine Brits, they say, don't eat locally enough: instead, they nosh oil, and consume too many calories in everything they do. To make things worse, they have *the wrong attitude*.

Not to be outdone, UK health secretary Alan Johnson takes a similar line. He casts obesity as a potential crisis on the scale of climate change, and insists that Britain's new towns should be designed so that people are forced to exercise. [17]

For Whitehall, as for thin, ascetic British Greens, fat motorists are the lowest form of life. For us, by contrast, all three groups remain human beings.

Veblen and conspicuous consumption

Malthus wasn't as misanthropic as today's Greens. In old age, he thought that humanity had risen to 'eminence' and might yet 'rise higher by the same means'. [18] Later, the Norwegian-American sociologist Thorstein Veblen (1857-1929) was more cautious.

In *The Theory of the Leisure Class* (1899), Veblen famously satirised the wealthy of America's Gilded Age for their conspicuous consumption. For example, when the rich favoured certain foods, or intoxicating 'beverages and narcotics', they were, for Veblen, engaging in the 'ceremonial differentiation of the dietary' – in other words, showing off their good taste and connoisseurship. [19]

Today Tim Jackson, a top British Green, invokes Veblen, but is more pessimistic. For Jackson, the environmentally damaging pathology of consumer society is imbued in humans, because evolution has encoded, within human genes, the desire to use consumer goods to 'advertise ourselves to our competitors, to the opposite sex, to any number of our fellow human beings'. Implicitly suggesting that the rapid consumption of energy can also be a form of social and sexual selection, Jackson asks: 'What is it with young men and fast cars?'. [20]

Too many environmentalists make glib, sub-Malthus critiques of personal transport and mating patterns. One leading British environmentalist thinks there's a better way to cut CO_2 emissions than reducing food miles: reducing the flights made by men to stag parties held in Tallinn. [21]

For many environmentalists, energy use isn't so much economic or political as *biological*. And if consumption, energy use and personal transport are indeed founded on deep-seated competitive and sexual drives, you will need even more awareness of your personal culpability for climate change – and even more advice about cutting down your travel.

Consumer excess as a deep, animal urge

To say that excessive personal consumption of energy is a deep, animal urge demeans people. It suggests that man's use of energy has deflowered Mother Earth of its energy resources. And it implicates women as sinners, too. Asked by a young woman what she could do about CO_2 emissions, Sir David King, for seven years the UK government's chief scientific adviser, admonished her to 'stop admiring young men in Ferraris'. He later explained:

> 'What I was saying is that you have got to admire people who are conserving energy and not those wilfully using it... young women think it is sexy to see men driving Ferraris. That is the area where a culture change is needed.' [22]

In their quarrel with excessive personal use of energy, environmentalists are fond of *psychobabble* – and much fonder of it than their intellectual predecessors.

In Chapter 1 we told how Alfred Marshall set a value on psychological 'externalities' such as the *goodwill* surrounding a business. Veblen also played up psychology. He emphasised the *motive* of what he called 'pecuniary emulation', or the drive to amass riches so as to gain respect from others, and thus more self-esteem.

Veblen was polite about Marshall, but attacked him for his static view of how a modern economy works. For Veblen, Marshall's theories of the normal case, equilibrium, and limits which were held to apply for all time, failed to account for 'developmental sequence' in economics. [23]

Unlike Marshall and today's Greens, then, Veblen didn't just indulge in psychological speculations. He also tried to integrate *technological advance* into his political economy. [24]

That's an example that a thoughtful politics of energy supply should follow.

Hobson's focus on consumers at home and oil abroad

Both consumption at home and oil abroad obsess radical folk. When environmentalists *do* choose to broaden their critique of energy use beyond the individual consumer, they move, in impressionistic style, to scarcity of world energy resources. They think they're doing something really new here, but there are two strong precedents for their thinking – one at the turn of the last century, and one at the apogee of the Cold War.

In 1889 two left-leaning English economists, John Hobson and Alfred Mummery, mounted their own critique of inequality. Writing in an age when mass poverty was still very real, they contended that insufficient consumption was economically harmful. [25]

Then in 1902, shocked by his experiences in the Boer War, Hobson ventured a very important thesis. The disparity between high production and weak public consumption at home was the 'taproot' of imperialism and of 'militarism, war, and risky, unscrupulous diplomacy'. Britain's economy depended on the tropics for food and raw materials, while limited consumer markets at home made America's industrial and financial 'princes' in oil, among other commodities, go 'seeking investments outside their country'. Indeed the latter trend, Hobson stressed, was what was responsible for Theodore Roosevelt and 'the adoption of Imperialism as a political policy and practice by the Republican party'. [26]

Later in this chapter we'll see how the simplistic axis of home consumer spending and iffy foreign oil reappeared in American thinking during the 1960s. Here we merely note that while Hobson at least wanted the masses to be able to consume *more*, modern Greens won't grant them that privilege.

Keynes shows a special haughtiness about consumption

During the Depression, the Liberal English economist John Maynard Keynes traced his famous *General Theory* back to Malthus on consumption, and held that Hobson's *Physiology of Industry* had marked 'in a sense, an epoch in economic thought'. For Keynes, more consumer demand would bring Britain much-needed economic benefits. [27]

So did Keynes, whose largesse extended to the Bloomsbury Group and the *New Statesman*, take an equally generous line on consumption? No. Indeed, without referring to Veblen, Keynes had earlier stolen his theme of pecuniary emulation and given it a distinctly misanthropic twist.

In his *Economic Possibilities for our Grandchildren* (1930), Keynes suggested that some consumer needs were 'non-economic'. For him, needs fell into two classes:

1. Absolute – felt regardless of the situation of others
2. Relative, in that satisfying them makes people *feel superior* to others.

Keynes concluded:

> 'Needs of the second class, those which satisfy the desire for superiority, may indeed be insatiable; for the higher the general level, the higher still are they.' [28]

For all his concern to raise consumer spending, Keynes anticipated today's Green critique of energy use. Condescendingly, he dismissed some needs as competitive and therefore insatiable – even if he wasn't Green enough to describe them as *selfish and illegitimate*.

Always satisfying the 'desire for superiority' in his personal life, Keynes forgot that *the simple effort to better oneself in relation to one's peers isn't wrong.* Like fire, competition has its dangers – but insofar as it drives people to aspire to more, it has its benefits as well.

Keynes also forgot that insatiable desire, like insatiable curiosity, isn't wrong either. The desire to achieve, aided by copious amounts of energy, is entirely human – and entirely commendable.

Maslow: mankind as needy, but also in search of knowledge, truth and wisdom

The Keynesian focus on popular consumption was buttressed during the privations of the Second World War. In 1943 the American psychologist Abraham Maslow codified the distinction between the basic human need for food and safety, and the higher, relationship-based needs for love, esteem and self-actualisation. [29] Today, no human resources or marketing slideshow comes without a pyramid diagram of Maslow's famous *hierarchy of human needs*.

Wrongly, Maslow thought man was a 'perpetually wanting animal.' Yet he was right in his conclusion that any theory of human motivation should be 'human-centred rather than animal-centred'. Maslow postulated a human 'desire to know, to understand, to systematise, to organise, to analyse, to look for relations and meanings'. [30]

For him, people weren't just consumers with cravings, but also *active protagonists* shaping the world. He'd see a 21st century not just of energy users with needs, but also of *energy scientists* and *technologists with talents*. How sad that over the decades, millions of marketing buffs have invoked Maslow

---

Inspiration to Obama: John Maynard Keynes.
'To dig holes in the ground', Keynes said figuratively,
'will increase, not only employment, but the real
national dividend of useful goods and services'
(*The General Theory*, Chapter 16). Maybe; but
Obama's programme to help the US private sector
create five million new green jobs could amount
to make-work more than efficient energy creation

without reading him, thus preparing a culture in which people are apprehended primarily as needy consumers!

Maslow didn't mention energy when he discussed basic needs. But for Canadian energy economist Peter Tertzakian, he should have included energy alongside food, water and shelter as a 'primary need'. [31]

In fact, Maslow's basic needs are ones that are immediately felt – needs that our earliest ancestors would recognise. Energy isn't like that: the modern idea of it only emerged in the 19th century. When individuals use energy today, they generally want convenience more than they want to show off. Their needs are both basic and 'higher', quite legitimately insatiable, and definitely *not* worth worrying about all the time.

Galbraith and the doctrine of consumer dependence

Today, environmentalists see people as unconscious, needy consumers who are also unhealthily dependent on energy. This condescension is nothing new: in fact, it dates back to the height of the Cold War.

In *The Affluent Society* (1958), another patrician economist, John Kenneth Galbraith, convinced millions of Americans that many consumer needs had come to be contrived, and were therefore not really urgent. First, following Veblen and Keynes, one man's consumption had become 'his neighbour's wish' – pecuniary emulation again. Second, in their quest for more and more production, corporations *created* wants, through 'advertising and salesmanship'. In what Galbraith damned as the Dependence Effect, many wants depended on the very production that satisfied them. So if those wants were bizarre, frivolous or immoral, no case at all could be made for them. [32]

Galbraith's analysis was very partial. Although 'keeping up with the Joneses' did go on in his day, the desire for convenience also informed Americans' purchases.

Galbraith was right in asserting that capitalism creates new and surprising wants. Indeed, it's partly through that process that it has brought about a certain amount of progress.

In the past, even Karl Marx had celebrated the creation of new wants as a dynamic aspect of capitalism when compared with feudalism. [33] But now Galbraith, a Democrat critic of the post-war boom and of growth itself, said that new wants *indicated consumer excess and dependency*.

As the London-based economics writer Daniel Ben-Ami notes, it took time for Galbraith's arguments to enter the mainstream. [34] Yet it wasn't long before another US bestseller made energy use a special target for vitriol.

Packard: industry makes consumers wasteful – and America dependent

In 1960, the muckraking American journalist Vance Packard was already famous for his assault on US advertising and its manipulation of consumers. [35] But in *The Wastemakers*, published that same year, Packard tipped his hat to Veblen and Galbraith and redoubled his attack, indicting industry for promoting *wasteful consumer behaviour* in order to sell its 'ever-mounting stockpiles' of products. [36]

In a key chapter titled 'The Vanishing Resources', Packard also developed themes first set out by Hobson:

* the US economy had become 'more vulnerable to a cut-off' because of its growing dependency on foreign raw materials
* in domestic oil, production would be 'peaking out' sooner or later
* America's growing need for foreign oil would put it 'deep into the hands of Arabian and Latin American politicians.' [37]

From Hobson around 1900, through to Packard in the Cold War, industry's reliance on consumer spending, America's lack of energy security and its resort to imported oil gradually emerged as the mud to throw at rich corporations, rich oil sheikhs and general inequality. Yet this radicalism, whose contemporary ideas on oil we will again consider in Chapter 5, was always too flimsy a platform upon which to build a thoughtful energy politics.

So long as political economy remains populist and focuses, mistakenly, on consumption, what begins as a right-on attack on Big Oil ends in *conservationist and conservative disdain for the lifestyles of human beings.* While ordinary men and women have better things to do than spend each day fretting about their pollution of the planet, environmentalists now denigrate them as guilty of the ultimate kind of ignorance, thoughtlessness and dependence: of being *consumers who are addicted to oil.*

A stupid but unquestioned metaphor: addiction to energy

After 9/11, *The Economist* denounced what it called the world's 'dangerous addiction' to oil. Later, the Californian environmentalist Richard Heinberg wrote that America, 'an energy-addicted society', would find it 'hard to wean itself from the habit.' In 2007 Thomas Friedman, a Pulitzer prizewinning writer, made two further and equally astonishing discoveries: the Soviet Union had died, in part, because of its oil 'habit'; and in the future Iran could succumb to 'the same disease'. [38]

The metaphor of addiction dominates today's critique of energy use. It

- steers debate away from energy supply
- makes people feel guilty about what they do with energy
- helps Greens order you to cut back on energy.

Yet nobody has a compulsion to inject oil. Regularly visiting your grandmother by car cannot be compared with heroin use. Indeed, given these facts, the popularity of the addiction metaphor deserves some historical unravelling.

The early 1970s was when the idea of addiction to energy first gained public credence, being popularised by the former priest Ivan Illich (1926-2002). Today, Illich is a largely forgotten figure; yet in 2006 George W Bush, in his State of the Union address, proclaimed America addicted to oil. In June 2008

Gordon Brown took up the same refrain. Finally, in August 2008, Barack Obama denounced America's addiction to foreign oil as a threat responsible for high petrol prices, redundancies, Middle Eastern terror, and 'the rising oceans and record drought and spreading famine that could engulf our planet'. Indeed, Obama felt that addiction to oil went to 'the heart of what we are as a nation, and who we will be'. [39]

We turn now to look at how, over 30 years, the concept of energy addiction moved from the pen of an early Green crank to the mouths of world leaders.

Environmentalism's climacteric: the early 1970s

Toward the end of the Vietnam War, Western states encountered recession and a loss of popular legitimacy. Rising private consumption appeared to undermine the Protestant work ethic, causing the state to try and modernise itself by finding new codes of conduct for people. In the process, the state did not dissent from the growing view that a dangerous *moral depletion* of society brought about by *unbridled consumerism* had also resulted in a dangerous depletion of natural resources. [40]

As the post-war boom drew to a close, Western governments tried to shore up their authority by outsourcing some parts of public policy to a wider group of 'experts' – especially experts in the environment. In 1970 Edward Heath's Conservative government attached a new and powerful Central Policy Review Staff to the Cabinet, with the job of independently assessing policy. In America it was a similar story. President Nixon established the US Environmental Protection Agency (1970), and saw the Endangered Species Act (1973) pass into legislation. Meanwhile in 1972, the UN set up the UN Environment Program (UNEP).

As the Green consultants SustainAbility usefully note, the first 'wave' of environmentalism peaked between 1969 and 1973. [41] Yet this wave was less a mass movement, and more a sub-elite's *concern about limits*. Modern environmentalism began as anxiety that the state, wanting to stabilise politics in the face of considerable unrest, had every interest in sponsoring. And once the diffusion of governmental authority toward experts began, environmentalists who held personal energy use to be dependency – an addiction contrary to *limits imposed by nature* – found they could gain a hearing. This was particularly true during and after 1973-4, when Arab oil producers ran an embargo on oil exports to the West.

Responding to the 1973-4 energy crisis, Heath's Cabinet considered that petrol rationing would provide few benefits. Nonetheless, it would 'have a marked effect on public opinion, and would underline the gravity of the crisis.' [42] In the event, Heath opted for 'Save it!': a propaganda drive pressing millions to recognise that personal energy conservation was a public duty. In the US, too, the Advertising Council ran a nationwide campaign on energy in 1975, with the pun-tastic tagline 'Don't be Fuelish'.

In the 1960s, two historians point out, the New Left in American had attacked consumption as 'seduction, a form of captivity'. [43] Then, in 1972, the Club of Rome's computer model of what it called the future 'predicament' of mankind, *The Limits to Growth*, was published. So was Victor Papanek's art school bestseller, *Design for the Real World*. In 1973 Ernst Schumacher's *Small is Beautiful* also gave Green politics a mass readership. [44] Yet the legitimacy crisis of the state, government attempts to counteract this, and the upset over oil together did more than Reds or Greens to institutionalise the idea of cutting back on home heating and road transport.

Since the stagflation of the early 1970s, the passing of the Cold War visions of Left and Right has opened up still more space for the state to characterise personal energy use as the eighth deadly sin: addiction. In putting the accent on personal use, the state merely deepens the tendency to outsource responsibility beyond itself.

With indecision the hallmark of many Western governments, playing up personal energy use absolves the modern state from making tough and costly choices in energy supply.

Green misanthropes (2): Ivan Illich casts driving as an addiction, and praises the 'psychic powers' of human feet

In his anti-car tirade *Energy and Equity* (1974), Ivan Illich argued that human beings were energy addicts. 'Beyond a certain threshold,' he asserted, 'mechanical power corrupts.' He continued:

> 'Even if nonpolluting power were feasible and abundant, the use of energy on a massive scale acts on society like a drug that is physically harmless but psychically enslaving. A community can choose between Methadone and "cold turkey" – between maintaining its addiction to alien energy and kicking it in painful cramps – but no society can have a population that is hooked…' [45]

The 'habitual passenger,' Illich added, was addicted to being carried along, and had 'lost control over the physical, social, and psychic powers that reside in man's feet'. A worldwide class structure of 'speed capitalists' had emerged, ensuring that more energy meant less equity. Indeed, once public transport offered speeds beyond 15mph, even non-automotive power added to inequality. By contrast, people on their feet were 'more or less equal'. [46]

For Illich it wasn't history, economics or politics that brought about a gap between rich and poor. Rather, technology, mechanical power and excessive energy created a distasteful technocracy and a reckless jet set. But just so the poor knew their place, the post-industrial, low energy, high equity society favoured by Illich would also be *labour-intensive*. [47]

For all his disgust with inequality, Illich had total contempt for his fellow man. Turning Veblen's pecuniary emulation to perverse ends, he announced: 'In a consumer society there are inevitably two kinds of slaves: the prisoners of addiction and the prisoners of envy.' Illich's whole purpose was not liberation, but rather 'a political process that associates the community in the search for limits'. [48]

Illegitimate arguments, subterranean influence:
Ivan Illich. Born in Vienna, Illich became a Catholic
priest in New York in the 1950s, as well as a virulent
critic of consumption. The opening lines of his essay
Energy and Equity, published just before the 1973/4
oil crisis, turned cutting back on energy into a moral
virtue. 'High quanta of energy,' Illich wrote, 'degrade
social relations just as inevitably as they destroy the
physical milieu.'

Said with a sneer: technology as a 'fix'

When Greens use the metaphor of addiction, they add the sneer that this or that energy innovation is just a 'technical fix'. In Chapter 1 we argued that Greens want people to change more than the energy supply. In this chapter we argue that the addiction metaphor lets Greens present technological innovations as being as stupid and dangerous as a heroin user's needle.

In a report on geo-engineering, or solutions to climate change that are conducted on a planetary scale, the London *Observer* calls them 'the ultimate technological fixes'. It continues:

> 'Opponents to such schemes point out that it is technology that got mankind in its current fix. An even bigger dose of technology is therefore the last thing the planet needs.'[49]

So technology comes in doses, and further doses will just make things worse.

As the journal *Nature* points out, critics of geo-engineering say that it is:

> '... a way to feed society's addiction to fossil fuels. "It's like a junkie figuring out new ways of stealing from his children", says Meinrat Andreae, an atmospheric scientist at the Max Planck Institute for Chemistry in Mainz, Germany.'[50]

For many Greens, the resort to ambitious technologies reveals the compulsive thoughtlessness of an addict. Human beings must kick two habits: that of using too much energy – especially oil – and that of looking to technology, rather than their own behaviour, as a way out of the mess they've made. They must leave the high-tech, cheap-energy party, and go on a low-tech, character-forming detox programme.

The metaphor of energy addiction has a strongly therapeutic character. Indeed, wanting technology to mend the world's energy supply and its climate is widely held as *a mark of mental illness*.

In her fascinating collection *The Technological Fix* (2004), Lisa Rosner, of the Richard Stockton College of New Jersey, suggests that the US nuclear physicist Alvin Weinberg, director of Oak Ridge National Laboratory, Tennessee, was first to advocate 'cheap technological fixes' as solutions to social problems in his 1967 book *Reflections on Big Science*. But by 1970, Rosner notes, the tide was turning. [51] Britain's Michael Gibbons, later an international doyen of science policy, acclaimed a book by René Dubos, presciently observing:

> '... alas, there is much work to be done before the ecological thinking of Dubos permeates the intellectual structures of Western society and becomes an effective alternative to the one dimensional "technological fixes" that society has so far provided to solve its problems.' [52]

Today that work has indeed been done. As an essay in *Technological Fix* on artificial hearts says, 'in a society less enthralled today with technological fixes than a generation ago', *healing*, not 'a mechanical response to a biological set of problems,' is felt to be the right way to deal with difficulties. [53] Indeed the healing, 'change your mindset' approach is what governments and Greens recommend for energy.

'Technological fixes,' the distinguished American historian of science Thomas Hughes sums up, 'leave us in a fix.' [54] What's missed here is the therapeutic way in which 'fix' is used to hint that advocates of technology are themselves *addictive personalities* – part of the problem, not part of the solution.

'Technological fix' is a charge made against intra-uterine devices, against the 2003 US invasion of Iraq and – significantly – against nuclear power. [55] But in energy as elsewhere, objections are not so much to this or that innovation, as to the whole idea of human endeavour.

The psychobabble approach

Mankind isn't addicted to hydrocarbons. The need for more energy is about human progress, not beastly longings. Nowadays, however, all kinds of activities are portrayed as addictive.

In the early 1980s, health insurance schemes started to cover US employees for addiction, while the media, therapy lobbyists and bestsellers in self-help began to treat gambling, shopping and sex as addictions. [56] By 1989, too, information technology (IT) was being classed as addictive: social inadequates, it was said, used IT too much. In 2001 it was claimed that the computer game EverQuest was addictive. Nowadays,

- the use of Facebook at work is compared with crack cocaine
- IT-based forms of gambling, shopping and sex are thought to be especially habit-forming
- Greens indict Britain's rulers as 'addicted to road building.' [57]

Today, cheap phrases stolen from psychology have become the main way in which politics, society, and especially *consumption* are understood. Tabloid newspapers and celebrity magazines are replete with addictive personalities, revealing their tendencies toward *denial* that they have a problem. That's one reason why anyone who cavils at environmentalism's authoritarian programme of social reform around personal energy use is denounced as a climate change 'denier.'

In fact, people are no more habituated to energy than they are to oxygen. Dictionaries merely define addiction as the persistent, compulsive use of a substance known to be harmful; but in popular parlance 'addiction' to energy suggests a *physiological* tie that's not just personally harmful, but also *socially reprehensible*. The tag is insulting.

Green critics of energy use love talking 'addiction': that way, they can tap into today's mainstream psychobabble. But they're also keen on *adding things up*. Nowhere are these two affections clearer than in their ideas around energy and *happiness*.

Happiness and the legacy of Jeremy Bentham

For all their dislike of consumer society, many environmentalists still take it as their starting premise on energy. They've plenty to say about energy corporations, government policy, and companies keeping their lights and PCs on after hours. But they go even further than this, and focus above all on consumption.

Greens are quite right about two things:

1. By itself, consumption brings little meaning to life
2. Alfred Marshall's neo-classical economics – which portrays consumers as always making rational decisions – doesn't stand up.

Nevertheless, because their analysis begins from consumption, environmentalists share the myopia of the free-market economists they despise. In particular, their view that more energy doesn't guarantee more happiness misses the point.

Despite recent economic growth, experts observe that 'life satisfaction has been kind of flat'. [58] In the UK, Strathclyde University professor Michael Common and Sussex University senior research fellow Sigrid Stagl pursue this idea in the realm of energy. They form their *alternative to GDP* by dividing

• Number of Happy Life Years – the product of average Happiness and Life Expectancy

by

• E, or energy consumption per head.

Trying to *measure happiness* as accurately as they do tonnes of oil equivalent consumed, the authors conclude from international data that developing countries generally beat developed ones in getting a lot of happy life years out of modest energy use. [59]

Number-cruncher: Jeremy Bentham. His *Introduction to the Principles of Morals and Legislation* (1791) tried to bring accountancy to ethics

In fact, as many critics of the Greens have said, to argue that economic growth and growing energy use in the West have failed to bring greater happiness is to take the correlation of two trends as causation.

That's a mistake. The purchase of more fuel does not lead to mental depression.

For Hamburg University researcher Dr Katrin Rehfanz and Southern Denmark University professor David Maddison, it's not energy but *climate* that explains 'differences in self-reported subjective well-being'. Drawing again on international data, the two authors daringly suggest that:

- people would prefer higher mean temperatures in their year's coldest month, and lower temperatures in its hottest
- global warming might improve winters in the North, but might make the those living in the tropics less happy. [60]

In corporations and government, bean counting and the target mentality have long been a way of life. [61] Green accountancy exercises, though, add up happiness, reducing temperament to temperature. Here there's no trace of the world's need for lots more energy, energy innovations, and convenience. Instead, energy use is ridiculed as bringing about unhappiness.

In adding up happiness, environmentalists revive the English utilitarian, Jeremy Bentham (1748-1832).

Like Malthus starting from the need for food and sex, Bentham began with nature, which he believed had 'placed mankind under the governance of two sovereign masters': pain and pleasure. To 'take an exact account' of the moral worth of any act, Bentham proposed summing up 'the values of all the pleasures on the one side, and those of all the pains on the other'. He also wanted to sum up 'the *number* of persons whose interests appear to be concerned'. [62]

Although Greens do like media stunts as a means of pressuring politicians, they generally prefer to view democratic action on climate change in Benthamite terms. For them, such action isn't about rallying behind a programme for more and better energy supply, but rather about adding up the sum total of ritualistic, moral and happiness-inducing energy conservations made by individual consumers.

A metaphor even sillier than energy addiction: your carbon footprint

The key way in which Greens push their diminished conception of democracy is to use a metaphor even sillier than energy addiction: that of your personal carbon footprint.

Right away, this second metaphor portrays human beings not as active and sociable, but as *clumsy consumers engaged in contaminating an otherwise pristine planet*. Yet it's the metaphor that's clumsy, not human beings. CO_2 emissions, after all, go upwards to the atmosphere, and don't relate to walking on soil.

So why *foot*print? In fact the metaphor of the carbon footprint gains its weird, upside-down incongruity from an earlier, land-orientated metaphor: that of the *ecological* footprint.

Between 1990 and 1994, University of British Columbia professor William Rees, a doctor in population ecology, supervised a PhD thesis by a Swiss, Mathis Wackernagel. Through a jointly published book in 1996 and later through Wackernagel's Global Footprint Network (GFN), they updated the idea of the *carrying capacity* of land, or its capacity to carry more or fewer people.

That Malthusian idea was first used by the colonial authorities in Northern Rhodesia to speculate about and warn against future population growth among black Africans – a particular concern among white settlers. [63] Now, with ecological footprints, GFN experts popularised carrying capacity by turning it into what they called a 'research and accounting question' about the past. For any city, the GFN says, an ecological footprint is how much land and water area it requires 'to produce the resources it consumes and to absorb its wastes'. [64]

Not content with computing ecological footprints in the manner of Bentham adding up pain, the GFN calculated that

in 2007, the world went into *ecological overshoot* on 6 October, when its demands on cropland, pasture, forests and fisheries exceeded the ability of these ecosystems to generate resources and deal with humanity's detritus by the end of that year. Over time, the GFN adds, ecological overshoots accumulate to create a *global ecological debt*, which in turn is bequeathed to future generations. [65]

Ecological footprints, overshoots, debts: what is forgotten in these pseudoscientific catchphrases is that to the extent that there's enough technological progress around to raise the efficiency of energy supply and lower its cost, so the consumption of energy is likely to rise. Indeed, that's one reason why the demand for energy is insatiable, and also why such demand can so easily be derided as an addiction. Yet people need more energy simply to do what they want to do – and the amount of energy available to the world faces, in principle, no limits.

The ecological footprint is a ghastly entry in the Green accountant's ledger book of past crimes. And that's true of the carbon footprint, too.

A recent and sympathetic review of the concept concedes that it covers everything from direct on-site emissions of CO_2, to all the GHGs emitted not just on-site, but also in upstream production processes. The units for measuring carbon footprints are unclear and, despite the metaphor's ubiquitous deployment:

> 'There is an apparent lack of academic definitions of what exactly a "carbon footprint" is meant to be. The scientific literature is surprisingly void of clarifications...' [66]

The carbon footprint, then, is a metaphor without merit. It distracts the mind from energy supply and energy innovation.

Carbon footprints as a Green 'Bootprint' on the brain

Adding up a footprint can only be a vague exercise, because the metaphor has a primarily moral intent. That's why you constantly have to undergo the following interrogations:

- Has *all* of your CO_2 been included in the scales of justice?
- If you've bought local food, you've saved food miles – but what kind of cooking are you doing with it?
- How much CO_2 does your Toyota Prius really emit in use, and how much in manufacture and disposal?

You could spend a lifetime accounting for the CO_2 in your every breath, but in practice you couldn't add it up. More importantly, the carbon footprint idea strips each individual and social activity of its merits and dissolves all goals into one: add up your carbon impact and reduce it.

In this scheme the value of regularly visiting your ailing grandmother by car is of course not comparable with the value of flying an artificial heart to save someone's life. Such things can never be quantified. All that is given a bogus quantification is Bentham's 'pains' and Marshall's externalities.

From the Manifesto Club, a campaigning network of humanists, Josie Appleton notes that, in ethical terms, people are now judged by the trail they leave behind. The question is not what their activity adds to the world in human terms, only the resources it takes away. [67]

Hating the consumer, but still taking consumption as its Alpha and Omega, environmentalism is forced to ignore all the great things people can do, the productive lives they can lead, and the impact they can have as citizens banding together around a new politics of energy supply.

Even more lamentable is just how many people buy into the idea of carbon footprints. Altogether, the footprint idea acts as a Green 'Bootprint' on the brain. Under its influence, dozens of physical, conversational and mental rituals around carbon footprints have emerged.

Designed to make one feel good, these rituals turn out to be exhausting exercises in *not being good enough*.

With carbon footprints, as with religion, we must know right from wrong, publicly admit to sin, do our bit, obey priests, and conform to sacred texts. But for all its similarities with religion, environmentalism does differ from it. While followers of religion at least want humanity to *transcend* the here and now, Greens want people merely to *react* against it.

But human use of energy isn't the hateful tread of a dinosaur, leaving its grubby but unwitting mark everywhere it goes. Individuals don't plague the Earth with soot. People are right to 'colonise' the planet's energy resources, because in doing so they're just trying to go about their business.

But many people, if by no means all, are also at risk of having their minds colonised.

In 1936, in the concluding paragraph of his *General Theory,* Keynes famously proclaimed:

> 'Practical men, who believe themselves to be quite exempt from any intellectual influences, are usually the slaves of some defunct economist.' [68]

Wrong about economics, Keynes was right to underline the long-term influence of ideas on society. Indeed, powerful ideas issued from *beyond* the domain of some defunct economists can have an enormous contemporary effect, given the right social conditions.

The 21st century isn't, as Illich said in 1972, enslaved by consumption as addiction and as envy.

But for as long as millions of people see themselves as consumers with a carbon footprint, they will indeed be slaves of a defunct idea.

Summing up this chapter

When pressed, many might agree that consumer actions have little effect on climate. They might also agree that talk of energy addiction, technical fixes and carbon footprints has gone too far. After all, alarmist popular literature on contamination, such as *High-Tech Holocaust* and *The Coming Plague*, long predates today's Lady Macbeth-like desire to expunge our CO_2. [69]

Yet many, too, would argue that contamination exists – and that, more importantly, you've got to start somewhere: with your home, your car, whatever. Some might add that, although consumer conservations of energy are mostly a ritual, you should still show you care about the planet. Finally, in the intervals between national elections and, thus, changes in government policy on energy supply, the daily chance to make even a tiny difference to the world appears valuable to many.

In fact, however, to continue with Edward Heath's Save It! and Jimmy Carter's sweater campaigns into the 21st century is to:

- accept personal culpability for climate change
- absolve the political classes of the responsibility of organising a bigger and better energy supply – before, during and after elections
- beckon the state to take an even greater interest in our private lives than it does already.

To conserve energy *and* perhaps save money in your personal life is all very well. But it isn't really an 'and' on top of improved energy supply. In national and international politics, it works out as an 'or' – a deluded alternative to a rational policy.

What, in energy, people need to add to is not the sum total of consumer conservations, but the sum total of *human knowledge and power*.

Altogether, the Green critique of energy consumption is an indictment of humanity – and especially of American men. But is that repulsive species really still in a macho 1950s time warp, where each lords it over his neighbours because of how many cylinder heads he has underneath his bonnet? Do American men really need rehab from their addiction to cars? Americans, it's well known, take short holidays and suffer from stagnant incomes; so are they really just leisurely junkies, always driving to places they've no business driving to?

In energy as elsewhere, Americans get a bad rap. But British Greens subject Britons, too, to outrageous charges. In 2004 the government-funded Sustainable Development Commission (SDC) warned then Prime Minister Tony Blair that 'unsustainable' economic growth – the kind that led to 'substantial' increases in GHGs – was as undesirable as crime, drug trafficking, sexual exploitation, and pornography. [70]

It's a poor social solidarity that Greens offer: because everyone shares a need to consume, everyone can come together… in feeling guilty about their crimes against the planet.

What the SDC forgot to mention in its repudiation of 'unsustainable' economic growth was rather important. *Who decides* what is sustainable and what is not?

In this chapter, we have seen that, for environmentalists:

- democracy has become the adding up of millions of feelgood personal energy conservations
- politics has become the adding up of your personal carbon footprint.

There is, however, one other abacus that Greens like to brandish. They claim that:

- science – especially climate science – is the adding up of worthy scientists' opinions into a consensus that can and should brook no denial.

Let's now turn to the issue of climate change.

Climate change: No need to panic

Science now dictates that the state gets tough about our profligate ways in energy, right? Wrong.

Politicians and celebrities

want consumers to conserve energy now. This demands no investment, and boosts self-satisfaction amongst *carbonistas*. Unlike the building of a new energy supply, this is also a measure that has immediate effect.

But just how urgent is the situation? How, to what extent, and how fast is mankind changing the climate?

Many factors drive climate in complex interaction, which is what makes climate science so complicated. However, simple concepts explain the key driving forces; and – whatever climate sceptics say – the biggest driving force for *change* in the climate turns out to be *mankind*.

The good news? Whatever Greens tell us, we believe that *the human origins of climate change form a basis for optimism, not breast-beating.*

Yes, mankind unwittingly made a mistake. It wasn't the first time and it won't be the last. But immediate atonement through lifestyle change is the wrong solution.

Without increasing energy, society will face real problems. But with new energy supply, there's also a chance to tackle climate change at source.

Overturning the personal habits that have evolved with modernity is different. Conserving energy and lowering one's personal carbon footprint aren't just ineffective: they're also by no means dictated by the real pace of climate change.

The future of the planet is not at stake in the next few years. There's time to make big changes in energy supply – and there's no need to panic.

This chapter first distinguishes climate from weather. It then surveys global warming, global cooling and – a key concept – radiative forcing.

That the origins of climate change are man-made, we suggest, doesn't make Greens right to moralise about man's past 'misdeeds'. In fact, humanity's continuing domination of the natural world means that it should be able to solve climate with new feats of achievement.

Much is certain about climate, but it's the *unknown* that rules many minds. The Precautionary Principle suggests that Anything Could Happen At Any Moment.

In fact, climate is unlikely to make quantum leaps in ferocity. Feelings that it might just do that, we show later, owe much to the Second World War and the Cold War.

While environmentalists play up mankind's intrinsic *uncertainty* about nature, they always stress the absolute, finished quality of scientific consensus on the *certainty* of climate danger. When it isn't being dumbed down, then, climate science is falsely elevated into the *New Scientism*: a technocratic and unanswerable demand for people to change their behaviour in a conservative direction.

The conditions surrounding the discovery of global warming in 1956-7 memorably contrast with those of today. The discovery was a tribute both to R&D, and to the role of chance in R&D. In the mid 1960s, however, the view grew that climate could be fundamentally indecipherable. By the time the IPCC was born in 1988, the zeitgeist in the West was not about R&D, but a growing politics of anxiety.

Those politics *predate* the end of the Cold War and 11 September 2001. Today, they demand harsh consumer penance. More broadly, the policy of *insuring future generations* is used to play down innovation in energy supply, and instead uphold more national and international state regulation.

Climate is what you expect; weather, what you get

According to the UK Meteorological Office, southern England experienced, in the 29 Novembers between 1971 and 2000, an average maximum temperature of 10.0° C, an average of 5.4 days of air frost, and an average of 77.4mm of rainfall. [1]

That's climate – a long-term average.

For England as a whole, the November of 2007 differed from the average November of 1971-2000. The Met Office reported that England was more anti-cyclonic during that particular month than during a 'normal' November, and that it also saw a couple of notable unsettled periods either side of mid-month. Temperatures soared to 18.8° C at Wiggonholt, Sussex, on the first day of the month, and reached 18.2° C at Portland, Dorset, on the second. Although England had some chilly nights in the first week, conditions were generally too mild for frost. [2]

That's weather – the circumstances that obtain at a particular place and moment in time.

Climate sceptics say that, since it's hard to make reliable forecasts of the weather one week ahead, nobody can predict

A whole lot of energy: Hurricane Katrina makes its second and most damaging landfall, southeast Louisiana, 29 August 2005

how climate might change in 100 years. Yet they miss the point. Hard though it is to make a long-term climate forecast, it's an easier exercise than making a long-term weather forecast. For example, it's pretty certain that January months in the Northern hemisphere over the period 2060-2070 will on average be colder than August months. But it's not at all certain what the temperature will be in London on 3 January 2065.

Climate science is about looking at long-term driving forces. If the atmosphere traps more heat, then, other things being constant, there will be a long-term rise in the average temperature. Just how and where the rise emerges – that's weather.

Journalists err when they state that climate change causes a particular weather incident such as Hurricane Katrina (2005). The long-term driving forces of climate change only make such events *more or less likely*.

What are these forces? Energy from the sun drives both climate and weather. Sunlight heats the Earth, but doesn't illuminate it uniformly. Sunlight is more intense at the equator, and disappears at night. It's absorbed mostly at the surface of the Earth, rather than higher in the atmosphere.

These imbalances drive the weather. Wind and water carry massive amounts of heat toward the frozen poles, forming gigantic eddies and swirls as warm and cold fronts collide, tumble over mountains, and move from sea to land. A famous example of a major current is the Gulf Stream, which carries heat from the Gulf of Mexico toward Western Europe.

The sun's heat is redistributed over the Earth, but the never-ending input of energy must ultimately be balanced by an output of energy from the Earth back into space. The Earth gets rid of its energy in the same way as the sun: it glows with heat. The difference is that the Earth is cooler than the sun, and so glows not with visible light, but with radiation at the infrared end of the spectrum.

In principle the Earth should radiate enough infrared light to balance the input of solar heat, and in this way the sun would regulate its temperature. That's the big picture. But to understand climate change, and how humans are influencing it, we need to go a lot deeper.

Warming, cooling – and radiative forcing

Two key natural effects, both modified by human activities, now need considering. The Greenhouse Effect warms the Earth, while the Earth's less publicised *albedo* cools it.

Though the atmosphere is transparent to sunlight, it contains GHGs which absorb the infrared light that the Earth radiates outward. To return to radiative equilibrium, therefore, the Earth has to give off more infrared rays for enough of them to escape into space; and that can only happen if it heats up. The higher the concentration of GHGs, the more they trap infrared radiation, and the more the Earth warms.

The most important GHGs – including water vapour, CO_2, methane, nitrous oxide and ozone – occur naturally. Without them, the Earth would be about 30° C cooler. The big concern over climate is that human activities will raise the concentration of these gases, and especially CO_2, enough to boost warming by several degrees.

So what about albedo? This is a percentage measure of how reflective the Earth is. Where it's very reflective, for example over its white ice sheets, most sunlight isn't absorbed, but rather reflected straight back out into space. In such areas, the Earth isn't heated up and so isn't made to re-emit incoming light as infrared.

By contrast, areas of low albedo, such as dark green forests, absorb almost all the sunlight that falls on them, heating the Earth.

Changes in humanity's use of land alter albedo, which today stands at about 30 per cent. Dark tarmac reduces albedo, while replacing dark forest with cropland often increases it. Through mechanisms not yet fully understood, human activity can also change albedo through influencing cloudiness, which is an important source of the Earth's overall reflectivity.

To quantify how both human and natural forces have changed climate, scientists use the concept of *radiative forcing* – the difference between the energy leaving the Earth and that entering it. A Christmas tree fairy light emits 1 W, but a radiative forcing of just a few watts per square meter over every part of the Earth, land and sea, 24/7, all-year-round, could warm it by several degrees.

Exactly by how much the Earth will warm, and how fast, depends on factors harder to calculate than the forcing itself. How radiative forcings translate into temperature is summed up – as we saw in Chapter 1 – in a number known as *climate sensitivity*. We'll come back to this question.

The chart overleaf is adapted from the IPCC's 2007 assessment. It shows the change, from 1750-2005, in the radiative forcings associated with different contributors to climate change. [3] To the IPCC's credit, it also highlights the *status* of different aspects of climate science.

The forcings are collected from many observations, and interpreted using several different theories. *Both observations and theories are associated, as always in science, with a changing mix of certainty and uncertainty.*

Uncertainty, certainty and the level of scientific understanding

As science progresses, uncertainty and imprecision tend to give way to certainty and growing precision – although relapses can often follow unexpected discoveries. Thus, the chart's horizontal *error bars* show the IPCC's best assessment of the values within which it's 90 per cent confident that the true value lies. That is, on each error bar, the IPCC claims that there's only a five per cent chance that the true value lies to the left of the bar, and only a five per cent chance that it's to the right. In addition, different aspects of science progress at different speeds. To register this, the IPCC rates, in the final column of the chart, the level of scientific understanding that's so far been reached in relation to each contributor to radiative forcing.

Radiative forcing of climate between 1750 and 2005

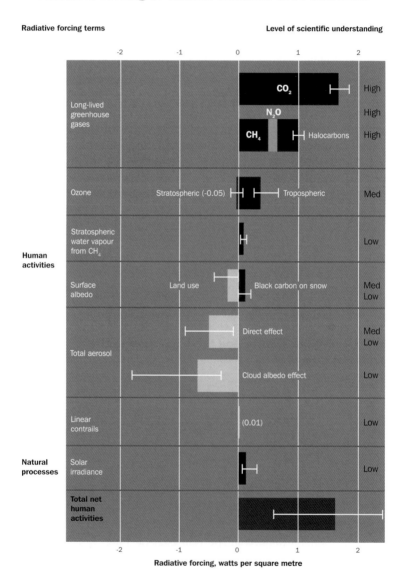

Radiative forcing terms

Level of scientific understanding

Human activities

Long-lived greenhouse gases — CO₂, N₂O, CH₄, Halocarbons — High, High, High

Ozone — Stratospheric (-0.05), Tropospheric — Med

Stratospheric water vapour from CH₄ — Low

Surface albedo — Land use, Black carbon on snow — Med, Low

Total aerosol — Direct effect (Med Low), Cloud albedo effect (Low)

Linear contrails — (0.01) — Low

Natural processes

Solar irradiance — Low

Total net human activities

Radiative forcing, watts per square metre

CO_2 main man-made component has been from the burning of fossil fuels. There are other significant sources, such as the cement industry

N_2O main man-made component has been from the use of fertilisers in agriculture

CH_4 main man-made components have been from paddy fields, landfill, ruminant animals such as cows, and leaks surrounding the use of natural gas.

Halocarbons – (F-gases) are almost entirely man-made, principally for refrigeration. Now regulated, to lower damage to the layer of ozone that's to be found in the Earth's stratosphere

O_3 ozone, a relatively unusual form of oxygen, is also a greenhouse gas. Halocarbons have destroyed enough stratospheric ozone – ozone higher than about 10km – to bring about a small fall in forcing. But the fall has been more than outweighed by a rise in the ozone that cars and other machines generate at ground level. This is known as tropospheric ozone

Stratospheric water vapour – when aircraft put out water in the Earth's stratosphere, which is normally extremely dry, they create a disproportionate warming effect

Surface albedo – soot from industrial activities such as burning coal has blackened the surface of snow, and so led to warming. Changes in the use to which humans have put the land have increased albedo, and so led to cooling

Total aerosol effects – burning coal, wood or dung causes the emission of small particles. In what is called the direct effect, these particles reflect sunlight, increase the Earth's albedo, and so have significantly cooled it.

Since the small particles in aerosols form points around which water can condense into droplets, aerosols also increase cloudiness, and so make a second contribution to albedo – the indirect, cloud albedo effect

Linear contrails – the trails of cloud that can be seen behind aircraft engines have a slight net warming effect, reflecting more light back down to Earth than up into space

Solar irradiance – although sunlight is by far the most important factor in determining the temperature of the Earth, it has only changed by a small amount. This is a long-term effect, on top of the 11-year 'solar cycle', over which there is a regular rise and fall in the sunlight reaching the Earth, but no net change

Given these significant reservations about certainty and the level of scientific understanding, what then does the chart convey?

It's pretty certain that climate change is man-made

First, humanity's total net and positive impact on radiative forcing has exceeded the sun's, by a large margin. Within 90 per cent confidence limits, that margin could lie anywhere between 0.25 watts per square metre and more than two. A low level of both scientific certainty and understanding still surrounds the cloud albedo effect. [5] As a result of the large error bar around that effect, there is a lengthy error bar about humanity's total net impact. According to the IPCC, then, uncertainty still surrounds that total net impact.

Second, over the past 250 years, the use humans have made of land has done something to offset the increase in overall radiative forcing brought about by man's production of GHGs. Man's production of aerosols has done even more. Yet even when put together, these two tendencies have not been enough to counter the impact of human-created GHGs. Man-made warming has exceeded man-made cooling.

Climate sceptics portray humanity's warming of the planet as a deeply uncertain affair. But a high degree of scientific certainty attaches to the positive radiative forcing brought about by man-made GHGs, and a low-to-medium degree of certainty attaches to the negative radiative forcing brought about by land use and aerosols. In Chapter 7, we show that much of today's 'natural' landscape has, for some centuries, been strongly contoured by man. What the chart shows is that, *in relation to both its heating and its cooling, the whole of the Earth – seas and climate included – seems to have become an artificial place.*

That climate change is man-made shows how fixing energy supply could make a huge difference

What's known about radiative forcing is a rebuke not just to climate sceptics, but also to climate zealots.

The chart suggests *a reality to man-made climate change more radical than climate zealots allow*. It shows that humanity's contribution to climate change:

- has been large
- has cut both ways.

Until the discovery of global warming, that contribution was made *thoughtlessly*. But today, through climate science, humanity *knows* about global warming. In technology and operations on a global scale, humanity is now also strong enough to control global warming. It must just thoughtfully invest in a new round of energy supply.

Mankind is ingenious enough – and has time enough – to fix global warming during a wider endeavour: to give itself enough energy to grow. That climate change is man-made shows how the right, ambitious choice of technique in energy supply can now make a massive difference.

There's no need to be frightened. Moreover, there's no need for mankind's *past mistakes* in climate to be labelled *misdeeds*.

Climate as a moraliser's murder mystery

Climate zealots are wrong to argue that humanity has been evil in sullying and thus heating the planet. Humanity has also blackened the planet, made it more physically reflective and therefore cooled it.

These are facts, not ethical judgments. In a recent book on climate change, Gabrielle Walker and Sir David King flippantly ask, 'Whodunnit?'. [6] They forget that human 'responsibility' for climate change is a scientific matter, not an intentionally murderous act.

Similarly, government documents lapse into criminology when they portray man-made contributions to climate change not just as *footprints*, but also as *fingerprints*. [7] Why not use the less freighted metaphor of *signatures*?

To moralise about human agency in climate change can only be right if industrialisation, agriculture, land use and aviation are deemed repugnant practices, because they oppress a conscious thing, *nature*. Greens often interpret climate change in such a way as to advance this point of view; but that doesn't make it the right one.

Humanity's mixed, unknowing impacts in the past can now be rectified by determined, aggressive and thoughtful action on GHGs, beginning with CO_2. But to be effective, and to bring real gains for humanity, the action must be around a bigger and better energy supply, *not* around imposing parsimony on the individual consumer.

Man-made climate change does not equal imminent catastrophe

Climate change will mean fewer cold snaps, more heat waves, rising sea levels, changing patterns of rainfall and storms, and changing conditions for agriculture and disease. But environmentalists claim more than this. They exaggerate the:

- speed of climate change
- conclusions of climate science
- precision that surrounds these conclusions
- scientific certainty that surrounds these conclusions.

They also ascribe many weather incidents, and even *wars*, to climate change. [8] To make themselves heard, environmentalists will often do anything to represent climate change as *an imminent catastrophe*.

Perhaps, in their urgent tones and their scientific exaggerations, some environmentalists seek, through alarmism, more government funds and more government kudos. But when climate sceptics attack zealots with the cry 'Follow the money!', we're no more satisfied than when zealots find Exxon dollars behind every sceptic. Nor, for us, does the liberal, sound-bite-orientated, hysterical character of the media really explain the popular resonance of the zealots' vision.

What's actually going on in this fast-forward vision of planetary disaster is environmentalists surfing on a much wider social culture of fear, distaste for man's works, and disgust with his wastes. [9] It's as if geniuses like Michelangelo or Einstein are unremarkable in comparison with the havoc, pollution and destruction that mankind has visited on Noble Nature.

What Kent University sociology professor Frank Furedi calls 'the expanding empire of the unknown' weighs on the environmentalist's mind as much as it does on the rest of society. [10] That's why, when climate sceptics pick on particular exaggerations of climate change and accuse environmentalists of abusing science, they're obtuse.

Environmentalists don't just deal in science. They invoke the Precautionary Principle – an argument that is legal in form, though moralistic in content.

Precaution: what diplomats and international Non-Governmental Organisations love

As the Hull University geographer and ethicist Sonja Boehmer-Christiansen notes, the Precautionary Principle is said to have made its way into the English language during the early 1980s, as a poor translation of the German term *Vorsorgeprinzip*. Literally, that term means 'prior care and worry'; but it's also readily linked to state-influenced planning or provisioning for the future. Applied to environmental matters, *Vorsorgeprinzip* emerged in the early 1970s, with German clean air legislation. [11]

The first treaty to refer to the Precautionary Principle was the 1985 Vienna Convention for the Protection of the Ozone Layer. After appearances in several subsequent international conventions, it then achieved a breakthrough in 1992. In establishing what was called the Single European Market, the Maastricht Treaty of that year amended article 130r(2) of the EEC Treaty: 'action on the environment', it proclaimed, 'shall be based on the Precautionary Principle' – although the Treaty declined to offer a definition of the Principle itself. [12]

Also in 1992, in Rio, civil servants and international Non-Governmental Organisations (NGOs) succeeded in pressing the UN Environmental Programme into adopting the Principle – not least, in relation to climate change. [13] In the Rio Declaration, Principle 15 reads:

> 'In order to protect the environment, the precautionary approach shall be widely applied by States according to their capabilities. Where there are threats of serious or irreversible damage, lack of full scientific certainty shall not be used as a reason for postponing cost-effective measures to prevent environmental degradation.' [14]

Clearly, a lawyer could have drafted this. Usually incanted as if it were an axiom of natural science, the Precautionary Principle is nothing of the sort. It's an edict of international *law*, concerned with procedures and the burden of proof.

That isn't so terrible. But in fact the phrase 'lack of full scientific certainty shall not be used' downgrades the role of science in decision making. To understand why, it's vital to distinguish between *risk* and *uncertainty*. [15]

Technically, *risk* exists when man can calculate – more or less accurately – the probability of different outcomes. When insurers use statistics for life expectancy, fires or floods, they're in the business of managing risk. Particular events cannot be predicted; but the statistics are quite reliable.

Uncertainty, by contrast, occurs when probabilities are unknown. What are the chances that a revolution in physics will solve the world's energy problems? Perhaps the probability is very low. But there's no real way of knowing.

Crucially, *it's through science that uncertainty can be transformed into risk*. This allows risk to be debated in a clear manner, and from different angles – technological, economic, political and environmental.

Officialdom uses the Precautionary Principle to play up *lack of full scientific certainty*. With the Principle, there's never a scientifically quantifiable risk, but always an infinite amount of uncertainty. As a result science and scientists must take a back seat. They have no role transforming uncertainty into risk. As two enthusiasts for precaution around climate change say, it 'entails a greater degree of humility or realism over the role and potential of science in the assessment of risks'. Indeed not just the setting of policy, but also the assessment of risks should encompass public agreement and participation. [16]

That all sounds very democratic. But take genetically modified (GM) foods: almost all scientists say that enough is understood about GM foods to treat the problems they bring as specific risks. Greens counter with the argument that mankind still has much to learn about biology – so nobody can be certain what might happen with GM foods. Scenarios which scientists hold unlikely must, in this framework, nevertheless be taken extremely seriously. Greens also invoke *the opinion of the European public* in their assessment of GM risks as being not worth the candle.

But who is the public, whose appraisal of risk is as worthy as that made by scientists? Could it by chance consist of 'the interests, including beliefs and political tactics, adopted by those who claim to speak in the name of the public, society or even future generations'? [17] In environmental matters, 'the public' often amounts only to a group that has promoted an issue according to its own beliefs or interests. [18]

That was what happened in Rio in 1992. There, state bureaucrats and international NGOs managed to fix juridical uncertainty and fear as *the* operating concept in all future official assessments of climate change.

Denied data on probabilities, it's natural for the real public to focus on worst-case scenarios. Give up trying to quantify how likely a disaster is, and there's no place to stop, and no way to prioritise what you value and what you don't. If you dread some circumstances that are highly unlikely, it's simple enough to envisage others that are even worse. What if there's a design flaw in the next generation of nuclear power stations – one that brings multiple Chernobyls? What if the next generation of solar cells turns out toxic? What if a world dependent on wind power encounters a volcanic eruption big enough to disrupt global wind patterns?

To begin from what *isn't* known is to let the flesh creep. Yet US Defence Secretary Donald Rumsfeld did exactly that on 12 February 2002, when he famously publicised what elites had long been discussing: 'unknown unknowns'. [19]

Our response to this phantasm is: so what? The unknown is a large, real and ever-present realm. But what can ever be said about it, other than that it is unknown?

What governments and Greens are really up to with the Precautionary Principle is smuggling in presumptions about how human society is highly vulnerable. That might or might not be true; but the starting point for assessing vulnerability, as anything else, must be what is *known*. Greens often underestimate how much is known. And where it really is the case that little is known, precious knowledge must still form the basis for decision-making.

When applied to climate change, the Precautionary Principle ensures that the future can only be a lurid journey into a Dantean inferno. Stern's very own PowerPoint – complete, in the original, with orange tones fading into hellish reds – confirms this. [20]

Climate portrayed as leaping about

Let's now move from legal principles to the physics of climate change. Attacking Bjørn Lomborg, a sophisticated critic of climate alarmism, Cambridge economics professor Sir Partha Dasgupta shows how the empire of the unknown now dominates mainstream thought. Here's the (shortened) conclusion of his polemic:

> 'If there is one truth about Earth we all should know, it's that the system is driven by interlocking, nonlinear processes running at different speeds. The transition to Lomborg's recommended concentration of 560ppm would involve crossing an unknown number of tipping points... We have no data on the consequences if Earth were to cross those tipping points... Even if we did have data, they would probably be of little value because nature's processes are irreversible... [Estimates] of climatic parametres based on observations from the recent past are unreliable for making forecasts about the state of the world at CO_2 concentrations of 560 ppm or higher. Moreover, the nonlinearities mean that doing more of a bad deal [Kyoto] may well be very good... These truths seem to escape Lomborg... [He] believes we shouldn't buy insurance against potentially enormous losses resulting from climate change.' [21]

On climate change, then, what is fashionable is to highlight the *almost complete absence of certainty.* The only certainty allowed is that planetary behaviour is profoundly non-linear – that it can leap about at bewildering speeds and unpredictable rates. Multiple interlocking non-linear processes make the Earth a deeply unstable place. Thus *immediate* personal conservations of energy, together with Kyoto-style agreements, CO_2 taxes, trading or rationing, represent sensible, precautionary insurance policies, which will save much more money later.

But is such a fast-moving programme really the right one?

Lord Stern's 'projected impacts of climate change'

Global temperature change (relative to pre-industrial)

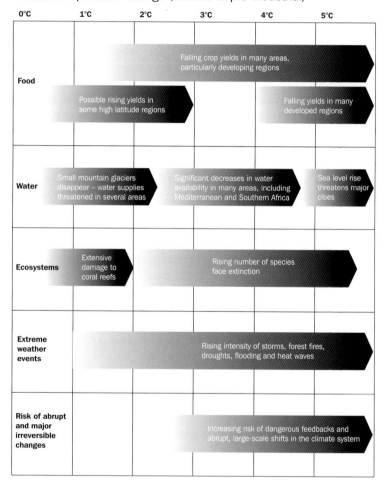

	0°C	1°C	2°C	3°C	4°C	5°C
Food			Falling crop yields in many areas, particularly developing regions			
		Possible rising yields in some high latitude regions			Falling yields in many developed regions	
Water		Small mountain glaciers disappear – water supplies threatened in several areas		Significant decreases in water availability in many areas, including Mediterranean and Southern Africa		Sea level rise threatens major cities
Ecosystems		Extensive damage to coral reefs		Rising number of species face extinction		
Extreme weather events			Rising intensity of storms, forest fires, droughts, flooding and heat waves			
Risk of abrupt and major irreversible changes				Increasing risk of dangerous feedbacks and abrupt, large-scale shifts in the climate system		

Exponential growth, feedback, non-linear behaviour, and chaos

When discussing climate, Greens love to throw around modern mathematical jargon. Sadly, they often have little respect for the precise meaning of the concepts they use. In a fearful variant of Orwellian Newspeak, terms such as *exponential growth, feedback, non-linearity* and *chaos* have become, in Green commentary, little more than code for *the planet spinning out of control.*

The first worrier about exponential growth was Malthus. He fretted that while 'the means of subsistence, under circumstances the most favourable to human industry, could not possibly be made to increase faster than in an arithmetical ratio', population was different. Food production could not keep pace with population. 'When unchecked', Malthus warned, population 'goes on doubling itself every twenty-five years, or increases in a geometrical ratio'. [22]

Exponential growth is defined as a process that doubles (or halves) over a period of time that remains constant. In the case of Malthus, this period was every 25 years. Closely related to exponential growth is *feedback*, which comes in two variants: positive and negative. In positive feedback, a change creates a greater change, magnifying the original effect. In the case of Malthus, the feedback is simple – more people breed yet more people. In negative feedback, a change provokes an opposite change, stabilising things.

CO_2 added to the atmosphere raises temperatures, which leads to the evaporation of more water vapour, which – being a GHG – raises temperatures further. This is a positive feedback. Less well understood is that more water, by forming clouds that reflect sunlight, may also make for a negative feedback.

The most significant uncertainties in climate science surround feedbacks. Without feedbacks, a doubling of CO_2 concentrations would produce a temperature change of only 1.2° C – which is nothing really to worry about.

The difficulty is that there are undoubtedly many feedbacks. These are very likely to be on balance positive, increasing the rise in temperature into the range of 3° C in the century 22nd century. Greens worry that feedbacks may produce even more warming than that.

Non-linear is a broad adjective. The mathematician Stanislaw Ulam has been credited with the remark that 'non-linear science' is like 'calling the bulk of zoology the study of non-elephants'. [23] Broadly, in a linear equation the cause is *proportional* to effect. If the cause doubles, then the effect also doubles.

For scientists, linear equations are attractive because they can be solved. Solutions can also be added together to produce new solutions. Any solution can be broken down into simpler, more tractable components – a fact that allows linear equations to be solved systematically.

Linear equations describe many situations. The fundamental equations of quantum mechanics, describing the motion of atoms and molecules, are linear. They're also good approximations to many situations for small motions – the swing of a pendulum when it's not too large, a sound wave when it's not too loud, or a ripple on the surface of water.

Non-linear simply describes all other situations – situations in which the effect is *not* simply proportional to cause. If, say, the cause is doubled, and the effect is quadrupled, that's an example of non-linearity. Rivers, electrical circuits, car engines, the sun and most other phenomena in nature all exhibit non-linear behaviour.

Tipping points are also examples of non-linearity. Beneath a certain threshold, change in a cause produces little or no effect; but a small change that crosses the threshold creates a large effect. Clearly, the effect is not simply proportional to the change.

Non-linear equations cannot usually be solved exactly. They can be solved approximately by a mass of arithmetic, nowadays run on computers. This is how computerised models of climate are built and operated.

Feedback is often thought of as non-linear, and sometimes

it is. But often feedback can be described by a linear equation. When a change in a quantity is proportional to itself, the result is exponential growth. That's the case with growth of bacteria: the increase in the number of bacteria is proportional to the number of bacteria that are already present. This is an example of feedback described by a linear equation.

More important is that non-linearity creates the possibility, although not the necessity, of *chaos*. In chaotic systems, states that begin close to one another can rapidly diverge. This is *sensitive dependence on initial conditions*. Very small changes – perhaps changes that are too small to measure – can have very large consequences.

But chaos requires more than a simple explosion in which points that were close together move rapidly apart. It also requires that states starting far apart come close together.

Weather is a chaotic system. Patterns of wind and cloud that look similar will evolve into very different states over a few weeks. That's why weather forecasting is hard. On the other hand, weather patterns continually fall into familiar regularities – coming back closer again.

Neither non-linearity nor chaos should be interpreted as closing off human understanding. Newtonian gravity, for example, is non-linear. Considered over billions of years, the motion of the planets is chaotic. [24]

Newton's step forward was once hailed as the foundation of the Enlightenment. Though today's insights into non-linearity and chaos in fact build on Newton, *the expanding empire of the unknown in today's human imagination ensures that they are interpreted as revealing the limits of science.*

Even if it's impossible to predict the exact motion of a chaotic system, it's often still possible to understand its properties. The pressure of a gas, for example, arises from the molecules that make it up chaotically colliding with the walls of a container. Many molecules hit those walls every millisecond; but such chaotic behaviour averages out, so that, at human scale, pressure is predictable.

In the same way, it may be that even if we cannot predict the weather in a week or a month, we have good prospects for understanding the climate system on longer timescales, including hurricanes, heat waves, ice sheets and the rest.

Climate exhibits feedback, non-linearity and perhaps tipping points. But while feedbacks will, in aggregate, raise temperatures, the mere existence of non-linear processes cannot be pressed into the service of an accelerated apocalypse.

Even tipping points need not be uniquely frightening. 'Tipping Points in the Earth System', a workshop held at the British Embassy in Berlin in October 2005, brought together 36 experts. Continued GHG emissions, they said, might push humanity past nine possible tipping points over the next 100 years. Their examples show that tipping points can be managed similarly to other environmental questions. [25]

Later on, we examine the melting of the Earth's ice sheets, by way of a look at Greenland. Here we summarise the workshop's other examples.

The *melting of summer arctic ice* would not have serious consequences for human beings. It would put species such as polar bears under additional pressure and add to the overall pace of warming (dark water reflects less heat than white ice). But humanity should be able to get round these problems.

Disruption of El Niño, the circulation of warm water in the Pacific, could lead to broad changes in regional climate across the globe. Yet these changes would likely occur over a century or more, giving plenty of time for adaptation. The possibility is also remote: it might happen within a millennium, but 'the existence and location of any threshold is particularly uncertain'. [26]

Disruption of the *Indian summer monsoon* would be a problem, albeit one that could be adapted to with better water management. But though a possibility, it's more related to clouds of smoke from more traditional air pollution across Asia than to GHGs. If anything, additional greenhouse warming stabilises India's summer monsoon.

If global warming disrupted the *West African monsoon*,

that would also be a problem. But a side effect would be a greening of the Sahara and its surroundings – 'a rare example of a beneficial potential tipping point'. [27]

Serious damage to the *Amazon rainforest* has been predicted if temperatures rise by 3° C or more. Yet again the story is not simply one of climate change: the fate of the Amazon 'may be determined by a complex interplay' between changes in land use and climate change. [28]

For *forests in Canada and Russia*, rises of 3° C or more are again projected to be problematic. But it's less clear why transformation of such forests to grassland would be catastrophic. In any case limitations in existing models and physiological understanding make such a transformation a matter that is still 'highly uncertain'. [29]

No doubt our reading of the facts put forward by the Berlin workshop is much more sanguine than that made by those who participated in it. But whether the facts drive an alarming analysis, or a calm one, relates more to differing approaches to precaution and uncertainty than to the facts themselves.

Behind environmentalism's accelerated Apocalypse

It's glib to describe environmentalism as a religion. While Green images of burning heat and a second Great Flood do recall religious faith, they're more *symptoms* of environmentalism than premises. Still, if radical climate change isn't an instantaneous matter, why do so many environmentalists want to believe it is?

There are several reasons. In Chapter 1, we discussed the fear and lack of investment that today surround technological innovation and R&D in the West. Chapter 2 showed how, in the early 1970s, environmentalism won a legitimating status for itself. Altogether, then, contemporary techno-fear and the statist origins of modern environmentalism make Greens' views of the future highly conservative. In general, environmentalism interprets the future as *something that happens to you*, rather than *something that you make happen*. That explains why action on climate change is framed at the modest level of individual consumption, rather than organised at the ambitious level of global energy supply.

Beyond that, however, two major historical experiences have shaped the particularly fast-moving character of the doom envisaged by modern environmentalism: the Second World War, and the Cold War.

History has always contoured forecasts of the future. In particular, the speed of the Second World War's onset in Europe following the Munich Conference of October 1938, the drama of Japan's attack on Pearl Harbour and the brevity of the Pacific War's conclusion in Japan have had an enduring impact on the Western psyche. Environmentalism, too, has absorbed this impact.

And when political, business and environmental leaders talk about climate change, they very often search for epic effect by bringing up the Second World War. [30]

The Second World War as the template for lightning change

Addressing the UN in 2007, Gordon Brown upheld Nicholas Stern's view that the likely costs of climate change would compare with those of the Depression and the Second World War put together. [31] After that, Richard Branson, the head of Virgin Group, told the UN that people needed to take global warming as seriously as the British did the last war. [32]

Greens love that war. They long for:

1. Rationing through personal carbon allowances [33]
2. A 'supreme effort of national mobilisation' [34]
3. Climate sceptics to be given the same short shrift as those who would deny the Holocaust. [35]

Above all, though, references to the war assist Greens in drumming up a sense of *urgency* about climate change.

After 1945, Presidents Truman, Eisenhower, Kennedy and Johnson pressed the lessons of Munich into the service of the Cold War. To make peace with aggression was seen as an error. Post-war 'security' in the West meant that world-shattering turbulence, now led by communists rather than fascists, must be *nipped firmly in the bud*. Even in the US presidential campaign of 2008, George W Bush felt the same way, hinting that Barack Obama's foreign policy amounted to 'the false comfort of appeasement, which has been repeatedly discredited by history'. [36]

The imperative to compress the likely evolution of climate change is really a Green shadow, in the world of nature, to the bad faith still felt about appeasement in the world of politics. From Hitler at Munich, through Admiral Yamamoto at Pearl Harbour, to Osama Bin Laden and climate change today, 'evil' has been assumed typically to gain such a swift dynamic that only super-urgent, monumental and unanimous action can reverse it.

The idea of an *unstoppable chain reaction* first entered popular consciousness with Hiroshima; and today environmentalism loves that idea. In 2005, the UK government-funded Carbon Trust ran a series of television commercials. Each drew on the ancient, mystical Hindu text, the *Bhagavad Gita*, which Robert Oppenheimer, the father of the Bomb, had ruefully recalled during the Trinity atomic test, conducted in the US on the eve of Hiroshima (see Chapter 4). To background footage of a mushroom cloud, the commercials proclaimed: 'I have become the destroyer of worlds'. [37]

The alarmist approach suggests that *when mankind plays with the fundamentals of nature, a conflagration will follow in no time at all*. In this sense, your carbon-profligate lifestyle helps pulverise the planet, and must cease forthwith.

Rapid infection, falling dominoes, ladders of escalation: the influence of the Cold War

The next experience shaping modern thought about doom tomorrow was the Cold War. To begin with, the West developed a *biological* metaphor for Apocalypse. Thus, preparing top Congressmen for the Cold War in 1947, US Under Secretary of State Dean Acheson used the language of contamination and epidemiology to emphasise that one bad thing can quickly lead to another. Speaking of pressure by the Soviet Union on the Near East, he advised:

> 'Like apples in a barrel infected by one rotten one, the corruption of Greece would infect Iran and all to the East. It would also carry infection to Africa through Asia Minor and Egypt, and to Europe through Italy and France…' [38]

The fears that once accompanied the spread of the 'communist menace' today attend global warming. A paroxysm knowing no national boundaries is felt to be imminent.

The 21st century already dreads international contaminations and geometrically multiplying viruses. Little wonder that climate cataclysm is felt to be possible at once, everywhere.

In 1954, when Vietnam began to best France during the battle of Dien Bien Phu, Eisenhower added another fearful, affecting image of communist expansion – the falling domino:

> 'You have a row of dominoes set up, you knock over the first one, and what will happen to the last one is the certainty that it will go over very quickly. So you could have a beginning of a disintegration that would have the most profound influences.' [39]

Soon, under Kennedy, the possibility that *the world could be destroyed through a rapid, uncontrollable and irreversible trail of human-initiated events* became further enshrined in the US doctrine of nuclear deterrence through Mutually Assured Destruction (MAD).

As we showed in Chapter 2 with Galbraith and Packard, the height of the Cold War engendered critiques of consumerism and energy use. But the Cold War's peak also moulded futurologists, climate scientists, and environmentalists.

Notoriously, the Pentagon corralled emerging, mathematics-based disciplines – cybernetics, game theory – into its cause. [40] After the advent of the integrated circuit in 1957, computers were also used, in practice and in propaganda, to make military manoeuvres more respectable.

In 1960 Herman Kahn, arguably the inventor of modern-day forecasting, used his background in the highly computerised RAND Corporation to predict, with assiduous calculations, the death counts for future nuclear conflicts. [41] Then, in 1965, Kahn developed a metaphor for what he called the 'coercive aspects of international relations': complete with 44 rungs and six 'firebreaks', it was that of a *ladder of escalation*. [42]

What for Kahn was a metaphor became, in the imagination of many, a very real ladder. The future looked different. In a nuclear world, it could well be a succession of discrete catastrophes, linked over shorter and shorter intervals, each magnifying the last.

Sound familiar?

The Old Scientism and the New Scientism

To meet, Kahn was so big, he even looked like a think tank. [43] As so much the expert, his views could *not* be contradicted. Perversely, Kahn used computers to predict devastation in the future, and to back the most aggressive postures in the present. His approach was but one example of the wider Cold War phenomenon of scientism: modish, computerised, cool, 'independent', unanswerable. [44]

Kahn used 'systems analysis' to draw up digitally-based models of nuclear war. Once popularised, his outlook suggested that the future could move in quantum leaps of lethality; so it didn't prove hard, in the 1960s, for some scientists to interpret the future of climate as a set of lurches toward hell. After all, computer models were used to make the new nightmares especially authoritative.

In today's war against global warming, people are once again told that what they're up against isn't smooth, graceful, geometrical progressions. In a kind of subconscious residue of Cold War fears, a series of rapid, bucking, ever more disruptive changes, or *switches*, is held out as jeopardising the very existence of the Earth. [45] And more than ever, computer models of everything are invoked to smudge over the difference between natural science and the social sciences, between science and its interpretation, between science and policy proposals about what to do.

In the process, science is perverted – so much so, indeed, that Lord David Sainsbury, Tony Blair's adviser on science and technology, once described it as a tool of British foreign policy in dealings with the Chinese. [46] And while the old scientism of the Cold War had its critics, today's New Scientism runs pretty much unquestioned.

In 1973, when *The Limits to Growth* came out, Christopher Freeman, director of the Science Policy Research Unit at Sussex University and one of the world's top technology policy gurus, satirised the approach as 'Malthus with a computer'. [47] The Fall of Man, some felt, could not be verified by the movement of electrons around printed circuit boards.

Today things have changed. Unchallenged, the Stern report referred more than 500 times to 'models' of

- climate change and its monetary cost
- hydrology and crop growth
- risk and uncertainty
- innovation, technology and energy.

Stern's opening words give *science* an all-determining position:

'It is the science that dictates the type of economics and where the analyses should focus, for example, on the

Finger on the dark button of fate:
Dwight D Eisenhower (1890-1969)

economics of risk, the nature of public goods or how to deal with externalities, growth and development and intra- and inter-generational equity.' [48]

Similarly, while the IPCC' s Working Group I confines itself to the *physical science* of climate change, Working Group II, focusing on the impacts of climate change and *human adaptation and vulnerability* in the face of those impacts, makes a mish-mash of monolithic computer simulations around disciplines quite separate from climatology. [49] The same is true of the Working Group III on the *mitigation* of climate change. Here models of climate merge into free-market models of economics and into projections of demography. [50]

Given its willingness to mix up natural science with social forecasts, the New Scientism, like the old, is not actually very respectful of science. In fact, the New Scientism is about *deifying nature*. Once nature is put before humanity, science becomes merely the winged messenger for nature, there to tell a dumb human species that it must have more 'awareness' of how dumb it is.

Mitigation, Adaptation and Transformation

Official answers to climate change suggest that it can either be averted, or lived with.

Slowing or stopping climate change is known as **mitigation**. In practice, this generally means cutting the net levels of GHGs added to the atmosphere each year, by:

- conserving energy
- decarbonising energy supply
- conserving, enhancing or fireproofing *carbon sinks* – large features of the planet, whether natural or artificial, that absorb CO_2. [51]

In our view, the first option is the wrong way, but the second is the right way to go. The third is fine, although when there are more appropriate uses for natural sinks such as forests and oceans, then developing new, bigger and better sinks may be preferable to simply conserving, enhancing or fireproofing the ones that already exist.

An alternative to mitigation is **adaptation** – measures that the IPCC says 'reduce the vulnerability of natural and human systems' to climate change. [52]

Bad weather already causes damage, especially in the Third World. Therefore adapting both landscape and settlement to handle climate change makes sense. Better roads and telecommunications, for example, could speed the pace of evacuations.

Yet the world needs better roads and IT networks regardless of emergencies. It needs to do more than just 'ruggedise' its cities against the immediate effects of climate change.

For Greens, adaptation represents too much, not too little. They prefer mitigation for three reasons:

1. By contrast with leaving future generations to adapt to the consequences of today's errors, conservation brings benefits now.
2. Adaptation might be a Band-Aid; taking precautions through mitigation is a dead cert. Carry on adding GHGs to the atmosphere? That's like poking an 'angry beast'. [53]
3. The poor simply can't afford to adapt to the effects of climate change.

We don't agree. We've already dealt with precaution; but there are two other arguments to be made here.

First, today's legacy to future generations isn't just a burden to be lightened. Through innovation and progress, humanity can leave its descendants a much more vibrant bequest.

Second, it's true that the poor cannot afford to build flood defences. But to prefer mitigation to adaptation because the Third World will never be strong enough to withstand harsh weather –

that's a circular argument. No doubt achieving economic takeoff in Mozambique and Bangladesh will be hard. But will it be any harder than a crash cut in GHG emissions?

Our programme of **transformation** goes beyond mitigation and adaptation. In our view,

1. The business of mankind is not just to slow or stop climate change, leaving climate in a more 'natural' state, but also to take control of as much of the environment as is possible.
2. Rather than just adapting human arrangements to deal with climate change, both the energy and the non-energy aspects of the environment merit a transformation to meet human needs.

Transformation is about making the planet a more human kind of place. It goes beyond energy and GHGs, even if human place-making is one reason why the world needs more energy.

As hinted earlier and more fully developed in Chapter 7, the beginnings of transformation are to be found everywhere. Human beings now live not in raw conditions, but in a built environment. They have altered between a third and a half of the Earth's land, and used more than half of its accessible fresh surface water. [54] More nitrogen, which is crucial for all life, is now fixed by human industry for use in fertiliser than by the entire natural biosphere. [55]

Transformation has tended to be the rule in the past, and should definitely be the rule of the future.

Global warming's discovery, 1956-7, and the conditions that allowed it

The height of the Cold War didn't just nurture a computerised sense of urgency about the future. It also installed a science regime committed to fundamental exploration. Despite the military origins of much – though not all – of US research into weather and climate over 1956-65, both fields enjoyed fundamental breakthroughs. [56]

In the 1930s the American oceanographer Roger Revelle researched ocean chemistry, and, among other topics, its carbon dimension. In the 1950s he won funds from the US Navy to measure radioactivity and ocean mixing. He concluded that radioactive wastes introduced into the upper layer of the ocean might stay there for many years.

Then, in 1957, Revelle published a paper with Hans Suess of the US Geological Survey. [57] It included perhaps the world's most famous paragraph ever to have been Scotch-taped to an original draft. As the brilliant US science historian Spencer Weart summarises that paragraph, seawater

> '... needed to absorb only about a tenth as much gas as a simple-minded calculation would suppose. While... most of the CO_2 molecules added to the atmosphere would wind up in the oceans within a few years, most... would promptly be evaporated out.' [58]

In 1956 Revelle said that the Greenhouse Effect might bring harm by 2000; in the following year, he warned that the effect might turn Southern California and Texas into 'real deserts'. [59] But the confidence of US science in his day, reflecting America's economic boom and its overall military superiority to the Soviet Union, tempered these kinds of fears.

Concluding his 1957 paper, Revelle famously wrote that mankind was performing an unprecedented and unrepeatable 'large scale geophysical experiment' with climate. But as Weart perceptively remarks, the word experiment:

> '... sounded benign and progressive to Revelle as to most scientists... he only meant to point out a fascinating opportunity for the study of geophysical processes. People's attitude toward the rise of CO_2, he would write in 1966, "should probably contain more curiosity than apprehension".' [60]

Relative to the economy, Cold War R&D in the US was broader and more intensive than it is today. A military but expansionist context buoyed up science, and allowed it to progress – sometimes through a kind of organised serendipity.

Society's mood, then, can profoundly affect the interpretation of science.

When environmentalists bang on about consensus, they miss how science, despite its enormous recent progress, has yet to reach agreement on the detailed mechanisms of climate. But even if consensus *is* eventually created on areas of scientific uncertainty, environmentalists would still be wrong to interpret it as a directive for you to *minimise your carbon footprint their way right now*. That would be an outrageous distortion of science, undertaken for decidedly political ends.

In its emotional claims to objectivity, the New Scientism deflects society's focus, and the focus of science, right away from energy supply.

Today, the world should revive Revelle's emphasis on experimentation and curiosity – and uphold the vital role of well-funded serendipity in science. Luck can never replace thoughtfulness in energy supply. But investment *and* luck in energy R&D would today be wiser than piling on more apprehension about the rapidity of climate change.

Such apprehension began with Revelle himself, in 1956. Meanwhile, general Cold War jitters also reached a pitch. Soon humanity came to be blamed not just for global warming, but also for dangerous climate change.

By 1960, rises of atmospheric CO_2 were found consistent with Revelle's line on weak sea absorption of CO_2. Then, after the organised serendipity that had attended Cold War oceanography, a more accidental serendipity in Cold War meteorology allowed non-linear planetary behaviour and chaos theory to be discovered.

The man who first alerted the world to climate change: Roger Revelle (1909-1991)

Difficulty in predicting the weather turns into the impossibility of understanding climate

In 1961, a meteorologist at the Massachusetts Institute of Technology, the late Edward Lorenz, found by a chance computer simulation that weather exhibits such sensitive dependence on initial conditions that its long-term behaviour was impossible to predict. Given the basic distinction between weather and climate, Lorenz was correct.

Lorenz's pioneering researches revealed the new power of computer models – though his were models of *natural*, not *social* phenomena. But in 1965, concluding his opening address to a major conference on climate change held at Boulder, Colorado, Lorenz turned *chaos in weather* into something much broader – *human uncertainty about climate*. He said:

> 'Climate may or may not be deterministic. We shall probably never know for sure.'

In just four years Lorenz had moved from the intrinsic *unpredictability of weather* to mooting an intrinsic *incomprehensibility of climate*. This was a mistake, reflecting the uncertain times.

As Revelle summed up the Boulder conference, minor and short changes in the Earth's past behaviour might have been enough to 'flip' its atmospheric circulation from one state to another. [61] But as Weart shows, it took another 30 years for measurements of past climatic flips to narrow their duration from thousands of years to decades or less. [62] Nevertheless, apprehensions that human beings could cause future flips in climate quickly and to disastrous effect grew up as hastily as Lorenz had put the whole idea of climate science into doubt.

When Lorenz invoked uncertainty about climate science, going on to develop the concept of chaos, he was not alone. At the Cold War's peak, Western culture was deeply uncertain about the future.

America's reaction to the Soviet launch of Sputnik in 1957 showed how panicky things could get. In seminal articles for the *New Yorker*, Rachel Carson contended that, along with the possibility of the extinction of mankind by nuclear war, 'the central problem of our age' had become pesticides and insecticides. [63] Among historians of science, the effect of the ambiguities in Thomas Kuhn's *The Structure of Scientific Revolutions* (1962) was to throw the whole idea of *progress* in science into doubt. [64] So in the tense and anxious late 1950s and early 1960s, it was easy for climatologists to leap, as Tony Gilland of the Institute of Ideas puts it, 'from rudimentary findings to cataclysmic worst-case scenarios'. [65]

It wasn't possible to detect anthropogenic warming at least until 1980. [66] Yet that didn't stop the 1965 Boulder conference from agreeing that the climate system 'showed a dangerous potential for dramatic change, on its own or under human technological intervention, and quicker than anyone had supposed'. [67]

The formation of the IPCC

The final years of the Cold War supplied a second episode shaping environmentalism. SustainAbility describes the period 1988-1991 as the second wave of environmentalism. [68] In those years, a second absorption of environmentalism into statecraft took place – an absorption firmer than that which befell environmentalism in the early 1970s.

The formation of a functioning IPCC in November 1988 occurred in a much larger context than that which greeted the first heyday of environmentalism. This time, the outsourcing of policy was extended to an intergovernmental panel of government experts. In addition, the IPCC was born during a resurgence and multiplication of social and environmental fears.

In 1985, UNEP, the World Meteorological Organisation (WMO) and the International Council for Science (ICSU) sponsored a conference of 89 scientists – including biologists and engineers – working in a personal capacity. Held in Villaich, Austria, the conference proved a key ramp toward the formation of the IPCC. Urged by the three sponsors to make policy recommendations, the conference concluded:

> 'the rate and degree of future warming could be profoundly affected by government policies on energy conservation, use of fossil fuels, and the emission of greenhouse gases.' [69]

This was an inauspicious prelude to the IPCC. 'Independent' scientists were invited by inter-state bodies to step beyond climatology and make political proposals. And – surprise – the proposals were not about reducing the carbon intensity of energy supply, still less about increasing that supply. Instead, they were about… energy conservation and use.

The 1980s produced anxiety on a grand scale. There were fears around:

- cruise missiles, Ronald Reagan's Strategic Defence Initiative, and the possibility of 'nuclear winter' [70]
- the likely future incidence of AIDS among Western heterosexuals
- the extent of child abuse in families
- a repeat of the explosion at Chernobyl, Ukraine, in 1986.

In 1986, the German sociologist Ulrich Beck published *Risk Society*, which suggested that the big problems human beings face followed from the unforeseen consequences of past technological developments. In 1987, the American journalist James Gleick published *Chaos*, which popularised Edward Lorenz's doctrines. Then, in 1988, Margaret Thatcher came out in favour of sustainable economic development and issued a warning about climate. Echoing Revelle in her own admonitory style, she said that mankind was engaged in 'a massive experiment with the system of this planet itself'. [71]

In these years, the IPCC was formed out of a convergence

between

- scientists, both in and well beyond climatology, who had by this time become fearful of climate change and politically active around it
- UNEP, WMO and ICSU, which wanted to build on their past successes in the control of ozone
- above all, the Reagan administration, which sought to restrain UNEP and settle sharply differing views among various US government agencies. [72]

The IPCC is mostly a US government creation, and wholly a political body. Its three Working Groups are mandated to assess not just climate change, but also its *social impact* – and what to do about it.

So when environmentalists say that IPCC pronouncements mean that 'The Science' has spoken, they misrepresent it. And through this device, they give politicians a nice, neutral, high and mighty way to attack you for your supposedly profligate lifestyle.

Green misanthropes (3): Achim Steiner and Rajendra Pachauri insist you change your habits

Achim Steiner knows how to do a press launch. He is Executive Director of the United Nations Environment Programme , a co-founder of the IPCC. On 2 February 2007, the IPCC published a 'Summary for Policymakers', anticipating the full report of its Working Group I on the physical science basis of climate change. [73] Speaking at a press conference to mark the event, Steiner said that the evidence for human beings causing climate change was 'on the table, and we no longer have to debate that part of it'. For Steiner, 'the science' should not disempower individuals: rather, it meant that 'every individual can today walk out of their front door and cut their emissions by more than what Kyoto had ever envisaged'. [74]

Steiner's biography boasts of his track record in 'sustainable development policy and environmental management', as well as his 'first-hand knowledge of civil society, governmental and international organisations'. Before joining UNEP, he ran the International Union for Conservation of Nature, managing 1000 environmentalists. [75] A German born in Brazil, Steiner was educated at Oxford, and spent time at Harvard.

None of that, however, gave him the right to claim that scientific debate on the human input to climate change had been concluded. None of it gave him the right to say that science dictates that you cut your GHG emissions.

Nevertheless, Steiner plunges on, insisting that the

Spinning science their way: United Nations dignitaries Achim Steiner and Rajendra Pachauri

143 CLIMATE CHANGE: NO NEED TO PANIC

world has 'less than seven years' to stabilise GHG emissions. [76]

Rajendra Pachauri represents a similar story of scientific sobriety and independence. An Indian economist and vegetarian, he is chairman of the IPCC and, in 2007, saw it awarded – along with Al Gore – a Nobel Peace Prize.

Pachauri isn't backward coming forward about climate change. The prospect of cars in India costing just 100,000 Rupees (£1300), he told Indian industrialists, meant that

'I am having nightmares, I don't know what will happen then.' [77]

But it's not just around driving that Pachauri exhibits strict scientific neutrality. His kinds of worry about the effect of cattle farming on climate change has made him advise readers of the London *Observer*: 'Give up meat for one day [a week] initially, and decrease it from there.' [78]

After the Cold War, a fad for tipping points

In retrospect, the Cold War was a reasonably stable era. [79] Yet even before 9/11, post-Cold War visions of mass disaster stayed as strong as they were in the 1980s. Indeed, from Russians with nuclear suitcases in the early 1990s to bird flu today, fears of a conflagration have grown.

Today, more than any other contender, climate change has come to embody and concentrate risk consciousness. And it's around climate change, more than any other issue, that computer forecasts of society have had a baleful influence. In the UK, officialdom now goes to market with models of the future by:

- starting with the headlines it wants to generate
- mixing in rich diagrams to show The Indisputable Science
- adding grave, illustrated Days-in-the-Life-of-Daisy
- alluding vaguely to an Annex, somewhere, that contains hard number-crunching around some obscure algebraic formulae.

Et voilà! The future is laid out for everyone to worry about. It can be about climate, but can just as well be about obesity. [80] Once again, this New Scientism *looks* hip and incontrovertible. But in fact it's deeply fatalistic. IPCC Working Group II recommends 'altered food and recreational choices' and planning regulations. [81] Working Group III wants people to adopt an 'efficient driving style'. [82] But just like climate sceptics, both groups miss the point.

The consequences of climate change are like climate itself. They depend little on personal consumption and driving habits, and a lot more on each particular society's level of economic development, and, not least, on the state of its energy supply. In the Third World as elsewhere, these things *can* be improved – and that's a policy proposal determined not by climate science, but by respect for *human talents*.

A new century has made a fad of tipping points. After all, the 'millennium bug' in computers was supposed to lead to a world standstill. More importantly, a very popular book on the power of word-of-mouth communications in modern society proved a gift to climate zealots.

Malcolm Gladwell's *The Tipping Point: How Little Things Can Make a Big Difference* was published in 2000. Using the language of *epidemiology* to track such trends, it suggested that the emergence of social trends was best understood if they were thought of as *viruses*. [83] Gladwell held that 'little causes can have big effects', and that human beings had 'a hard time' with *geometrical progressions* 'because the end-result – the effect – seems far out of proportion to the cause'. He went on:

> 'We need to prepare ourselves for the possibility that sometimes big changes follow from small events, and that sometimes these changes can happen very quickly.' [84]

Gladwell pointed out that viruses transform themselves and so can become much more deadly. But later, in an Afterword, he noted that people develop *resistance* to viruses. [85]

Despite the commendable balance Gladwell showed on viral growth, environmentalism quickly grabbed hold of his elegy to the power of small events. As Spencer Weart notes, 'Around 2005 the phrase "tipping point" appeared in both scientific and popular climate reports, an admission that change could be not only rapid but irreversible'. [86]

In 1972, Edward Lorenz gave a talk titled 'Predictability: Does the Flap of a Butterfly's Wings in Brazil Set off A Tornado in Texas?'. [87] Today, this metaphor for chaos is largely forgotten. Instead, environmentalists refer to tipping points around future climate change with a knowing air, confident that nobody will object to the concept.

Melting ice as a tipping point: the example of Greenland

For climate alarmists, melting ice is an iconic image. The world boasts a lot of ice – in glaciers, floating at the North Pole, on Greenland and on the Antarctic. Lots of ice is also melting. But like climate, it's never a cut-and-dried affair.

Take Greenland's ice sheet, remembering that the issues it raises are similar to those raised by Antarctica. Future rises in sea levels will not only, or even mostly, come from it melting. According to James Hansen, whom we cited in Chapter 1 and who is very alarmed about climate, the seas are rising at about 3mm per year – but the melting of Greenland's ice makes up only about 10 per cent of that. About 50 per cent of the rise is from expansion of water in the oceans as they warm. [88]

Greenland's ice sheet has become the focus of concern because of the fear that it may pass some sort of *tipping point*. Indeed, though it's only slowly melting at present, the whole thing could begin rapidly to slide into the sea. And once that happens...

Hansen writes that 'reticence may be a consequence of the scientific method', but that 'in a case such as ice sheet instability and sea level rise, there is a danger in excessive caution'. Addressing the possibility of climate tipping points, he concedes that the non-linearity of the ice sheet problem makes it impossible to accurately predict the sea level change on a specific date. But he continues that the threat of a large change in sea levels is a 'principal element' in his line that additional global warming must be kept at less than 1° C above the temperature in the year 2000 – and that 'even 1° C may be too great'. [89]

So would it be right to panic about Greenland's ice? Well, if we leave out lots of other complications, the mass of that ice is in fact determined not by *one* process, but by *the balance between two*.

Warming is thickening the ice at the centre of the Greenland ice sheet. Why? Because today's warmer temperatures mean more evaporation over the world's oceans. In turn, that means more precipitation – adding to Greenland's ice.

At the same time, warming has led to more melting at Greenland's edges. That leads to more meltwater run-off, and more 'calving' of icebergs.

If all Greenland's ice were completely to melt, or slide into the sea, sea levels could rise by seven metres, drowning most of the world's coastal cities. When Al Gore was challenged that his Photoshop images showing this were in fact worst-case scenarios, he invoked what he termed two 'wild cards': Greenland and West Antarctica. He continued: 'Greenland is the wilder of the two.... It's undergoing a radical discontinuity'. Gore said that scientists, when asked off the record if Greenland could break up this century, 'cannot rule that out and privately will not'. [90]

In 2007, however, the IPCC painted a rather more sober picture.

To start with, it's still not entirely certain how much Greenland's ice is changing. The IPCC is probably fair in pointing out that 'Lack of agreement between techniques and the small number of estimates preclude assignment of statistically rigorous error bounds' – and in adding that 'the short time interval covered by instrumental data is of concern in separating fluctuations from trends'. [91]

The IPCC's best estimate is that the annual change in Greenland's ice ranged from growth of 25 billion tonnes to

Melting icebergs in Disko Bay, off of the western coast of Greenland, 2006

shrinkage of 60 billion tonnes, 1961-2003.

Since 2003, it's true, Greenland's ice has been shrinking. Indeed shrinkage now stands at more than 100 billion tonnes a year. Yet that rate translates into an annual rise of sea levels of only 0.3mm. In turn, over a century, this amounts to just a few millimetres. [92]

That's not such a big deal. As Bjørn Lomborg has pointed out, the last 150 years saw sea levels rise at 3mm per year. [93]

The Greenland ice sheet has been melting since the end of the last Ice Age, and mankind has pushed that process along a little faster. But even with a global warming of 3° C in the 22nd century, models show most of Greenland's ice remaining intact by the year 3000.

It's true that models have sea levels rising by more than two metres by AD 5000. [94] By then, though, humanity could well have radically tamed the climate, preserved Greenland or – more simply – moved coastal cities inland.

Environmentalists claim that the IPCC was wrong to exclude poorly understood non-linear processes. [95] But the IPCC was only doing its job – laying out the relatively well-understood science. If Greens want to worry about things that are not yet established, they're free to do so. What they cannot do is claim the mantle of established scientific consensus.

It's also true that seven metres of sea rise all at once, or even all in a single century, would be bad news. And as we've said, there remain important uncertainties in climate science.

New processes have been discovered. Instead of flowing to the sea across the surface of ice, melt water can cut a crevasse downwards, going on to flow to the sea either through or under the ice. [96] In the words of Al Gore, the ice could be 'like Swiss cheese, metaphorically, and vulnerable to a sudden breakup'. [97] The worry is that this could lubricate the flow of ice, so that large parts of the Greenland ice sheet could simply slip into the sea.

Perhaps environmentalists are right that Greenland's ice will melt sooner rather than later. But a recent study suggests that the IPCC was right to be conservative. Using the Global

Positioning System, the study's authors measured the movement of Greenland's glaciers over time. They concluded:

> 'it has been suggested that the interaction between meltwater production and ice velocity provides a positive feedback, leading to a more rapid and stronger response of the ice sheet to climate warming than hitherto assumed. Our results are not quite in line with this view... the internal drainage system seems to adjust to the increased meltwater input in such a way that annual velocities remain fairly constant.' [98]

This, of course, will not be the last word in a rapidly advancing field of research. But two points have been established.

First, fears about Greenland melting are not based on settled science. On the contrary, they are based on fear of the unknown. Second, as science develops, worst-case scenarios don't always play out. In the 18 months since the IPCC published its last assessment, catastrophic scenarios have become less realistic, not more.

Everyone will agree that more study is needed. But as for what else mankind should do, it's clear that social attitudes toward risk and the Precautionary Principle will determine outcome more than further scientific insights.

Insuring future generations

Even if climate is not a guaranteed catastrophe, economists argue that spending a lot of money now to cut emissions is justified as a form of insurance. After all, people don't refuse to spend money on fire or health insurance on the grounds that that the worst will never happen.

For economists, the question of how much to invest in, say, clean energy involves the same sort of calculations that an investor makes when looking to maximise return. Those calculations work reasonably well for deciding, say, at what interest rate it's worth borrowing money to invest in a factory, or how quickly to pump oil out of a field to maximise profit.

However, over climate, economists have produced some peculiar results: first, when trying to factor in the Precautionary Principle to their calculations, and second, when trying to apply insurance to the long term future of the whole of society.

Stern acknowledges that most economic models show that 3° C of warming would be far from catastrophic. Indeed, he acknowledges that such a rise in temperature may even have beneficial effects:

> 'Up to around 2-3° C warming, there is disagreement about whether the global impact of climate change will be positive or negative. But, even at these levels of warming, it is clear that any benefits are temporary and confined to rich countries, with poor countries suffering significant costs.' [99]

SInce the IPCC's central projection is that a doubling of greenhouse gases will most likely produce a warming of 3° C, discussion might reasonably now focus on ensuring that *all* countries become rich – and on how to make benefits permanent. However, Stern uses 'uncertainty about the shape of the probability distributions for temperature and impacts, in particular at their upper end' to justify focusing on worst-case scenarios and magnify the costs. [100] He proceeds by expanding uncertainty along two dimensions, warning about 'surprises' in climate, and warning, too, about climate bringing about 'conflict, migration and flight of capital investment'. [101]

While the IPCC gives little basis for considering the kind of climate surprises Stern raises, he arbitrarily adds 'amplifying natural feedbacks in the climate system' to the IPCC assessment. [102] There have been remarkably few objections to this cavalier approach.

Another precautionary economist, based at Harvard, is Martin Weitzman. He also runs into problems when trying to convert the Precautionary Principle into an insurance premium. Paul Krugman, who won the Nobel prize for economics in 2008, described Weitzman's paper of the same year as 'driving much of the recent high-level debate' on climate. [103]

In that paper, Weitzman tries to juggle vanishingly small probabilities of climate catastrophe against costs that verge on the extinction of human life. [104] Showing himself to be a better mathematician than economist, he concludes that the costs of climate catastrophe are... infinite. Indeed, Weitzman muses that the industrial revolution – let alone future emissions – may have not been worth the GHGs it has created. He wonders whether conventional economics is simply not equipped to deal with the type of risk raised by climate. Finally he believes that nobody has the answers he is looking for. [105]

Weitzman's derives his pessimistic results on the basis that extremely large climate changes are possible. He justifies this partly by the Precautionary Principle, and partly on the grounds that science cannot tell us even the *scale* of likely change – whether it is likely to be 0.3, 3, 30 or 300° C. But here Weitzman is wrong. Science has established the scale of climate sensitivity at 3° C. It may be half, twice or conceivably three times that. But it is not 10 or 100 times as large.

More attention to what is realistic would dramatically bring down the costs of climate change projected by both Stern and Weitzman.

A second problem arises from the attempt to take a long-term view. The American economist William Nordhaus points out that Stern's projected costs of climate change – a 20 per cent cut in consumption per head, now and forever – are not what they seem. That's because Stern counts as costs today problems that will not arise for a long time to come. As Nordhaus puts it:

'the relatively small damages in the next two centuries get overwhelmed by the high damages over the centuries and millennia that follow 2200. In fact, if the Stern Review's methodology is used, more than half of the estimated damages "now and forever" occur after 2800.' [106]

Stern justifies his approach here by claiming that to neglect such long term costs would be a betrayal of future generations. In fact the approach is simply absurd. The costs that Stern imagines are from the spread of disease, effects on agriculture, flooding and so on. In reality even a discussion of 2100 seems highly speculative.

In the 22nd century and beyond, there will be new technologies, settlements in new locations, and no doubt new diseases and problems to deal with. But to imagine we can today anticipate the extent to which warming will aggravate – or relieve – these problems is to underestimate how far society is capable of progressing.

To imagine that we can sensibly discuss the consequences of warming in the centuries after 2800 is to detach oneself still further from reality.

For Stern, the relationship between present and future generations – like the possibility of human extinction – comes down to actuarial calculations and assumptions about discount rates. His only question is who inflicts damage on whom, and how much should the culprit should be made to pay. That's why he misses entirely the prospect that, through innovation now and in the future, the world will be made better and better for succeeding generations.

The mindset that begins and ends with insurance policies is a poor one with which to negotiate climate change. As an individual, you can insure your house – and if it burns down, the insurance company can compensate you.

But the long-term future of the world will not work out quite as simply as that.

Regulation cannot be a force for energy innovation

If – and it's quite a big if – international diplomacy and the Precautionary Principle win a new agreement on climate change, succeeding the Kyoto Protocol for the year 2012 and beyond, it will actually be nothing to celebrate. Regulators may pass laws and set targets for reductions in CO_2, but these guarantee nothing.

In 2005, the consultants McKinsey, often described as as the Jesuits of capitalism, pronounced that regulation, for decades the *bête noir* of free-marketeers, was a good thing. Acutely, McKinsey wrote:

> 'For companies in many nations, regulatory policy increasingly shapes the structure and conduct of industries and sets in motion major shifts in economic value. In network industries such as airlines, electricity, railways, and telecommunications, as well as in banking, pharmaceuticals, retailing, and many other businesses, regulation is the single biggest *uncertainty* affecting capital expenditure decisions, corporate image, and risk management. In the electric power industry, for example, the smallest price revisions can have a dramatic impact on corporate profits.' [107]

Regulation, McKinsey insisted, should become a core element of corporate strategy. Inside companies, a high-level executive with easy access to the CEO should run 'the regulatory function'. That function should expand way beyond the traditional role of compliance and periodic interaction with regulators. Instead, savvy companies should aim to be 'thought partners' for regulators. They shouldn't only manage regulatory risk, but also shape their industries and create potential opportunities for themselves. [108] In short – though McKinsey was only implicit about this – *regulation, properly handled, could be a force for innovation*.

In fact, regulation *cannot* be a force for innovation. Innovation in energy supply must take precedence over new rules, because realities will be determined by innovation, not by legislators.

In December 2007 the revered consultants in innovation, Arthur D Little (ADL), came to conclusions similar to those of McKinsey. Carbon – defined, worryingly, as 'greenhouse gases that include carbon dioxide, methane, nitrous oxide and ozone' – was helping to rewrite the rules of competition in business, locally and globally, by 'creating new opportunities for competitive advantage'. The thing to do was to go about 'creating a carbon-integrated strategy; and ADL hinted this should be done at board level. But in that strategy, innovation should *not* mean technological innovation. ADL said:

> 'Innovation plays a key role in carbon management for business protection and business creation. However, senior executives need to recognize that their business will gain most benefit from innovation that goes beyond exploiting carbon markets and new technologies. What's needed is innovation based on understanding how markets will look in a low-carbon economy in 2020; understanding your core competencies, now and in the future, as part of an effective partnering strategy; and understanding new routes to market.' [109]

McKinsey wanted the regulatory function to be a 'thought partner' of the state. For its part, ADL wanted the carbon function to go forecasting, checking competencies, partnering, and thinking about mechanisms of distribution.

These are not programmes of innovation; rather, they show the dominance of the regulatory mindset.

Regulation today is driven not by a commitment to genuine innovation, but by a political crisis of legitimacy and a strong aversion to risk. It is less direct than in the past, and less driven by the excesses of the market. But it is more pervasive, and – as we shall see in Chapter 7 – more perverse in its consequences than in the past.

For environmentalists and governments, 'doing something' means passing laws and agreeing treaties. For us, 'doing something' means taking action on the ground, and in the real world. We hold that more energy for the world will require each and every one of the key technologies discussed in the next three chapters to be planned, implemented, evaluated and improved.

In the global politics of regulation, the role of the EU is notable. With 27 members, the EU is now, as the *Financial Times* observes, the world's biggest economy. It is also the world's biggest single trading bloc, 'setting many of the world's de facto regulatory standards'. [110]

In energy, the EU's benign interventions include proposals that:

- manufacturers of domestic appliances cut, by 2020, the power that their machines use while on standby by 73 per cent [111]
- carmakers cut the average carbon emitted by new cars from 160g/km in 2006 to 120g/km by 2012, and to 95g/km by 2020. [112]

These measures seem innocuous enough. But they are far from free of problems.

Through regulation, the EU hopes to improve the energy efficiency not just of domestic appliances, but also of every kind of consumer product – lights, air conditioning units, PCs. Yet even if the EU never regulated these things, firms, in the 21st century, already compete to improve the energy performance of their goods. Sometimes it is cheaper to manufacture appliances that use a little more energy; but there's rarely a reason not to consider running costs.

Mobile phone operators are worried about the 'mobile footprint'. [113] Much of the international design community, including thousands of design students, is imbued with the idea that the new, right and proper mission of product design is to minimise carbon footprints. But as regulations multiply, the focus of the corporate innovation will tend to shift away from all-round improvements toward compliance with decrees. In particular, complying with regulation takes time – something that small and medium enterprises, often a source of innovations, have little of.

Strikingly, when carmakers fail to make the EU's regulatory targets, they will be fined – and *the fines will fund innovation*. Errant manufacturers, the European Parliament suggests,

> '... should pay an excess emissions premium in respect of each calendar year from 2012 onwards. The premium should be modulated as a function of the extent to which manufacturers fail to comply with their target. It should increase over time... to provide a sufficient incentive to take measures to reduce specific emissions of CO_2 from passenger cars, the premium should reflect technological costs. The amounts of the excess emissions premium should be considered as revenue for the budget of the European Union and used to increase support for CO_2 reduction research and innovation activities in the automotive sector.' [114]

It all sounds great, doesn't it? The Brussels Commission can cane manufacturers into behaving, and make regulatory fines a force for even more innovation.

The trouble is that heroic targets, as Joseph Stalin found out, don't necessarily make for heroic results. California's attempt to legislate zero emission vehicles is instructive here. Without the technology becoming a reality, the legislation had to be scrapped.

In 1990, the California Air Resources Board (CARB) passed the Zero Emission Vehicle Mandate, requiring a rising percentage of California's cars to be free of emissions. Not too long after that, however, US car manufacturers negotiated a Master Memorandum of Agreement with CARB making the mandate require them to build the electric car only to the extent that there was consumer demand for it. By 2003, the date at which 10 per cent of new vehicles were meant to be zero emission, CARB chairman Alan Lloyd ended the Mandate. [115]

The lesson is that passing a law or making a regulation demanding higher environmental standards can lead nowhere.

There is a place for regulation. Regulation can ensure that businesses stick to standards that have been agreed upon as socially acceptable. The public needs to be protected from serious hazards and businesses need to work on a level playing field. But regulation is most effective in codifying the status quo. It's too blunt to be a consistent agent of change. If the demands of regulation are too far out of line with what's technically possible, or with how people behave, then regulation is experienced as *diktat*, and is actively resisted where it is not simply ignored.

Socially acceptable standards deserve full political debate before regulations are adopted. That much is confirmed by the example of the EU creating regulated markets for CO_2. These were certainly not subjected to popular European debate – and the results have been decidedly mixed.

Capping and trading CO_2 cannot be a force for innovation

We saw in Chapter 1 how Ronald Coase effectively preferred the *state putting a market price on the right to pollute* to pollution taxes. The Kyoto Protocol gave much impetus to this second strategy.

The Protocol demands that developed economies cut GHG emissions by five per cent, 2008-12. Since January 2005, the EU's Emission Trading Scheme (ETS) has tried to meet Kyoto's provisions by setting a limit on the aggregate CO_2 emissions made by 11,500 energy-intensive industrial facilities – including power stations; together, these are responsible for nearly half the EU's CO_2 emissions and, thus, 40 per cent of its GHG emissions. [116] The ETS has accounted for more than 80 per cent of the world's market for CO_2. [117] It can therefore be taken as the prime example of the world's attempts to cap and trade CO_2.

How is the ETS supposed to operate? At the end of each year, firms in the EU emitting less GHGs than the amount allowed under National Allocation Plans (NAPs) are able to sell their excess allowances on exchanges, while those emitting more than their quota must either clean up their act, and/or buy the extra allowances they need on the market. The hope has been that, by engineering a scarcity of allowances, NAPs would force cuts in emissions.

It has not worked out that way.

The first trading period of the ETS ran from 2005 to 2007. Here, a number of EU member states gave away, for free, too many allowances – each of which gives the right to emit one tonne of CO_2. That lowered the price of allowances, adding to difficulties already experienced in verifying data and harmonising allocations between different member states.

The EU's top bureaucrats failed to anticipate this development. As they blithely put it,

> 'The Commission has no view on what the price of allowances should be. The price is a function of supply and demand as in any other free market.' [118]

How a market created by Brussels could function like 'any other free market' seemed to escape the collective eminence of the Commission.

In the second trading period, running from 2009 to 2012, the EU, now supplemented by Norway, Iceland and Liechtenstein, intends to cap national emissions at an average of about 6.5 per cent below 2005 levels. It also intends to give fewer allowances away, and instead auction more. For the third trading period, running from 2013 to 2020, the EU intends, in 2009, to

- establish one EU-wide cap instead of 27 NAPS
- keep cutting this cap till past 2020, at the rate of 1.74 per cent each year

- auction 60 per cent of allowances in 2013, and higher percentages in later years
- extend the ETS to new sectors (petrochemicals, ammonia and aluminium), and to GHGs beyond CO_2 (N_2O emissions from the production of acids, and perfluorocarbons from the production of aluminium).

Altogether, and without including all the modifications of the second and third trading periods, these measures are meant to cut EU CO_2 emissions from about two to about 1.7 Gigatonnes. [119]

The best-laid regulations, however, can go awry. In July 2008, a one-tonne CO_2 allowance cost €29.33; by November 2008, it cost merely €18.25. As recession hampered economic activity and so diminished emissions, so firms needed to buy fewer allowances to prove themselves clean. As Carl Mortished, world business editor of the *Times*, commented, the ETS had made 'a mockery' of Europe's 'stumbling attempts to lead the world in a market-based carbon strategy'. [120]

In revising its own regulations, the EU's administrators ask themselves no fewer than 34 questions about it. [121] That just might suggest that regulation cannot dynamise innovation. Yet back in 2005, McKinsey was emphatic that it could. Kyoto's implementation was reshaping international energy markets. The ETS had created a multibillion-euro market for CO_2 emissions certificates. It had

> '... reshaped the incentives for electricity production as generators switch from coal-burning to natural gas-fired plants to achieve lower levels of CO_2 emissions, for example. The strategic landscape is being redrawn as a result.'

A Europe-wide electrical utility, McKinsey reported, had brilliantly modelled how best to allocate CO_2 emissions certificates before the Protocol's implementation. It had used that magical thing, *a holistic perspective*, on different national markets within the EU. And so? Its final allocation plans 'highlighted arbitrage possibilities' in:

- replacing capacity or building new capacity in neighbouring markets
- re-importing electricity through the European power grid. [122]

So the reality of the nation state within the schemas of international energy regulation makes for innovation – but innovation in the sense of *arbitrage*, or profiting from the differences in prices between different markets; or in the sense of *re-importing* electricity.

How terrifically innovative!

From Kyoto to Copenhagen

In 1997 the Kyoto Protocol committed developed nations to a five per cent cut in emissions below their 1990 levels, as measured over the period 2008-2012. At the end of 2009 talks are set for negotiations in Copenhagen to replace the Kyoto Protocol.

First, it is worth assessing Kyoto. From 1990 to 2006, EU emissions fell by 4.6 per cent, with the record of individual countries varying from the UK's 15.6 per cent cut to a rise of 53.5 per cent in Spain. Over the same period, Japan's emissions rose by 5.8 per cent. In the US, which signed but did not ratify Kyoto, emissions rose by 14 per cent, while in Canada, which did ratify the treaty, emissions rose by 54.8 per cent – mainly due to the development of oil sands.

In total, countries committing to cuts under Kyoto, including the US, reduced emissions between by 4.7 per cent. But most of that was due to the collapse of industry in Eastern Europe following the end of communism. For the Eastern European 'economies in transition', emissions fell by 37 per cent. The remainder of the developed world increased emissions by 9.9 per cent. [123]

Eastern Europe shows that economic collapse is a route to reduction of emissions, albeit a destructive one. In the case of the UK, reductions were achieved by a shift from coal to gas. Many of the easier cuts in non-CO_2 GHGs, for example in agriculture, have already been made; so future cuts are likely to be harder and concentrated around energy. Investments in clean energy will begin to have a greater impact over the coming years – but falls in emissions are unlikely to meet Kyoto targets by 2012 unless the economic downturn following the Crash of 2008 proves to be prolonged and deep.

To meet their targets, countries in the developed world are relying on the Clean Development Mechanism (CDM). Instead of setting the developing world targets, Kyoto set up the CDM as a scheme that allows the developed world to meet targets by paying for emissions reductions in the developing world. But the record of the CDM is not very encouraging.

In principle, investments under the CDM must be in projects where the emissions would not otherwise have been cleaned up, known in bureaucrat-speak as 'additionality'.

In practice, as with emissions trading generally, the CDM has been surrounded by suspicion of scams and corruption. In some cases it appears to have stimulated emissions for the sole purpose of cleaning them up to gain credit, as in Chinese and Indian factories producing hydrofluorocarbon refrigerants. [124]

The CDM has not brought much genuine clean technology to the developing world. Instead, it has acted as a means for the developed world to avoid the real challenge: using innovation to develop cheap clean energy.

No doubt Copenhagen will attempt to correct what are seen as the weaknesses of Kyoto. In some cases, however, the mistakes could be magnified. If conservation of forests is written in as an alternative to technological innovation, then investments from the rich countries in preventing deforestation in poorer countries that want to develop their land may provoke far more tension than has been the case with the CDM.

The biggest source of tension generally will be between

the developed world and the rising economies led by India and China. From 2000 to 2008, European smugness had an outlet in the figure of George W Bush. With Obama in the White House, it is likely that the East, and China in particular, will loom larger as a target for Green opprobrium.

It also seems likely that the developed world will try to agree a global cap and trade system. But how the countries of the East will fit into that system is unclear. At present, their negotiating position is a robust one: that they cannot compromise on growth. Whether that position will shift remains to be seen. But China has already in some respects gone further, demanding that the developed world spend at least one per cent of its GDP on technology transfer. [125]

Two key factors will be more important than paper targets and markets for CO_2. First: innovation to cheapen clean energy. Realistically, it is this that will set the pace at which emissions are cut. The biggest danger at Copenhagen is that cuts will be agreed that try to go too fast, too soon, by relying on efficiency and cutbacks instead of ambitious, long-term R&D.

The second factor is the overall pace of economic growth in the world economy. A slowdown may, at first, appear as good news for emissions, which will fall for a year or two. But aside from its disregard for economic welfare, that perspective is shortsighted. In the longer term, energy will only be cleaned up by new investment and replacement of the energy infrastructure. That demands economic growth.

Consider again China's demand for technology transfer. If the economy of the developed world were to grow for a single year at three per cent rather than two per cent, or two per cent rather than one per cent, then that extra growth would be sufficient to fund a clean-up of China's energy sector.

A small amount of extra growth makes a big difference.

Summing up this chapter

The balance of evidence suggests that man-made global warming has outrun man-made global cooling – and that man is the key factor in climate nowadays.

But *the balance between certainty and uncertainty in climate science* is also important. We've argued that the certainties are much greater than they were when Roger Revelle discovered global warming.

Today's remaining uncertainties deserve full debate. They don't, however, justify Greens in what might be called not a *ladder of nuclear escalation*, but a greasy *slide of climate categories*. On that slide,

1. Selfish consumption and untempered economic growth boost global warming
2. Feedback effects raise temperatures still further
3. Greenland's ice, or ice somewhere else, melts irreversibly
4. Sea levels rise, mass migrations to higher land begin
5. With climate as a whole, non-linear, chaotic and tipping-point behaviour grows – irreversibly
6. Climate proves capable of infinite surprises
7. The world goes to the dogs.

Yet enough now is known about climate, in fact, to say that melodramatic leaps, switches, or flips are unlikely. If they do happen, science is also unlikely to be taken by surprise. The picture with climate is evolution more than revolution. If global warming speeds up still further, an attentive science community is likely to spot the trend.

Mankind shouldn't lose its nerve. It has some years yet to develop a more rational energy supply. Here, it's time to ditch illusions in:

• science as consensual and precautionary
• taxes and permits as wise insurance policies for future generations

- regulation and international treaties as dynamos of innovation.

The New Scientism insists that climate science has met the end of its history and means you must change your habits. By contrast, we argue that climate science remains open. Our *interpretation* of it, however, suggests that *mankind can and should take the time to build a bigger, better and fundamentally cleaner energy supply.*

Has global warming stopped since 1998?

According to many climate sceptics, global warming stopped in 1998 – or perhaps in 2000. [126] Others who accept the IPCC's science as a basis of policy – Nigel Lawson and Bjørn Lomborg, for instance – also make the point that warming has stopped. They do that as a reminder that the science is not really settled. [127]

The issue is an interesting example of the distinction between climate and weather. Sceptics argue that the past 10 years are enough to characterise climate, while mainstream science sees variation over that period as more akin to weather.

Here's a chart of global average surface air temperatures over more than a century:

Global average surface air temperatures, 1880-2007: change from 1951-80 average baseline, º C [128]

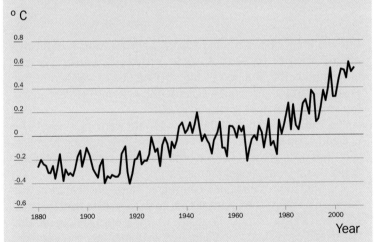

Now here's a close-up of the same temperature data over the years beginning in 1998. Choosing 1998, an exceptionally hot year, as a starting point disguises the long term trend... but that's the way climate sceptics make their argument:

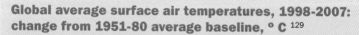

Global average surface air temperatures, 1998-2007: change from 1951-80 average baseline, ° C [129]

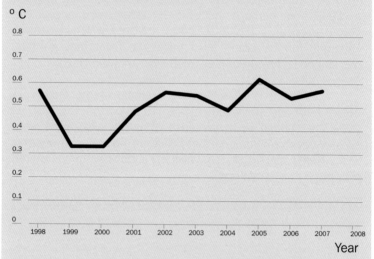

The first chart shows a trend toward warming, with fluctuations around it. The causes of some of these fluctuations are understood. The eruption of Mount Pinatubo, a volcano in the Philippines, brought about a dip in 1992. A strong El Niño caused the striking peak that appeared in 1998.

The acceleration of the trend toward warming since 1975 is clearly evident – and, what's more, the past decade is in line with that acceleration. On the other hand, recent temperatures could be taken as roughly constant.

Statistical techniques can refine these observations, but they don't change the basic picture. Looking at data over the past 10 years, it's consistent with global warming having stopped. But it's just as consistent with continued warming. Taken in isolation, it just isn't very informative.

Nobody would argue that a single cool day, month or year disproved warming. But what about two years? How many are needed?

Variability in temperature data is such that 10 years is just too short a period to comment on. To see whether or not the trend has levelled off or is continuing, meteorologists probably need to make observations till 2015.

If the theory of global warming rested on global surface air temperature series alone, it might be in trouble. It certainly couldn't be held with much confidence.

In fact, however, mankind's understanding of climate rests on a vast web of interlocking observations and theory. For example, there are observations not just of global temperature, but also of how temperatures vary

- in different regions
- across the oceans and in their depths
- through the atmosphere, between day and night
- between the seasons.

A particularly important observation is of sea temperatures. Heating water needs more energy than heating air (think of heating a pan on a stove). Although the oceans also show some variability, the warming is slower and steadier, and so more revealing of the long term trend.

The latest data on ocean warming go up to 2003, and show no slowdown in warming. [130]

There are observations of other variables such as precipitation and wind. And there are also connections to other parts of science, where both theory and experiment allow predictions to be made.

All these pieces of evidence underlie the proposition that CO_2 is having and will continue to have a warming effect. [131] There is no space in this book to explore most of that evidence. But given everything else that is known about climate, including the long-term trend in global temperatures, we think it very unlikely that warming has stopped.

However, the fact that on short time scales it is hard to make out the warming signal does underline that change is presently gradual rather than catastrophic. As Lawson and Lomborg suggest, it should also remind people to keep an open mind. New data needs to be kept under constant review.

Is climate science now settled?

When the climate protest group *Plane Stupid* attacked the expansion of Heathrow airport, it held up banners claiming it came 'armed … only with peer-reviewed science'. [132] Among campaigning journalists and Green bloggers, too, peer-reviewed science has become a totem for the ecologically correct. It's held to have confirmed the case for action, and in conclusive style.

Climate sceptics often agree that mainstream science presents a monolithic picture. For them, however, the granite-hard agreement is a result of a nefarious conspiracy, through which peer review crushes dissenting voices.

Neither of these positions holds up to scrutiny. An examination of the scientific literature shows that plenty of doubts are voiced.

The amount of research going into climate right now is vast. As a result, it's easy to find articles in leading journals proclaiming that:

- in ocean mixing, 'much remains to be discovered' [133]
- predicting how changes to the *stratosphere* will affect surface climate remains 'a substantial task' [134]
- the imprecision of computer models of atmosphere and oceans is 'irreducible' [135]
- with forests, global models of the biosphere-atmosphere system are 'still in their infancy'. Extrapolating from lab experiments or site-specific field studies to large scale climate models remains 'a daunting challenge' [136]
- models of climate addressing the next few decades differ in the regions for which their predictions are most accurate; there's still 'much to be understood'. [137]

Nor do many policy reports, when closely read, back up the impression often given by newspaper headlines – that 'the science is in'.

It's true that policymakers are fixated on science. But as this chapter's treatment of the Stern Review shows, what are really obsessed about are the *limits* to science.

Science, it's claimed, has shown the *possibility* of disaster. Therefore science just isn't strong enough to be able to rule disaster out, and often cannot even quantify its *probability*. In fact, though, Greens, governments and climate sceptics alike present everything as scientific so as to avoid political arguments.

For *Greens*, human actions are limited by nature. Since science is the study of nature, Greens are bound to elevate it so it becomes an all-powerful oracle.

For *governments*, science provides a basis for consensus at a time when they possess no big visions that can command mass loyalty. Not just in climate, but right across the board, government managers appeal to what's termed *evidence-based policy*. Old-fashioned ideologies no longer get a look-in.

For *climate sceptics*, emphasising science is a way of avoiding political combat. Sceptics find it hard to make a substantive reply to Green politics, often viewing it simply as a continuation of left-wing thought. Moreover, free-market sceptics are often suspicious of politics altogether, seeing it as little more than a grubby interference with 'natural' economic processes.

Of course any critic of climate alarmism needs to answer the scary scenarios put forward by the Greens. But the case cannot be won on technical grounds alone – and attempts to do so inevitably lead to a distortion of science.

More importantly, the argument that the science isn't settled is not the trump card that sceptics believe it to be. After all: if, like Greens and governments, you believe in the Precautionary Principle, then uncertainty in science is a reason to panic *more*, not less.

Nuclear Power: Forget Doctor Faustus

Late and uniquely artificial as a means of energy supply, nuclear power should no longer be demonised. It can help meet the world's need for energy in a big way, now

4

'We nuclear people have made a Faustian bargain with society.' So wrote America's nuclear chief Alvin Weinberg (Chapter 2), in 1972. Nuclear scientists, Weinberg argued, promised society cheap, inexhaustible energy which, 'when properly handled', was 'almost non-polluting'. However, in return, the need was to develop 'a vigilance and a longevity of our institutions that we are quite unaccustomed to'. The need for vigilance followed from the danger of a catastrophic nuclear accident; the need for longevity, from the spectre of long-lived nuclear waste. [1]

Societies had made similar choices before, Weinberg added. Humanity's move into agriculture, once accomplished, demanded that fields be tended forever; Dutch dikes required eternal maintenance. Weinberg concluded that 'society must then make the choice', a choice that 'we nuclear people cannot dictate'. For him, though, investing in nuclear energy seemed 'well worth the price'.

Since Weinberg, environmentalists have come to different conclusions. For them, the bargain post-war nuclear science made with society hasn't been worth the price at all. With nuclear matters, Greens worry about *safety, waste, terrorist attacks, proliferation, costs and secrecy*. Long able to slow the building of nuclear reactors, Greens have put the West's nuclear industry on the defensive.

The East is building nuclear power much faster than the West. But instead of forgetting Dr Faustus, the West now wants to control what it sees as the East's embrace of the nuclear bogeyman.

Legend has Faust selling his soul to the Devil in return for knowledge – and thus power. Traditionally, Faust was the scientist tempted to embark on reckless experiments in pursuit of learning. The legend captures how experimentation has always raised uncomfortable problems, even as it portends progress. But in the more optimistic tellings of the story, such as Goethe's, Faust in fact outwits the Devil. [2]

With Weinberg the story is changed. Nuclear scientists are portrayed not as Faust, but as *the Devil*. It's not them who have to make a fateful choice, but society. Altogether, Weinberg confirmed nuclear science and the energy available from the atomic nucleus as *demonic powers that are outside normal society*.

Weinberg's Faust fits with more general thinking since 1945. Alongside the idea of consumer society (see Chapter 2), technology has come to be seen as an unstoppable, alien force that is beyond the ambit of consumers, though exercising a great influence on them. Thus the Internet is a 'driver' of globalisation, and genetics imperil the family. Or so we're told...

Before the acclaimed Internet Age and the Biotech Century, mankind was supposed to be in the Atomic Age. Nuclear power, then, always bolstered theories that had society as determined *by,* not the determinant *of,* technology. In the flawed post-war framework of *technological determinism*, nuclear power in the end appeared as intrinsically so brutish and military in origin, its entire trajectory could only be downhill. [3] Mankind's nuclear dabblings were thus *scripted* to go awry, irrespective of social regime, economic conditions or political priorities.

In Chapter 3 we mentioned how Robert Oppenheimer, the father of Hiroshima, framed the power of the atom in fully apocalyptic terms.

But a chain reaction in the atomic world, or in a reactor, need not lead directly to a conflagration for society.

Uniquely artificial, with a uniquely high energy density

As a form of energy, nuclear is the latest on the world scene. It's also the one in which science and the military have been most involved. Nuclear fuels require a deeper interference with nature than fossil fuels or renewables.

For Greens, then, nuclear energy's uniquely artificial character makes it *sinister* – and its military associations confirm the point. Historically and logically, nuclear appears to have been Faustian from the outset.

The artifice of nuclear emerges in *the sheer quantity of energy on offer from very little fuel*. For those who want to conserve energy rather than meet new demand for it, that's bad. But handled with care, and often generating more than 1 GW in a single plant, nuclear materials show just how much mankind can transform energy supply.

Why does 'nucleonics' punch above its weight? [4] Chemical reactions, such as burning fossil fuels, concern the electrons that orbit nuclei within a tenth of a millionth of a millimetre. But nuclei themselves are a hundred thousand times smaller than that. [5] Now, it's a basic law of quantum physics that smaller distances mean higher energies. For this reason, the physics of sub-atomic particles is called *high-energy physics* – and the energy released by nuclear engineering is colossal.

In chemistry, nuclei are unchanged, and relatively low amounts of energy are released. But if, bombarded by neutrons, a nucleus shatters, it unleashes enormous amounts of energy. Thus nuclear fuels have a much higher *energy density* than chemical ones:

Energy densities of some non-nuclear and nuclear fuels, kilowatt-hours per kilogram [6]

Fuels	kW-h per kg
Coals	3
Oil	4
Uranium	50,000
Uranium after processing	3,500,000

Note that spent nuclear fuel can be *reprocessed*. In the reprocessing, unburned fissile uranium, newly created plutonium and newly created fissile uranium are drawn off and separated from high-level waste, and the bulk of the energy in the original fuel is recovered. This allows far more energy to be extracted from each original kilogram of uranium. Instead of a handful of kilowatt-hours per kilogram, reprocessed fuel can deliver millions.

Does the energy density of nuclear fuels, then, make them *diabolically powerful*? To argue this would be to scapegoat the subatomic world, nuclear physics and nuclear engineering for modern society's loss of political and moral direction. If society can regain direction, then nuclear energy would be less like black magic; it would be less enigmatic and less frightening.

People don't mind unseen electrons working for them in the realm of IT. Similarly, hundreds of millions of people have learned to live with unseen nuclear reactions delivering energy to their homes. Nuclear physics is no longer a mystery. *It's simply retrogressive to hold high-energy exercises in the sub-atomic realm as above and beyond man's capabilities.*

Why we favour nuclear power

The difficulties that undoubtedly surround nuclear power are neither technological, nor to do with physics. Much depends on how society chooses to handle those difficulties. What, then are the positive reasons *for* nuclear power?

First, *the world needs lots more cheap energy – and nuclear power stations can help meet that need quite rapidly.* It's a sad fact that renewable sources of energy cannot immediately meet the rise of energy demand, whether in the UK or in Asia (see Chapter 6). But after 50 years of mostly trouble-free operation, today's nuclear power station designs represent a mature technology that can be installed *en masse*.

Reactors are ready to go. As we shall see shortly, no fewer than 300 are under construction or planned. Whereas wind turbines can be rapidly deployed in numbers today, their capacities rarely extend much higher than 3 MW. And, given all their problems of intermittency and energy storage, 1000 wind turbines are needed to generate energy equivalent to that put out by a typical nuclear power station.

We deal with the economics of nuclear energy at the end of this chapter. But this much is clear already: *the unique energy density of nuclear fuels makes them providers of prodigious amounts of power.* In turn, high energy density ensures that *fuel costs with nuclear reactors form a very modest part of running costs.* This is our second reason for backing nuclear power.

Although uranium mining, like mining and drilling generally, is a hazardous business, to extract the nuclear ore necessary to generate a kilowatt-hour of electricity is a much less labour-intensive affair than extracting the coal, oil or gas needed to generate the same kilowatt-hour.

Ironically, mining nature for nuclear fuels requires much less human effort than the search for fossil fuels. Thereafter, energy density ensures that, all along the supply chain to reactors, the costs of manipulating and transporting enough nuclear fuel to generate a kilowatt-hour are low compared with the cost of manipulating and transporting fossil fuels. Tucked away in an annex of a UK government White Paper published in 2006, the chart below confirms that as an input to reactor operating costs, fuel counts for little:

Indicative composition of nuclear running costs over a year [7]

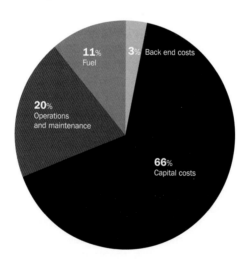

Because fuel costs with nuclear reactors form a very modest part of running costs, nuclear is ideal for baseload electricity generation. Running reactors flat out costs little more than running them slowly. While output from a gas-fired station can be racked up, rather expensively, to meet peak demand, a reactor is typically always on and delivering maximum output. Like coal (see Chapter 5), nuclear is the reliably powerful workhorse of future electricity generation.

There's another way of formulating the low running costs of nuclear. We uphold nuclear plants because we believe that the large-scale development of fixed capital testifies to the progress, such as it is, achieved by capitalist society.

It costs a lot of money to build a nuclear facility, and those costs are only amortised over a number of years. The relatively low running costs of nuclear power are something that it shares with renewables, and differentiate it, too, from fossil fuel plants. In the latter, coal and gas are continuously bought on the open market and respectively comprise a large and very large proportion of running costs. By contrast, nuclear embodies not so much repeated market transactions as a political and economic commitment to making energy for years and years. That kind of commitment is worth defending in its own right.

Thirdly and finally, we favour nuclear power stations because in operation they generate no CO_2. Famously, this is the line taken by James Lovelock, inventor of the Gaia theory. [8] It isn't our primary argument, but in the course of building a new round of nuclear power stations, it represents an advantage over plants based on fossil fuels, and especially over coal-fired power stations. [9]

Today's nuclear revival

In 2006, just 439 nuclear reactors supplied 16 per cent of the world's electricity, a quarter of OECD countries' electricity, and more than three quarters' of France's. Nuclear engineers could boast of more than 13,000 years of operational experience. [10] Despite fears, a nuclear revival is underway.

In a risk-averse world, however, the revival is modest by historical standards. The IEA estimates that the planet needs between 24 and 32 new 1000 MW nuclear power stations each year between 2010 and 2050 – in other words, for between 960 and 1280. But as the IEA has also noted, between 1977 and 1993 France alone brought an average 3.6 reactors into operation each year. Given that world economic activity in 2005 was about 30 times that of France in 1985, the IEA points out, scaling up France's old nuclear programme to the whole world today would mean building 100 reactors a year – equivalent to adding 160 GW a year if each had the capacity set by Finland's Olkiluoto project, a 1.6 GW Pressurised Water Reactor (PWR). [11]

One hundred reactors a year sounds like a lot. But in the 1980s, a new one opened almost every two weeks, until the 1986 explosion at Chernobyl slowed the pace. [12]

The good news today is that, for the first time in 30 years, construction has resumed in the West. The East is also seeing dramatic growth. After years in which uranium markets were glutted and many reactors could be run on material from old Soviet warheads, demand for uranium is picking up. New mines are being explored in Africa and elsewhere. In the US, construction of plants to enrich nuclear fuel has revived.

Aged Western nuclear firms now see opportunities in the East. America's Shaw Group and France's Areva have each won major Chinese contracts. In the East there's a chance for the West to build up-to-date designs that only exist on paper back home. Thus Westinghouse, which is owned by Toshiba, will see the first implementation of its new AP1000 reactor design not in the US, where 12 are planned, but in two pairs of reactors at Sanmen and Halyang, China. Electricity generation is set for 2013 – and 5000 US jobs will be created by the project. [13]

Much of Western nuclear innovation is today based on simplifying reactors so as to cut costs and increase safety. That's the rationale behind the fourth generation (4G) reactors of the future. There are also smaller designs, to suit rural electrification. Partly such emphases reveal diminished ambitions;

still, more people should now be able to gain their first access to nuclear electricity.

Significantly, the more serious nuclear innovations now come from the developing world.

In South Africa, Pebble Bed Modular Reactor (Pty) Ltd plans to build eponymous reactors – the biggest change in nuclear reactor design since its inception. The company hopes to encapsulate fuel in thousands of graphite spheres rather than rods, making refuelling continuous, easy and safe. Construction is set to begin in 2009; plans are for 4-5 GW to come from 20-30 pebble bed machines rated at 165 MW each. [14] South Africa is also collaborating with China on PBMRs, not least because China was first to develop an experimental plant.

Meanwhile India, which has little indigenous uranium, has for some time been developing 'breeder' reactors that convert thorium, which is thought to be three times more abundant than uranium, into usable uranium fuel. [15]

The East is where the action is

Among developed nations, France and Japan have few domestic fossil fuel resources, and have looked to nuclear energy as an alternative. Behind only the US, they lead the world in terms of numbers of reactors generating electricity. But the centre of new nuclear development has shifted East: to Asia, Russia and Eastern Europe. South Africa is the only African country with definite nuclear plans, although these are at a relatively early stage.

Overall, the picture for reactors looks like this:

The East is going nuclear more than the West: reactors operable, under construction and envisaged, and planned capacity increases, GW [16]

	Operable	Being built	Planned / proposed	New GW, date
USA	104	0	32	
France	59	1	1	
Japan	55	2	13	9 by 2015
UK	19	0	0	
Germany	17	0	0	
Sweden	10	0	0	
Brazil	2	0	5	
Russia	31	7	35	22 by 2020
India	17	6	19	16 by 2020
China	11	7	100	40 by 2020
South Korea	20	3	5	12 by 2017
South Africa	2	0	25	
Rest of world	92	10	76	
Total	**439**	**36**	**311**	

In contrast with vibrancy in the East, nuclear power in the UK is going nowhere fast. Once a pioneer in the genre, the UK today continues in the more recent tradition of governments outsourcing policy – miring nuclear in consultations, deliberative workshops and stakeholder events.

It's worth recalling that, as late as 2003, a government White Paper declared that the current economics of nuclear made new power stations 'unattractive', and so refused to support building them. [17] More than five years later, the formal policy has changed, but real practice is little different.

Defensiveness in the UK

In the UK, environmentalists are certain that the government is all in favour of nuclear power and, indeed, in hock to the nuclear lobby. At one level, official pronouncements on nuclear power do suggest that the authorities favour it. On close inspection, however, *support in principle turns out to be prevarication in practice.*

At his 'addiction to oil' press conference, Gordon Brown made a call for 1000 new nuclear power stations to be built worldwide. But his pro-nuclear rhetoric was qualified by the note that such a programme would have 'serious implications' for 'security as well as cost and change' (sic). He went on:

> 'While I know there are nuclear protesters who object to any nuclear power, they need to know if they had their way, the resulting energy crisis would bring less security, more instability, faster climate change and more poverty.' [18]

On the surface, this appears to be a confident, aggressive response to Green critics of nuclear. But in fact Her Majesty's Government shares the apprehensions of those who abhor nuclear power.

Given that regulation cannot be a force for innovation (Chapter 3), *Energise!* does not advocate nationalising nuclear energy in Britain. But the fact remains that the government has at no time said that it will fund, still less build, or operate, nuclear power stations. Equally, industry refuses any forthright defence of nuclear, preferring to wait, diplomatically, for the next move by the government.

The government will pay neither for recycling nuclear waste, nor for decommissioning nuclear power stations. Instead, on 10 January 2008, introducing its response to a public consultation on nuclear, John Hutton, Secretary of State for Business, Enterprise and Regulatory Reform (BERR), merely 'invited' energy companies to 'bring forward plans' to build and operate new reactors. [19]

The 186-page White Paper that Hutton published on the same day was just as indecisive. Still, it had a confident title: *Meeting the Energy Challenge.*

The White Paper deserves in-depth discussion here because it highlights the feeble character of nuclear advocacy today. Here's a guide to its overall flavour:

Key concepts and number of mentions in the UK government's January 2008 White Paper on nuclear power [20]

Concepts	Number of mentions
Security (of installations)	about 110
Risk	135
Decommissioning	137
Climate	142
Security (of energy supply)	about 150
Emission	209
Regulation	216
Safety	233
Cost (excluding cost-benefit)	383
Waste	444
Benefit (excluding cost-benefit)	54
Research, in the sense of R&D	23
Growth (economic/in energy demand)	10
Innovation	1

Climate change and energy security dominate the government's rationale for nuclear power, with the need to generate more electricity far behind these two considerations. Meanwhile, *safety, risk, physical security, costs, waste and decommissioning* weigh heavily with officialdom. With these obsessions as the ground rules, nuclear power cannot win. Only if society's need for more energy is put in the foreground can nuclear power's supporters expect to be victorious in arguments about it.

Exactly how many nuclear plants does the UK need? How many megawatts should they generate? How fast should they be built, and where, exactly? The government gives no answers to these questions, and for a reason. Seven times in its White Paper, it repeats this point:

> 'The fundamental principle of our energy policy is that competitive energy markets, with independent regulation, are the most cost-effective and efficient way of generating, distributing and supplying energy. In those markets, investment decisions are best made by the private sector and independent market regulation is essential to ensure that the markets function properly and in accordance with our wider social and environmental objectives, particularly tackling climate change.' [21]

This is a *hands-off* approach to the future of UK energy supply.[22] Shrugging off any need for leadership, the government leaves decisions on actually building nuclear reactors to the caprice of market forces. Still, it's deeply committed to regulation. BERR, DEFRA and the Environment Agency each have responsibility for nuclear affairs. Apart from, but sometimes within, these government departments, there are further regulatory agencies at work:

Just some of the bodies charged with looking after nuclear affairs in the UK

1. Committee on Medical Aspects of Radiation in the Environment
2. Committee on Radioactive Waste Management (CoRWM)
3. Nuclear Decommissioning Authority
4. Nuclear Directorate of the Health and Safety Executive Health Protection Agency Radiation Protection Division
5. Office of Nuclear Development

As recently as 2008, in an innovatory, single-minded approach to nuclear power, John Hutton announced the creation of two new bodies: the Nuclear Development Forum, a state/industry talking shop, and the Office of Nuclear Development, designed to improve interdepartmental collaboration across government on nuclear matters.

The civil servants who wrote the White Paper are bold enough to exclaim about 20 times that nuclear power *'cannot be excluded'*. Here's a typical formulation:

> 'Ruling out nuclear as a low-carbon energy option would significantly increase the risk that the UK would fail to meet its CO_2 reduction targets because we would be placing greater reliance on fewer technologies, some of which have yet to be proven on a commercial scale.' [23]

This isn't a positive argument for more electricity and more nuclear power. It's a logic based on fear and the Precautionary Principle. Nuclear is here presented not on its own merits, as a high-tech, low running cost form of energy supply, but rather as an insurance policy against climate change, and against external threats to Britain's energy security.

For the government, nuclear isn't essential to economic growth. Instead, growth is subordinate to fears of CO_2:

> 'Without a healthy economy, the UK would not be in such a strong position to play a leading role in helping develop the new, innovative low-carbon forms of electricity generation needed to tackle climate change globally.' [24]

This is the White Paper's single reference to innovation. Yet even here HMG won't argue for nuclear power on its own merits. Instead, it enters the barren terrain of the CO_2 emissions at issue in every stage in the life of a power station (see Chapter 1). It then concludes that those for nuclear 'are about the same as those for wind generated electricity'. [25] This battle over CO_2 ends in a draw. Nuclear cannot be declared the winner.

Cutting CO_2 is nowadays the only politically correct justification for nuclear power. Still, let's now consider the government's fallback line of defence: energy security.

In Chapter 2, we showed that fears of dependence on overseas sources of oil have a long history. In the White Paper, the words 'option,' 'diversity' and 'mix' get no fewer than 250 mentions. Partly the text concerns diversity of low-carbon technologies, but mostly it's about keeping options open in case of overseas supplies of energy being 'politicised'. Thus:

> 'Diversity of energy sources can help to reduce our dependence on gas as reserves fall in the North Sea and reduce the impact on the UK should prices for fossil fuels rise globally.' [26]

It's true that there's no silver bullet in energy supply. To build supply quickly, cheaply and cleanly enough to meet the needs of the British economy is something that can't be done using a single technology. However, the idea that nuclear is needed to safeguard security of inexpensive supply doesn't amount to a convincing case for nuclear power.

To its credit, the White Paper repudiates Green fears about 'peak' uranium, citing, among other sources, evidence from the IPCC. [27] But then it drops the ball:

> 'We also recognise that, with no significant indigenous source of uranium ore, we will have to import uranium fuel. However, uranium imports come from a range of countries that are not necessarily the same as those that supply other energy sources. Uranium is currently mined in 19 different countries and resources of economic interest have been identified in at least 25 other countries. This provides valuable diversity of supply.' [28]

That will not mollify Green opponents of nuclear power. In a neo-protectionist world, there's no guarantee that even Australia – the main exporter of uranium to the UK – might not, at some point, deny supplies. Unlike, say, the economy of Saudi Arabia, Australia's economy is not heavily reliant on the export of one commodity. Disputes with the UK, though distant, are certainly not ruled out, especially given the country's republicanism and increasing orientation to China.

Our point here is not that overseas supplies of uranium are likely to fall into jeopardy, but rather that the wind and the sun are subject to no export controls, and no insecurity of supply. Similarly, the White Paper gives another hostage to fortune when it says that measures around energy efficiency are 'amongst the most cost effective ways of reducing energy demand' and hence of reducing CO_2 emissions. Its only argument *for* nuclear here is that, after they've improved the energy efficiency of their lives, people might *still* turn up their central heating, or buy more energy-consuming products. [29]

Only by blaming consumer extravagance, it seems, can Britain's government make a case for nuclear. What a pity!

Greenpeace UK muddies the water

In 2008, Greenpeace UK pursued an advertising campaign – between CHP (good) and nuclear power (bad). [30] Now, we're all in favour of CHP, and especially the large-scale provision of heat to industry upheld by *Securing Power*, a report on CHP that Greenpeace commissioned from the consultants Pöyry. [31] Indeed, CHP will itself make use of the heat from… 4G nuclear power. [32]

If nuclear plants cost £3-4.8bn to build, as Greenpeace suggested, do CHP plants really cost just its quoted figure of less than £1bn? *Securing Power* mentions neither figure. Probably CHP heat to industry will be a little cheaper than heat supplied through nuclear-generated electricity; but with CHP heat for *homes and offices*, the limited scale of such heat could make it dearer than the nuclear alternative.

Next, Greenpeace argued that while CHP facilities 'can be built on existing industrial sites, close to the demand for heat', nuclear reactors 'can only ever be built where there is enough water for cooling, eg by the sea'.

It's true that reactors need a lot of cooling water. But if Britain ever runs short of clean water, the solution won't be fewer nuclear reactors, but more energy – to power desalination plants that can provide the country with more water.

Anyway, reactors don't need to be near the sea. Switzerland has no coastline, yet 43 per cent of its electricity comes from nuclear reactors (just five, in fact). [33]

Location is always an issue with large infrastructure projects. But nuclear reactors can be built in many places in the UK, because water is more plentiful in the UK than Greens imagine. Similarly, large-scale CHP requires concentrated clusters of industries as customers. That's fine too – although it's striking that Greenpeace's expert consultants, in identifying such sites in the UK, picked nine that are all… by the sea. [34]

Lethargy and bureaucracy in the UK

With officials' arguments for nuclear power so weak, it's hardly surprising that the pace of UK nuclear energy's future development appears so sluggish. In a section titled 'Why decisions on nuclear power are needed now', the government's White Paper spells out why urgency is needed. On the Paper's own, conservative estimates (which have high hopes in measures of energy efficiency), about 30-35 GW of new electricity generating capacity will be needed over the next two decades, and about two-thirds of this investment should be made by 2020. Of the 22 GW likely to close by 2028, just over a half consists of fossil-fuel generation, and about 10 GW is nuclear. Indeed by 2023, all but one of Britain's nuclear power stations look set to have closed. [35]

But wait – because the government surely will. The Paper states that the public must learn that

> '… it takes a long time to plan and build nuclear power stations. This means that new nuclear generation can make only a limited contribution before 2020.' [36]

Really? In Kashiwazaki-Kariwa, Japan, the world's first two advanced boiling water reactors started commercial operation more than 10 years ago, after just 62 and 65 months of building. [37] Given that still faster rates of construction ought to be possible now, delay in Britain would seem to come down to weak planning, and weak skills in nuclear engineering.

The White Paper concedes that nuclear planning has been inefficient, costly and lengthy in the past: Sizewell B, for example, took six years to secure planning consent, and cost £30 million. [38] So let's survey, as quickly as is possible, how the government proposes to start to commence to get ready to take steps to prepare a different kind of planning regime – maybe.

In November 2007, the government introduced a reforming Planning Bill. If passed into legislation, it will establish a new 'single consent regime' for nationally significant infrastructure, under which the government will produce a National Policy Statement (NPS) that will 'reiterate the government's policy on nuclear power' – by building on the foundations laid down in the January 2008 White Paper. [39]

The nuclear NPS will cover the criteria that the government considers should be used to assess the suitability of potential sites for reactors. It may give 'an indication of certain locations' that meet these criteria – following, of course, a Strategic Site Assessment (SSA). If the NPS does give such an indication, there would then follow 'a process of engagement and consultation with those local communities on which the NPS had a direct bearing, before this was finally adopted'. Planning reform would also create an 'active pre-application phase, during which potential [nuclear power station] developers will need to consult publicly and locally on their proposals and engage with local authorities, statutory bodies and other key parties'. Developers would do this before submitting their applications to develop to a new Infrastructure Planning Commission (IPC), which will decide on each application – in England and Wales, but not in Scotland. [40]

Deliberative and participative, if not at all representative, these arrangements all *sound* very democratic. But far from being a single consent regime, the White Paper gives the state the job of seeking out multiple consents from 'stakeholders,' who themselves get 24 mentions and, interestingly enough, include 'faith groups'. This consultation and deliberation, which relies partly on direct mail to 5,000 'grassroots and community organisations', is actually not at all democratic. [41]

Throughout all the consultations, Whitehall will have the final say.

Still more questions need to be asked. If the nuclear NPS will also 'set the policy framework' for decisions made by the IPC, what exactly is the IPC? Typically enough, its 20-30 members won't be elected, but will be appointed by ministers. And who will these appointees be? They will be

> '… well respected experts, drawn from a range of fields [which] might include national and local government, community engagement, planning, law, engineering, economics, business, security, environment, heritage, and health, as well as, if necessary, specialist technical expertise.'

The latter expertise will no doubt include that related to the nuclear sector. [42]

Contrary to what Greens like to believe, the IPC does not offer nuclear firms a green light to develop new plants. Rather, it nationalises and makes official the *resistance* to development that, for the most part, experts in many of the fields mentioned above already strenuously uphold in Britain.

In sum, after endless rounds of consultation, a body that promises to consolidate the environmentalist allergy to nuclear power will make the final decisions about it. The *planning* of nuclear infrastructure in Britain, like that of infrastructure generally, promises to become as Byzantine as the planning of new houses. In every case the end-result of planning and so-called 'democracy' is to arrest construction, to make it more expensive, to have an inquiry about it, and to seek judicial review about whether it should be allowed to go ahead.

No wonder that the White Paper's 'indicative timetable', showing what it says is the 'fastest practical' route to new nuclear power stations, begins to look as complicated as a nuclear reactor itself (see overleaf).

From the timetable, it appears, about five years will have to pass before nuclear operators make a 'full decision' to 'proceed and commit'.

More delay is also possible. The reason: there just might be a genuine shortage of skills in UK nuclear engineering.

Motivating a new generation of nuclear engineers

In 2008, British Energy, a nuclear electricity generator in which the British government had a 35 per cent stake, excited interest from corporate bidders. In this contest, the government favoured a takeover by EDF, hoping that Britain could benefit from the French generator's nuclear expertise. [45] Indeed, as the White Paper had earlier pointed out, failure to move ahead with nuclear could mean British skills in the sector being 'lost'. [46]

Eventually, EDF was successful in its bid. But other help was at hand: along with the nuclear industry, the Engineering and Physical Sciences Research Council had begun to fund masters-level and continuing professional development for the nuclear sector. And what was its cash commitment to this cause? A cool £1m – half the average price of a house in Mayfair, London. [47]

Right across UK electricity, concerns about an ageing workforce are aired. But the real worry is about how to convince young Britons that electricity is a useful career. Three specialists, for instance, suggest a TV series featuring 'heroic actions of electricity people', or even 'a reality show focused on an electricity industry company or team'. [48]

Time was when adult politicians and industrialists could give children a good reason to grow up. In a new century, it seems, nobody knows how to motivate a new generation of nuclear engineers.

The purpose of the electricity industry is to keep the lights on. Since Greens organise voluntary blackouts, it's clear that not everyone thinks that's worthwhile. [49] *Electricity in general, and nuclear electricity in particular, needs articulate defenders.* People need inspiring with a basic confidence in what human beings can do with very high technology – at the sub-atomic level as much as anywhere else.

Why nuclear remains on the back foot

No intrinsic technological defects explain nuclear power's Cinderella status. Reactors do generate electricity, after all. To recruit a new workforce to the nuclear industry, in fact, citizens need *new arguments* even more than reactors can still benefit from *new technologies*.

The energised citizen should be able explain why:

1. More energy for everyone is a good thing
2. Nuclear reactors deserve favour
3. Nuclear's safety record has become pretty admirable, and is a matter of running the right economic and social regime
4. Nuclear's chequered career in the past need be no guide to its prospects in the future.

We've already made the first two arguments. A panel at the end of this chapter deals with the third. But the fourth is the most important. To rephrase it: *just exactly why have politicians and nuclear corporations become so defensive and lethargic about nuclear power?*

Here historical insights are even more valuable than technological ones. The problems that pertain to nuclear reactors, insofar as they exist, are those of social regime, not intractable technology. More even than nuclear accidents, *it's the military origins of and more recent political dissent about nuclear power that have put its enthusiasts on the back foot.*

Today the history of nuclear power ought in fact to be over.

The British government's 'indicative pathway' to possible new nuclear power stations [43]

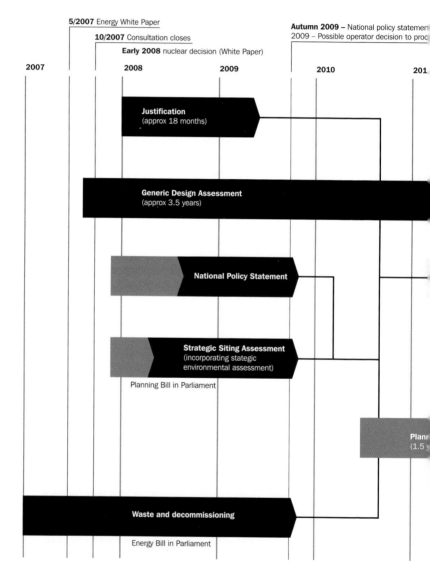

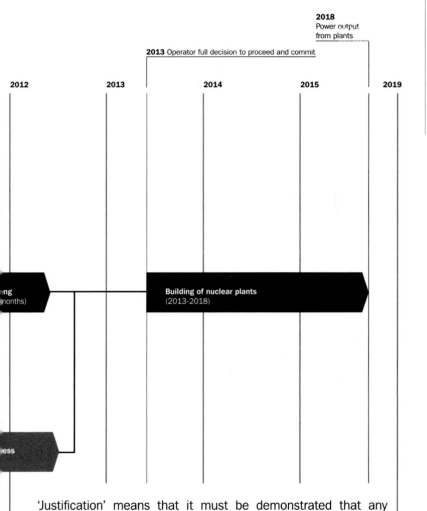

2018
Power output
from plants

2013 Operator full decision to proceed and commit

| 2012 | 2013 | 2014 | 2015 | 2019 |

Building of nuclear plants
(2013-2018)

ng
nonths)

ess

'Justification' means that it must be demonstrated that any benefits resulting from the introduction of nuclear power 'outweigh the associated health detriment'. The Generic Design Assessment assesses the safety, security and environmental impact of power station designs. [44]

Instead, it still hangs like a dark cloud over the industry. Why? Because nuclear's history

- begins with Hiroshima and Nagasaki
- developed through the perverse Old Scientism of the Cold War
- developed also through liberal and leftist dissenters from the Old Scientism.

Then, over the course of the 1970s, dissenters from nuclear war and nuclear weapons tests morphed into handwringers about the risks surrounding civilian nuclear power.

The 1970s: the decade when new doubts about reactors set in

For 30 years, nuclear power has made a historic retreat from confidence. So before turning to the earlier political history of nuclear power, let's just underline the significance of the 1970s.

During that decade, ironically, oil came to look like a bad bet. Ironically, too, nuclear reactor technique had already matured. Nevertheless, an exhausted New Left and a nascent environmentalism began to single civilian nuclear technology out for ridicule and fear.

Once again, however, it was not Reds and Greens that consolidated doubts about nuclear power so much as governments. The recession of 1973-4, which proved too profound to blame simply on OPEC, was a key factor in nuclear power's loss of impetus. Before that recession, demand for energy looked buoyant and nuclear power looked set for a great future. After it, many reactors were not built for years.

Before the 1970s, the main focus of establishment concern about nuclear power was its debatable economic performance against other forms of energy. After the 1970s, however, policymakers preferred to worry about risk. Institutions of the Old Scientism were wound up, and new agencies, with a more critical attitude to nuclear risks, were set up. So it was that Reds and Greens only buttressed the state's loss of faith in nuclear. Worried about risks of war and risks to the environment, Reds and Greens now merely converged with governments that, after OPEC's action and a major economic crisis, were already newly alive to risk.

Today's campaigns against oil and against coal-fired plants, it should be remembered, are relatively new. With nuclear power, by contrast, opposition began in the 1970s. Indeed, modern environmentalism can be said largely to have defined itself around nuclear power.

The defensiveness of nuclear's advocates today doesn't relate to flaws intrinsic to reactors. The difficulties around reactors can be and have been controlled. Rather, the evolving political culture around nuclear power, and especially its advocates' 30-year-long failure to counter critics, explain why it has moved ahead in a rather haphazard and fitful manner.

Beginnings: nuclear power bears the taint of violence

On 6 August 1945, the *Enola Gay* bombed Hiroshima and 140,000 people died. Then, in the Cold War, the Bomb seemed to threaten all humanity. Even today, when Hilary Clinton offers to 'annihilate' Iran in the event of it attacking Israel, fears of nuclear weapons run high.

Real dangers surround nuclear reactors. *But the passion that they elicit today can't be traced to these dangers alone.*

It's true that nuclear weapons haven't been used since Hiroshima and Nagasaki; that the Cold War is history; that perhaps 400,000 people have died in the Iraq war, but not through nuclear weapons. Nevertheless, civilian nuclear power still suffers from its:

- origins around the Bomb, and around national belligerence
- links to nuclear proliferation
- potential to act as a target for a military or terrorist attack.

Civilian nuclear power bears the taint of violence. Nuclear physics seems to have introduced

- a new and vicious logic to fighting, and perhaps to the fate of the Earth and to human actions generally
- a specially vicious and durable kind of contaminant: ionising radiation capable of causing genetic damage that carries through to future generations.

If ever there were a Faustian contract, nuclear war, beginning in Japan, would seem to embody it. As for the nuclear reactor, it would seem to be merely a codicil to that contract – a proposition complete with its own dangers.

Civilian nuclear power is seen as special. Nuclear physics *appears* to have led inexorably to Hiroshima and Nagasaki. Perhaps nuclear physics must have an almost equally remorseless and lethal momentum in the civilian domain.

For critics, civilian nuclear reactors can *only* foster the spread of materials that are lethal: lethal both immediately, in war, and lethal over time, though radiation. So it's little wonder that reactors have provoked dissent. In the public mind, their fuels have at least a historical association with violence toward nations, or at least the threat of such violence. Their fuels also have a historical association with civilian injuries and deaths caused by radiation, whether in nuclear weapons tests, or in nuclear accidents.

Yet the radiation around power stations is much weaker than that set off by a bomb (see panel below). What's more, that example reveals a general truth: the uses to which nuclear physics and nuclear reactor technology are put are to do with the decisions of men and women, and don't automatically adhere to uranium. Proliferation and breaches of safety in radiation are not determined by the structure of the nucleus. They relate to broader social, economic and political conditions.

Let's get precise about the dangers of radiation

In November 2006, the British press published striking photographs of Alexander Litvinenko, once a senior official in the KGB. Litvinenko was in a London hospital bed, dying of poisoning by polonium 210. Pale, wan and bald, he acted as a potent reminder about the effects of radioactivity. In high doses, such as he mysteriously received, radioactivity can kill you. In low doses too, it produces mutations in DNA that, building up over time, can produce cancer.

The dangers of radiation from nuclear waste provoke a special dread. But with a high level of ingenuity, mankind can deal with those dangers.

They should not be minimised. Nuclear waste gives out more radioactivity to those who work around it than do cosmic rays, to which long-haul jet travellers are strongly exposed. Nuclear waste is more radioactive than the soil. Only coal, once burnt into ash, is as radioactive as low-level nuclear waste.

Levels of radiation exposure can be measured by the quantity of energy absorbed. For purposes of calculating risks to humans, however, this quantity of radiation must be weighted for the effectiveness of each type of radiation in causing biological harm. The unit in which this is done is a **Sievert**.

One Sievert is a large unit. [50] For practical purposes, exposures are measured in milliSieverts, or thousandths of a Sievert:

Average annual exposure to radiation, milliSieverts [51]

Exposure	milliSieverts
From artificial sources, civilian	
Airborne discharges from nuclear power stations to the general public	0.0001
Liquid radioactive wastes discharged into the sea	0.0007
Workers in medicine/research industry	0.1
Workers in nuclear industry	0.4
Medical irradiation	0.41
Aircrew	2
From accidents	
Average dose within 5 miles of Three Mile Island	0.09
From artificial sources, military	
Post-war nuclear tests, average worldwide dose, 1963	0.113
Post-war nuclear tests, average worldwide dose, 1999	0.51
From naturally occurring sources	
Cosmic rays at ground level	0.392
Minerals (most significantly potassium and uranium)	0.765
Radon gas emitted from minerals, average dose	1196

From a rational point of view, doses of radiation from the routine operations of the nuclear industry, and even from accidents like Three Mile Island, are too small to be of concern. Notably, the artificial radiation dose from civilian nuclear power is hundreds of times less than the natural radiation dose from earth and space.

The fact is that for most people, the main artificial source of radiation likely to be encountered isn't a nuclear power station, but rather nuclear medicine. Here, as with other dangerous medical techniques, exposures are now both regulated and accepted as necessary.

The destruction of Hiroshima

Nuclear war suggests that
nuclear power is a race apart

When, after 9/11, anthrax was found in the US postal system, the world was briefly reminded that nuclear arms are not the only weapons of mass destruction. Nor are nuclear arms alone in causing genetic mutations. Those mutations brought about by US action in 1945 were terrible, but those caused by chemical weapons have since been termed 'comparable'. [52]

Some like to relativise the effects of nuclear weapons, which is an error. But many more would proclaim nuclear weapons a race apart – which is also not quite right. And so it is with nuclear reactors. They pose their own special technical challenges; but these challenges are not beyond human ingenuity to deal with, especially today and tomorrow.

Let's now see how reactors suffered from their association with military matters.

During the Cold War, nuclear weapons seemed to have rid human conflict of its few remaining rational aspects. Nukes made technology appear to have acquired a menace and momentum all of its own. In the age of the Enlightenment, Carl von Clausewitz had famously said that war was the extension of politics by other means. [53] But after 1945, in a nasty paradox, the rational calculations of nuclear physics, when applied to war, seemed to make it the extension of something less rational even than politics.

In nuclear weapons, pacifists and defenders of Stalin's Soviet Union felt they had alighted upon a technology of exceptional beastliness. They were optimistic that the fact of such weapons might shift popular feeling their way. By the late 1970s, however, the left was tired and increasingly desperate. But, in populist style, it had no compunction about adding civilian nuclear power to its list of easy targets.

Nuclear power certainly grew out of nuclear weapons. The first man-made nuclear reactor was built on the Manhattan Project, the $2 billion effort that led to Hiroshima. Nuclear-powered submarines wrote the book for US reactor design (see panel below). In Britain, Calder Hall, the first 'commercial' reactor, opened in 1956 by the Queen at what is now called Sellafield, Cumbria, was in fact chiefly designed to produce plutonium for bombs.

But it would be wrong to visit the sins of the father on to civilian reactors today.

Magnifying fear, governments and Greens insist that, like the prosecution of nuclear war, nuclear power can be subject to *rogue actors* bent on causing unimaginable disasters. This is a misanthropic vision. It's also one that, in a deterministic style, strips selected technologies of their human design and transforms them into a primordial force for evil.

Here the nuclear fuel cycle is inherently a tool for rogue states and wild Islamic fundamentalists. Just as inherently, the fuel cycle is the plaything of cowboy plant managers and rapacious nuclear shareholders, who are both only interested in a fast buck.

Now, there's certainly the *possibility* that such nuclear scenarios come true. They might come true for the first time (rogue states and fanatics), or repeat some of the industry's poor if comparatively rare episodes of the past (maverick managers and shareholders). But it's the reality of post-war American politics, not possible mishaps in the future, that explains the distaste which nuclear power arouses today.

Through its actions, the American state, run by both Republicans and Democrats, laid a real basis for modern suspicion of nuclear power. The state was the big 'rogue actor', if ever there was one.

At the same time the American left magnified the establishment's doubts about nuclear power.

The US Navy: basis for modern reactor design

Because nuclear energy doesn't consume oxygen, it can power submarines underwater for extended periods. In 1946, a US Navy sailor, Hyman G Rickover, started work on this principle. [54]

In the 'development' part of R&D, it's usual to manufacture components, construct a prototype and then organise a fully working system. Flouting this, Rickover concurrently built a prototype propulsion unit and a submarine. Ruthless on quality control, he launched the *USS Nautilus* in January 1954. [55]

Given a cover portrait in *Time* magazine, Rickover remarked that scientists could

'... think up thousands of reactors. But the Navy wanted a nuclear submarine... fast. We picked a simple type of reactor that we knew a lot about already. If we'd waited for the scientists, we'd still be fooling around.' [56]

Rickover's compact reactor used fuel forged into uranium rods. It was cooled with pressurised water. By a twist of fate, his PWR design, made by Westinghouse, had far-reaching significance.

Even before the launch of the *Nautilus*, Rickover was designing a nuclear powered aircraft carrier. But in 1953 President Eisenhower cut funds for the project. That made the US Atomic Energy Commission scramble to recast Rickover's design – for civilian purposes. [57] Four years later, the Duquesne Light Company of Pittsburgh began generating America's first 60 MW of nuclear electricity from a Westinghouse machine based at Shippingport, Pennsylvania.

Through Shippingport, engineers and contractors gained valuable early experience in reactors. Today, Rickover's PWR design remains the workhorse of the nuclear industry. Only now, with 4G designs, have engineers got round to thinking about which of Rickover's 'thousands' of other putative reactors are worth taking forward.

**The man who designed the first
Pressurised Water Reactor:**
Hyman Rickover (1900-1986)

B-Reactor, Hanford, Washington.
Completed in September 1944, it soon
began making the plutonium that was
eventually used on Nagasaki

The Manhattan Project and the bombings of Japan

The motives and need for the bombing of Hiroshima and Nagasaki have been widely debated. [58] The making of the Bomb, and the devastation it wrought, have also been widely documented. [59] Here, our focus is on three other crucial aspects.

First, 120,000 scientists worked on the Manhattan Project. Fatefully, America involved Britain, France and Canada in it. Very soon after the bombings, this shared work led to important tensions within the West over both military and civilian nuclear power, leaving aside the obvious East-West tensions of the Cold War. Thus as early as 1946, the US Atomic Energy Act, which established the military-civilian Atomic Energy Commission (AEC), ended West-West collaboration on nuclear research.

From the start, then, nuclear weapons raised issues of what became known as nuclear proliferation. Thus later, when civilian nuclear power developed, proliferation, nation states and conditions of secrecy were all prominent factors around it. We will come back to these issues.

Second, *the first and only use of nuclear weapons set the pattern for the Old Scientism*. Nuclear weapons scientists, including a majority of the staff at Los Alamos, were often politically keener on using the Bomb on Japan than much of the US military and President Truman appeared to be. By the time of Hiroshima, there was little left to bomb in Japan. [60] Conventional bombing nevertheless continued, planning for invasion continued, and US diplomatic feelers toward the Emperor of Japan continued. [61] The point of Hiroshima was to deliver what Henry Stimson, US Secretary of State for War, sought at a key meeting on 31 May 1945: a 'profound psychological impression on as many Japanese as possible'.

At the meeting, after it was suggested that one A-bomb wouldn't differ from 'any Air Corps strike of current dimensions', Oppenheimer interjected:

> 'The visual effect of an atomic bombing would be tremendous. It would be accompanied by a brilliant luminescence which would rise to a height of 10,000 to 20,000 feet.' [62]

In terms of international relations, then, Hiroshima and Nagasaki amounted to two *demonstration projects* – whichever country the demonstrations were aimed at. [63] The Pacific War also had a pronounced racial component. [64] Given this context, the US authorities tended to regard Japanese victims of the bombings less as humans in agony, and more as a source of military and medical information.

Catherine Caufield's political history of radiation, *Multiple Exposures* (1989), is exhaustive and illuminating, and we draw on it now. [65] Caufield confirms that the Old Scientism was insouciant, callous and dehumanising. It had no regard for Manhattan Project employees. University of California radiology guru Robert Stone, who led the Project's health division, frankly noted that clinical study of Project staff was

> '… one vast experiment. Never before has so large a collection of individuals been exposed to so much irradiation.' [66]

Finally, Manhattan, Hiroshima and Nagasaki prompted *dissent* among some of the very scientists involved with them. At Los Alamos, the chief metallurgical chemist called the Project's worker insurance 'inhumane, unethical, and unfair', and chemists refused to purify raw plutonium without extra insurance. [67] In June 1945, physicists at the Project's Chicago operation, led by James Franck, advised Stimson not to use atomic weapons against Japan, and went on to found the *Bulletin of Atomic Scientists* as a forum for all-sided debate. [68]

On 6 July 1945, dissent accompanied Trinity, the world's first atomic explosion – done, like Hiroshima and Nagasaki, in the air – at Alamogordo, New Mexico. In 1965, Oppenheimer said of the scientists who witnessed it: 'Some wept, a few cheered. Most stood silently'. [69]

After the devastation in Japan, dissent intensified. For about 40 years after Trinity, US nuclear weapons scientists were broadly split between 'arms racers' and 'arms controllers'. [70]

On top of that, the radioactive fall-out, residual radiation and genetic mutations around Hiroshima and Nagasaki gradually became controversial. Meanwhile popular murmurings grew, not just around nuclear war, but also around the fall-out that accompanied nuclear weapons tests.

Radiation and the Old Scientism

Manhattan was about unleashing a bomb, not worrying about radiation. Perhaps just 200 scientists active in the Project knew its ultimate purpose. The rem, a unit of radiation dose covering alpha, beta and neutron radiation and not just gamma and X-rays, was only developed during the Project (it is now superceded by the millisievert, where 1 mSv = 0.1 rems). [71]

By June 1945, however, new calculations showed that the Trinity test would be accompanied by serious radioactive fallout, as indeed it was.

Oppenheimer had medical reports on Trinity kept separate from other Trinity documentation, and made access to them subject to his personal approval. A war was on. Secrecy on the part of the US military was hardly surprising, and 50 years had to pass before there was any glimmer of what the New York City writer Sean Collins calls *the dogma of transparency* about institutions – the vapid, all-purpose call for secret conspiracies to be unveiled. [72]

Nevertheless, in its policy on radiation, the Old Scientism, as embodied in the post-war nuclear agencies of the American state, had nothing to be proud of.

The Old Scientism covered things up. It censored and harassed dissidents. And in its disregard for the effect of military-related radiation on human life, it sparked off irrational hostility to the radiation around nuclear reactors.

Despite the Old Scientism, a controversial culture surrounded radiation from the start. [73] Thus in 1947, worried about genetic mutations, America's newly rebranded National Council on Radiation Protection (NCRP) cut the limit of exposure to X and gamma rays from 1 to 0.5 mSv a day, renaming weekly limits of 3 mSv the 'maximum permissible dose'. [74] In 1948, too, the NCRP adopted the theory, contrary to the AEC, that no threshold *existed below which radiation was safe.*

Only in 1954, however, did the NCRP publish its dose limits. The first federal radiation regulations only became effective in 1957. [75]

The bombings of Japan set the framework for post-war dissent about radiation

From 1947 onward, funded by the AEC, the US Atomic Bomb Casualty Commission (ABCC) captured data on Hiroshima and Nagasaki. [76] In her comprehensive 1995 study of post-war debates around the radiation set off by the twin bombings, the Dundee University social scientist Sue Rabbit Roff is properly caustic about the ABCC. [77] It did not treat victims. Despite the incidence of radiation at the Trinity test, it refused for many years to countenance the idea that an airburst explosion, such as that over the two Japanese cities, could result in fallout and residual radiation. The ABCC also deliberately obscured the relationship between radiation and distance from the epicenters of the bombings.

In 1980, to cap everything, researchers at Lawrence Livermore laboratories discovered that the ABCC's estimates of radiation around Hiroshima had contained a major error. As a result, gamma rays are now thought to be 40 per cent more likely to cause cancer than previously imagined. [78]

Nevertheless, Hiroshima and Nagasaki gave America vital data on radiation's human impact.

Use of atomic bombs over Japan was always justified as a means of bringing about a swift and less bloody peace. However, just as this rationale tended to downplay the immediate casualties, so, for many years, did it provide a context in which US officials cynically and scandalously ignored, denied and underestimated the longer-term effects of radiation. Indeed, the shameful records of the AEC and ABCC on radiation, Hiroshima and Nagasaki were matched only by the technological determinism of nuclear dissenters.

Roff contends that with radiation today, 'the only certainty is uncertainty'. Yet while this shows her closeness to the New Scientism's conception of climate science (see Chapter 3), she's nearer the mark when she notes that 'all contemporary discussions about the types and extent of injury caused to humans by ionising radiation are grounded in the biomedical studies of the survivors'. [79] The reason for this state of affairs is worth examining (see panel).

Understanding low doses of radiation

If the dangers of civilian nuclear radiation should not be minimised, neither should they be exaggerated.

Radiation consists of fast moving subatomic particles, expelled by the decay of a radioactive atomic nucleus. It can cause disease by smashing through the molecules in a cell, in particular DNA molecules. Artificial radiation shares this mechanism of toxicity with natural radioactivity, such as cosmic rays. Chemicals such as tobacco smoke also work in a very similar way, chemically attacking DNA.

In all of these cases, *doses matter in determining biological outcomes*. If both strands of DNA's double helix are broken at a single point, that's enough, in principle, to start a cell on the path to cancer. A cancer could, therefore, be triggered by a single radioactive atom. Luckily though, the body is good at repairing DNA breaks. Natural radioactivity and chemical assaults result in many DNA breaks every day of your life. Either these breaks are repaired, or unlucky cells die.

Acute radiation sickness results from very high doses of radiation delivered in a short time. It's caused by the death of many cells at once. That's what killed Alexander Litvinenko. High, short doses are also how radiotherapy kills tumours.

For a few cells to die through low dose radiation, by contrast, is no problem. The body can replace those cells as part of normal cell turnover. When wrongly repaired or damaged cells survive after low doses, however, there is a long-term danger of cancer.

Each radioactive decay is another chance to create a cancer. In the long term, then, the effect of radiation is likely to be **additive**. Risk is in proportion to exposure. Since exposure to radiation from cosmic rays is 490 times larger than that from nuclear power station discharges to air and sea, the chance of developing cancer from cosmic rays is 490 times larger than the chance of developing cancer from nuclear power station discharges.

Proportionality of response to dose is known as **linearity**. At Hiroshima and Nagasaki, the ABCC's measurements of large doses among large numbers of victims confirmed the broad outline of linearity. They also provided a clue to a much more difficult question: that of small doses among small numbers of victims. In the absence of a large research sample of people exposed to low doses of radiation, however, just how right is it to extrapolate down from the Japanese data?

A number of hypotheses exist about the lower end of dose/response relationship. The simplest possibility is that proportionality continues right down to zero dose. This is known as the 'linear no threshold' model, or LNT, on the grounds that there's no threshold below which the impact of radiation on human beings can be neglected – even below natural background levels.

It's also possible that very low doses may be more or less harmful than claimed by LNT, or even beneficial. The details depend on complicated biology. What's clear, however, is that any such deviations from linearity are hard to detect. *If people were living much longer or shorter lives from low-level radiation, the phenomenon would show up in mortality statistics.*

The peak of Cold War starts the discrediting of civilian nuclear power

At Bikini Atoll in the Marshall Islands on 25 July 1946, in what was called Operation Crossroads, Baker, an underwater test, exposed 42,000 participants (including 400 US safety personnel unequipped with detectors of alpha rays) to copious amounts of radiation. Food on the participants' boats was affected too, but it was officially announced that nobody had suffered ill effects.

Negative publicity broke out about Crossroads and subsequent tests. [80] Then, on 1 March 1954, the US exploded its first full-scale hydrogen bomb at an islet near Bikini. Bravo was 1000 times more explosive than the bomb that incinerated Hiroshima; and it was Bravo that confirmed that fallout was really a problem for the world.

Nearly 90 nearby islanders, including 29 children under 10 years old, received 17.5 mSv. All 23 crew of a Japanese tuna trawler – *Fukuryu Maru*, ironically translating as *Lucky Dragon* – were also affected. Typically enough, though, the AEC said that the islanders were reported well, and its new chairman, a top Wall Street banker, pronounced that *the LD was* a 'Red spy outfit'. [81]

In the era of Joe McCarthy, all this was to be expected. Nevertheless, it's interesting that the Old Scientism anticipated, in the spirit of the Cold War, the New Scientism's criminalisation of dissent. Like those who attack climate change 'deniers' today, the AEC frequently intimated that the very idea of fallout was unacceptable. It was unpatriotic.

The hostility of Japanese to US nuclear tests grew. [82] In a significant marker for future East-West tensions, India called for tests to be halted. The AEC was hit by a series of embarrassments. [83] In 1956 the United Auto Workers tried to stop a new reactor being built in Detroit.

By 1957 the AEC had moved the monitoring of fallout out to the US Public Health Service, but found itself asked questions on the subject by the US Congress's Joint Committee on Atomic

Energy. Fallout from further weapons tests conducted in the US, together with other developments, now forced the AEC to get a bit more serious about radiation. [84]

In these crowded, fearful years, nuclear weapons tests constituted much of the substance of the Cold War: in 1958, more than 100 were made by the US, the Soviet Union and the UK. During the same year, however, a kind of epiphany occurred among US policymakers. Eisenhower rejected the recommendation of scientists that a massive programme of fallout shelters was the kind of measure that would see the US through a nuclear war. Technology, it was now felt, couldn't quite do that. Instead, for the US, it was time to draw breath, calm down, and in arms, negotiate from a position of strength. [85]

Together, the conduct of the Old Scientism, the dynamics of the tests, and a more sober and eventually more public discussion about nuclear weapons

- gave a Faustian tone not just to the radiation associated with tests, but also to nuclear radiation in general
- began to undermine the whole idea of conducting experiments in the nuclear realm.

At Easter in 1958 Britain's newly-formed Campaign for Nuclear Disarmament went on the first of its Easter marches in protest against nuclear weapons. In June, Dr Alice Stewart and others, having traced 1416 of the 1694 English and Welsh children who had died of leukaemia or cancer before their 10th birthday during the years 1953 to 1955, discovered that abdominal examinations of their mothers using X-ray radiation had been responsible for no fewer than six per cent of all fatalities. [86] In October, again in the UK, what had been Calder Hall (civilian nuclear) turned out also to be Windscale (military nuclear), which accidentally suffered a radioactive leak.

Radiation, it seemed, was everywhere a malignant force. Around disarmament, the moratorium on nuclear tests agreed between the US, the Soviet Union and the UK in November

1958 did not dissuade France or the superpowers themselves from going on to conduct tests. Later, in 1963, the Partial Test Ban Treaty was followed by more tests than ever. Tests were now simply carried out underground.

For many, weapons tests and the failure to regulate them did much to discredit all experimentation in the nuclear domain. In climate science, experimentation helped Revelle and Lorenz (see Chapter 3). At the same time, however, Packard and Galbraith were beginning their wide-ranging critiques of American capitalism's creation of new products based on dubious needs (Chapter 2). Politically, therefore, it was always likely that the bad smell surrounding nuclear weapons would spill over to civilian nuclear power.

After the Shippingport plant started in 1957, the US found that it had built only three more nuclear power stations in five years. Then, in 1962, conservationists in California began what proved to be a successful campaign: to stop a civilian reactor being built at Bodega Bay. [87]

Despite increased testing, fallout declined. In the spring of 1969, a celebrated dispute broke out, between dissenters both outside and eventually inside the AEC, on the genetic effects of radiation on American infants over the period 1950-65. [88] Meanwhile, the scale of proposed discharges from a reactor into the Mississippi ranged the AEC and the nuclear industry against the state of Minnesota.

By 1969, scientists were looking at the radiation around reactors more closely. Attention shifted from fallout to nuclear power.

In January and February 1970, the US Congress's Joint Committee on Atomic Energy, including Al Gore, held lengthy hearings on the Minnesota conflict, and radiation generally.[89] In a sign of the growing crisis of legitimacy for state institutions (see Chapter 2), the AEC revised, in 1971, its standards on radioactive releases from reactors, tightening controls on them. Meanwhile, developments on the left prepared the ideological foundations for still more opposition to civilian nuclear power.

The left and the nuclear

With dissent about nuclear war, a potentially useful attempt to understand technology as a symptom of a failing society soon turned into a selective kind of technological determinism. In the 1930s, the left had feared chemical weapons. Now, it highlighted nuclear weapons. Horrible technologies remained a defining feature of its critique of capitalism.

At first, some of America's brightest leftists focused on the economics of weapons. [90] Capitalism had changed, they argued: to capital goods and consumer goods, it had added 'the production of the means of destruction' – spending on which prevented an economic crisis. [91]

Though nuclear weapons turned out cheaper than conventional ones, Hiroshima orientated the left's discovery of a *permanent war economy* toward the Bomb. This, leftists thought, was immoral, but integral to the new capitalism. Indeed in 1955, two years after Eisenhower announced his 'Atoms for Peace' programme for developing *civilian* applications of nuclear technology, one far-left economist was still convinced that only socialism could bring atomic power to the US. [92]

Once the US began to boom, however, the left had to shift its critique from economics to sociology. In a 1956 bestseller, the radical sociologist C Wright Mills attacked US military men for the 'increased personnel traffic' between them and the corporate realm, and warned of the possible 'triumph in all areas of life of the military metaphysic, and hence the subordination to it of all other ways of life.' [93]

Mills was wrong: civilian control of Mills' American 'warlords' in fact increased from 1947 through to the 1960s. [94] Nevertheless, the personal connections between the military and industry, plus the unstoppable, immoral dynamic of military technology, became the degraded staples of post-war leftist thought. [95]

Here, if only in embryo, were ideas much favoured by later opponents of nuclear power.

From Eisenhower to Marcuse

Oddly, Eisenhower left office in January 1961 with a farewell television broadcast that systematised Mills' complaints against corporations connected with the new technologies of war. Public policy, Ike warned in a farewell address, might become the captive of a scientific-technological elite. The technological revolution, he argued, was largely responsible for the creation of a *military-industrial complex* that, reaching into 'every city, every statehouse, every office of the federal government', might endanger liberties or democratic processes. [96]

It took five years and a festering war in Vietnam for Ike's warning to seem 'more than the swan song of an old and naïve man'. [97] But much earlier, in October 1961, *The Nation*, weekly voice of leftish Democrats, reworked the song into a special issue on the loss of civilian control over the military. Fred Cook, an investigative journalist and a critic of Big Oil, wrote the issue, and then published a book on it, titled *The Warfare State*. [98]

Cook indicted both the logic of technology and human recklessness. The quick responses required by Cold War weapons meant that they had 'come to dominate the Military as the Military dominates the State'. Even and especially 'the lower echelons' of the military endangered the planet:

> 'Not only do faulty information, bad judgment, rash action become at some time and place fatally inevitable, but one psychotic or one fanatic devotee of the let's-blast-'em-and-get-it-over-with school easily could settle the fate of the world.' [99]

In stressing the connection between *nukes and nutters*, Cook was the first to popularise the theme of rogue actors – one that has run from Stanley Kubrick's *Dr Strangelove* (1968) through to the present day. [100] Soon Herbert Marcuse, a German leftist and the key intellectual inspiration of the 1960s counterculture, laid out a larger framework that was to have still more influence on nuclear energy's modern tormentors.

Acknowledging a debt to Mills, Packard and Cook, Marcuse's *One-Dimensional Man* (1964) followed form in distinguishing 'false' needs from 'true' ones. [101] Marcuse's departure was to overturn the Old Left's traditional support for technology. 'By virtue of the way it has organised its technological base,' he announced, 'contemporary industrial society tends to be totalitarian.' [102]

For Marcuse, science was no longer neutral, but now developed under a technological *a priori* that projected nature as merely 'stuff' for humans to 'control.' Instead of society, rather than technology, being the basic historical factor in the development of mankind, technology now circumscribed 'an entire culture,' including science. Both technology and science moved under the same logic – that of capitalist domination. [103]

In the Cold War and Vietnam conflict, then, a populist of consumerism fused with fear of Armageddon and distaste for technologies deemed sinister or inappropriate. *More generally, technology (bad) was counterposed to nature (good).*

It wasn't hard to recoil, impulsively, at the Old Scientism. In 1959, to turn nuclear fall-out shelters into good coin, Herman Kahn had recommended studies of human behaviour in other overcrowded environments: concentration camps, prisons, troopships. On 25 July 1961, in a televised address on the East-West crisis around Berlin, President Kennedy had called for fall-out shelters, saying: 'We owe that kind of insurance to our families – and to our country.' [104] Now Marcuse could fairly attack Kahn and the lack of feeling displayed by the Old Scientism, as well as ridicule shelters and the Scrabble they came with. [105]

But then Marcuse went overboard. What he called the 'defence structure,' he angrily proclaimed, 'makes life easier for a greater number of people and extends man's mastery of nature.' [106] *For Marcuse, it was not the particular exercise of technological power that was the problem, but technological power itself.*

Such one-sided reactions to the Cold War and its

technologies were not confined to the left. Contrasting the impetuous and heedless pace of man with the deliberate pace of nature, Rachel Carson complained that radiation was now 'the unnatural creation of man's tampering with the atom'. [107] Even with Galbraith, there was a trace of the Marcusean vision. In his *The New Industrial State* (1967), Galbraith's consumers were still the captives of corporations; but now corporations themselves had to obey what he famously labelled the 'imperatives of technology.' [108]

Before living standards began to stagnate in the early 1970s, their rise obscured hair-shirt, low-tech alternatives to consumerism. But even before Ivan Illich (see Chapter 2), the willingness to be transfixed by technology, the desire to draw unfavourable contrast between technology and nature – these things became the unquestioned foundations of Western dissent.

Later on, nuclear power, as the most profound of mankind's manipulations of nature, proved unable to escape a full and vindictive interrogation.

The 1970s: campaigns against weapons make way for campaigns against reactors

On 2 October 1969, the US tested a 1.2 Megaton nuclear bomb on the island of Amchitka, off the west coast of Alaska. Fearful that earthquakes and tidal waves might ensue, radicals from Vancouver, Canada – the nearest major city to Amchitka – stormed America's nearby border, closing off traffic for two hours. According to one account:

> 'In Vancouver at that time there was a convergence of hippies, draft dodgers, Tibetan monks, seadogs, artists, radical ecologists, rebel journalists, Quakers, and expatriate Yanks… Greenpeace was born of all of this.' [109]

Greenpeace began as a Canadian branch of the Sierra Club, an elite group of American environmentalists. It emerged as a protest against American nuclear weapons tests – but this time the protest was not so much against nuclear's impact on human beings, as against its impact on nature.

As its name suggested, Greenpeace mixed environmentalism into the old anti-war feeling. Protests by a succession of Greenpeace boats against subsequent nuclear tests, not least by France, threw the organisation further in the limelight. [110] By the time of the Arab oil embargo of 1973, then, Marcuse's legacy and the birth of modern environmentalism added to establishment hesitations about the most obvious alternative to oil: nuclear power. In these halcyon years for the British left, too, striking miners had leaders who were hostile to nuclear power; indeed, the most famous son of the National Union of Mineworkers, Arthur Scargill, was a longstanding member of CND.

Eventually, outcries against DDT faded. The US soon forgot about Agent Orange, the chemical defoliant it had used against Vietnam. Détente between the superpowers took attention away from nuclear weapons. But nuclear power lived on, and emerged as a focus for Green suspicions. For environmentalists, it could never be the beneficial application of the objective laws of physics.

In the summer of 1976, the UN Conference on Human Habitat, Vancouver, proved an opportunity for Greenpeace to hold a benefit concert and to launch another of the organisation's boats. The secretary-general of the conference requested a ride; but the most significant development was a delegate telling Greenpeace that the US authorities had enough missing plutonium for there to be a black market in the stuff. That broke a national news story in Canada. [111]

In April of the following year there was another sign of the times. More than 1400 anti-nuclear protesters were arrested at the site for Seabrook Nuclear Power Plant, 60km north of Boston.

Like the military, nuclear power was by now seen as *inevitably* secret, dangerous and run by macho men. Indeed by the early 1980s, the decade of Reagan and Thatcher, it was possible for the Washington correspondent of the London *Observer*, together with a lawyer and ex-member of Gough Whitlam's Labour government in Australia, to publish a 578-page exposé entitled *The Nuclear Barons – the inside story of how they created our nuclear nightmare.* [112] Similarly, one UK energy specialist detected the existence of a *nuclear-industrial complex*.[113]

In the 1980s, a pale residue of post-war radicalism made its own bid to rob nuclear power of credibility. However, policymakers had by then entertained second thoughts on nuclear power for nearly a decade.

Nuclear grows before 1973/4, slows after

Until 1973/4, nuclear reactors were broadly planned, built, and made operational. But then things changed. For elites, the recession was no time to continue the 1960s upswing in building reactors.

The downturn was especially evident in the US. In February 1974, the AEC confidently predicted that worldwide nuclear energy would grow from under 100 GW to more than 500 in 1985, roughly 1000-1500 in 1990, and roughly 2500-4000 in 2000. But during 1974, demand for electricity use in the US failed to grow at all, after a steady annual increase of six or seven per for many years. As early as the spring of 1975, 87 of the 180 reactors on order in the US had been deferred, and eight had been cancelled. [114]

Perhaps it was not surprising, then, that in 1975,

- criticism of the AEC reached such a crescendo, Congress abolished it, putting its civilian nuclear responsibilities in the hands of a Nuclear Regulatory Commission (NRC)
- criticism, budgetary pressures and a burgeoning Japan saw the ABCC replaced by a new, Japanese-American agency, the Radiation Effects Research Foundation (RERF).

Thus did the two US nuclear agencies guilty of the worst excesses of the Cold War meet their end.

Worldwide, the slowing of nuclear took longer, but completely undid the AEC's forecasts. Between 1972 and 2006, no fewer than 165 plants and more than 134 GW of capacity were cancelled *after* the granting of construction permits. Indeed on 62 cancelled plants, building had either begun, or was even complete. [115] World nuclear generating capacity would have been broadly 50 per cent higher, 1975-2005, had it not been for cancellations.

Worldwide nuclear capacity:
how cancellations soared after 1973/4 [116]

Generating capacity (GW)

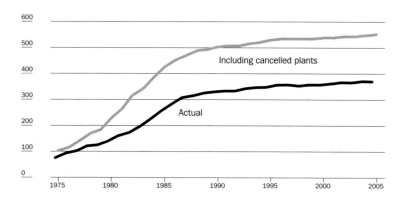

In Britain, a similar caution took a rather different form. Typically enough for the British, historians of nuclear power contributed to a broad disenchantment with it. In 1967, writing for Britain's right-wing Institute of Economic Affairs, the journalist and economist Duncan Burn had suggested that the 1945-51

Attlee government's fondness for nationalisation had been to blame for the anarchic trajectory of civil nuclear power in the UK. Burn wrote:

> 'The American type of organisation was not ruled out... by lack of resources or the smallness of firms or markets. The determining factor in the British choice is to be found in attitudes rather than circumstances; the choice reflects views on private enterprise, competition, central planning and administration and the competence of ministers.' [117]

However, in 1974, Margaret Gowing and Lorna Arnold's magisterial history of British nuclear affairs, *Independence and Deterrence*, confirmed to many opinion-formers that the genesis of atomic power had been much more about military dogmas and aspirations than economic ones. [118] Burn was wrong, Gowing and Arnold implied: post-war, Labour had not excluded firms such as ICI and English Electric from nuclear power. Such firms had been involved from the outset, but had concluded that nuclear work would drain precious manpower, turn out unprofitable, and lead to political hassles.

In the austerity of the 1940s and 1950s, Gowing and Arnold contended, Britain's commitment to nuclear matters had been more about delusions of imperial grandeur than about cheap electricity. In taking this view, the two authors helped bring about an important shift in British officials' opinions about nuclear power. Expert opinion now began to see it more as *a matter of security in the sense of military and political risk management*, and less as a matter of economic potential.

Also in 1974, a paper by Harvard's Irvin 'Chip' Bupp and MIT's Jean-Claude Derian began to change minds about nuclear. They estimated that light water nuclear reactors in the US were much dearer to run than had been previously thought, and that such reactors were likely to become more expensive still. Significantly, they put the increase in costs down to delays caused by public opposition. [119]

By 1976, Gowing, Arnold, Bupp and Derian were cited, not without glee, by the Canadian environmentalist Walter Patterson. His book *Nuclear Power*, a relatively balanced Green critique of the sector, became something of a primer on the field. [120] Patterson himself was a founder of Friends of the Earth in the UK. From then on, FoE played a leading role in undermining the case for nuclear power.

Proliferation adds to the risk-orientated conception of nuclear power

In 1968, the Nuclear Non-Proliferation Treaty (NPT) was signed in London, Washington and Moscow. [121] By then, proliferation had already become an international concern. First, the arms race between the superpowers was spiralling upwards. Second, the superpowers feared the spread of atomic weapons to countries that had not yet developed such a capability.

The NPT addressed both points. Articles I and II outlawed the transfer of nuclear weapons technology to non-weapons states, and obligated them not to seek it. Article VI, meanwhile, committed states possessing nuclear weapons to

> '… pursue negotiations in good faith on effective measures relating to cessation of the nuclear arms race at an early date and to nuclear disarmament, and on a Treaty on general and complete disarmament under strict and effective international control.'

For all the mangled prose, subsequent history was clear: the nuclear weapons states – America, Britain, China, France and the Soviet Union – proved far keener on keeping others out of the nuclear club than in pursuing disarmament themselves.

Altogether, the NPT ratified the status quo in the nuclear arms race, in the hopes of slowing it down. But since America had, apart from the upset of Sputnik, long led that race, the NPT amounted to *the Atlantic Alliance facing down Asia in the nuclear realm.*

Against Russia and China, which had exploded her first hydrogen bomb in 1967, the Treaty sought to freeze in a Western advantage. And had India and Pakistan signed the Treaty, they would never have gained the nuclear weapons they now have.

Most significant to the story of nuclear energy, however, are Articles III, IV and V. These guaranteed non-nuclear weapons states the benefits of peaceful nuclear technology. Thus weapons states undertook to transfer nuclear expertise to those states that did not yet have it. The NPT gave the UN's International Atomic Energy Authority, set up in 1957 to promote Eisenhower's Atoms for Peace, a new job, and one it still has today: to police nuclear matters in the developing world through an intrusive system of inspections.

Nevertheless, 40 years ago, the nuclear superpowers formally paid some respect to the ambitions and the sovereignty of less powerful countries. They also expressed some confidence that the development of nuclear technology could have benefits independent of its military uses. As the 1960s turned to the 1970s, however, these kinder sentiments began to vanish.

Under President Jimmy Carter, a former submarine nuclear engineer, US concern over proliferation came to a head, confirming that nuclear power was now more a matter of *risk* than of economics. In April 1977, Carter announced to the nation that there was no dilemma more difficult to resolve than that connected with the use of nuclear power. The world needed energy, but 'components of the nuclear power process' could be turned to providing atomic weapons. [122]

Carter announced restrictions. He continued with an embargo on the export of equipment or technology that would permit uranium enrichment and chemical reprocessing. In effect, he flouted the NPT's provision that the developing world be given access to nuclear technology.

Such *realpolitik* was hardly a novelty. What was much more remarkable was that Carter also scaled back and restructured *American* nuclear activities in line with his non-proliferation edict. [123] He ordered that reprocessing within the US be halted, on the grounds that, while it extracted the full energy from uranium fuel, it could also be adapted to extract plutonium or weapons-grade uranium for use in bombs. He worried that continued reprocessing in the US would both create more material that might be diverted to weapons, and encourage the spread of reprocessing technology around the world.

It's true that the halt to reprocessing came in the wake of recession. Cancelling new power stations also lowered future requirements for nuclear fuel, making reprocessing a much less attractive economic and technical proposition than it had been. Yet if Carter was right, the world's leading superpower could not trust its own institutions to look after nuclear materials within its own borders.

That went further than mistrust of the East. It marked a new level of unease about the whole idea of nuclear power.

Since Carter's era, America, Britain, the UN and the IAEA have used all aspects of uranium extraction and refinement, along with the production of fissile plutonium in nuclear reactors, as a means of restricting the sovereignty of Pakistan, India, North Korea, and of course Iraq. More recently Britain, France and Germany have tried to set controls on Iranian fission, while the UN has also upbraided Brazil and South Korea for their experiments. A discreet veil has been passed over Israel's 200 nuclear warheads, although even here there are fears of proliferation.

Operational risks: Three Mile Island and *The China Syndrome*, 1979

Early on the morning of 28 March 1979, there was a mechanical failure in the cooling system of Unit 2 of the power station at Three Mile Island, Pennsylvania. Overheating led to venting of radioactive gases to the atmosphere and, later, the release of liquid waste to the Susquehanna River. However, no workers were injured.

Among those living close by during the accident, a 1998 follow-up of more than 30,000 found 'no consistent evidence that radioactivity released during the nuclear accident has had a significant impact on the overall mortality experience of these residents.'[124] In 1945 alone, at Hanford, Washington State, America's plutonium production facility had secretly released more than 10 million gigabecquerels of radioactive iodine into the atmosphere – an enormous amount (one becquerel equates to one decay per second). By contrast, official estimates suggest that Three Mile Island released less than 550 gigabecquerels.[125]

After Governor Richard Thornburgh's advice to evacuate pregnant women and children within a five-mile radius, more than 100,000 people jammed the roads – 50 per cent of

Three Mile Island, Pennsylvania

the population of Three Mile Island. [126] In the aftermath, regulatory oversight of the entire US nuclear industry was restructured and expanded.

By coincidence, the movie *The China Syndrome* was released within two weeks of the accident. It starred Jane Fonda as an investigative journalist who makes contact with a whistle-blowing engineer at a nuclear power station. She goes on to show safety taking a back seat as short cuts are made to maximise profit, with the corrupt collusion of government regulators. The media's corporate backers, meanwhile, put on pressure to stop the story coming out.

That the film was so readily taken up as a means of understanding real events showed that Hollywood's account had successfully gripped the public imagination. The mass panic that greeted the accident formed a stark contrast to the reality, in which the accident caused no casualties. Just five years after the Watergate scandal of 1972-4 ended in Nixon's resignation, the conviction that *we are not told the truth* took firm hold.

Europe also saw a reaction against nuclear power, above all in Germany, where in the late 1970s opposition assumed the scale of a mass movement. In Germany, Green activists gained an international following for their yellow bumper-sticker logo of a smiley sun and, in various translations, their slogan *'Atomkraft? Nein, Danke!'*. In 1980 they formed a political party.

In later years, Germany's social-democratic party, the SPD, took on a Greenish hue, and diminished the influence of Green political organisation. Nevertheless, the international impact of German environmentalism in the realm of theory was considerable. Hans Jonas, a philosopher, attacked nuclear energy for helping our old friend, 'energy gluttony', while more broadly prioritising the *survival of future generations* over hopes in progress. [127] Meanwhile, the sociologist Ulrich Beck interpreted society as manufacturing 'invisible risks' – above all, those associated with nuclear power. [128]

On his own terms, at least, Beck was fortunate. The publication of his key work, *Risk Society*, occurred in the same year as a reactor blew up in the north of the Ukraine.

Chernobyl, 1986: too little technology, too little economic development

At 1.23am on 26 April 1986, reactor number four at the Chernobyl Nuclear Power Plant, three kilometres south of Pripyat, Ukraine, exploded. The accident happened as new emergency procedures were being tested. For 10 days fire raged, releasing a cloud of radioactive material through the wrecked roof that spread across Europe, raining out caesium and iodine.

Chernobyl was without doubt the worst nuclear accident in history. It drove a third of a million people from their homes. But was it, as the British philosopher John Grey alleges, a warning against human hubris? [129] It's important to understand Chernobyl for what it was. What exactly happened?

According to Greenpeace, Chernobyl resulted in 200,000 deaths in the years 1990-2004 alone. The organisation links the disaster to respiratory disease, endocrine problems, premature ageing, and psychological disorders, among others. Significantly, it also claims that 'the worst is yet to come'. [130]

The Chernobyl Forum, a group of experts brought together by UN agencies and the governments of Ukraine, Belarus and Russia, came to a different conclusion. It noted 50 deaths amongst workers and among 'liquidators' brought in to clean up the accident. The Forum estimated that there would eventually be another 4,000 fatal cancers in the most exposed population. In less exposed populations the rate of cancer increase will, at less than 0.01 per cent, be undetectably small. Still, there are also non-fatal consequences: liquidators suffering from cataracts, and about 4,000 thyroid cancers amongst children. The latter are highly treatable, albeit with severe long-term side effects. [131]

Chernobyl was a very serious disaster. Things could have been much worse: immediately after the explosion, winds blew radioactive material away from Pripyat. [132] But before examining the matter more closely, we should ask: what accounts for the differences in the deaths that are quoted – between Greenpeace's

200,000 and the Chernobyl Forum's 50?

The Chernobyl Forum started with what's known about radiation and the exposure of the local population. Rightly, however, it also factored in:

- the rise in diagnoses of cancer following from intense post-accident surveillance
- the effect that the subsequent collapse of the Soviet Union in 1989 had on local health.

A helicopter decontaminates the
disaster site of the Chernobyl explosion

Greenpeace also noted the uncertainty inherent in estimating health impacts. But it then used the concept of uncertainty as an excuse to blame every 'unexplained' problem on Chernobyl. Instead of proceeding on the basis of science, it began from the premise that the explosion explained every ill.

What should people make of the real Chernobyl? Its consequences were typical of a disaster in a poor country. When a plant making pesticides in Bhopal, Madhya Pradesh, India, released a toxic cloud over surrounding slums in 1984, 3,000 people died *overnight*. A further 15,000 may have died in later years, and tens of thousands were affected. [133]

Mining in China also kills. In 2004, more than 6,000 people died in accidents. [134]

Chernobyl was less deadly than Bhopal or Chinese coal mining, but is analogous in that all three disasters would likely neither have happened, nor led to such grave consequences, had there been better control technologies based on more substantial investment. The loss of life happened in industries forced to function in cash-strapped contexts: they were not the result of inherent defects in the industries themselves. [135] Also important is the broader economic and political context. To protect their thyroids, children in Pripyat were given iodine pills within hours of the explosion. In Belarus, things happened differently. There the authoritieslet children drink milk contaminated with iodine-131 for a week. [136]

The consequences of a Chernobyl in the developed world would be limited by 'containments' – formidable concrete and steel walls surrounding reactors, giving them their characteristic spherical shape. Containments have always been standard in the West. According to physics professor at the University of Pittsburgh, Bernard L Cohen, accident analyses since Chernobyl show that if there had been a US-style containment, 'none of the radioactivity would have escaped, and there would have been no injuries or deaths'. [137]

The lesson of Chernobyl is the need for more technology and economic development, not less.

After 9/11, fear of terrorist attacks on reactors

Since its birth, nuclear technology has generated runaway disaster scenarios. After 9/11, however, policymakers began to take worst-case thinking about nuclear matters much more seriously than they had in the past. Fears of terrorist attacks on power stations rose up the agenda.

Curiously, a new consensus emerged. Anti-nuclear campaigners, once suspicious of government security measures, now worried as much as the state about terrorist attacks on nuclear reactors. 'Imagine a world without New York City,' instructs Robert F Kennedy Jr, Senior Attorney for the Natural Resources Defense Council in the US. He continues that terrorists have already done that imagining:

> 'In November 2002, the FBI warned that Al Qaeda sleeper cells could be planning attacks on US nuclear power plants near our largest cities to try to inflict "severe economic damage and maximum psychological trauma".' [138]

In the same vein, Helen Caldicott, president of the Nuclear Policy Research Institute, Washington, DC, claims that the design for an atomic bomb can easily be found on the Internet – and that a trip to a hardware store will allow miscreants to 'complete' nuclear arms production. [139]

Sensible security measures are just that: *sensible*. But, as with security checks at airports, people need to ask whether such measures do any real good, or whether they merely prevent people from getting where they need to go.

A terrorist attack on a nuclear power station is unlikely to be catastrophic. Of containments, a review published in the journal *Science* pointed out: 'No airplane, regardless of size, can fly through. This has been calculated in detail and tested in 1988 by flying an unmanned plane at 215 m/s (about 480 mph) into a test wall 3.6m thick'. The authors continued:

'To tell people that they and the Earth are in mortal danger from events that cannot cause significant public harm is to play into the hands of terrorists by making a minor event a cause for life-endangering panic.' [140]

The Devil's Trinity today (1): accidental releases of radiation

The key contemporary charges made against nuclear power concern *accidents, proliferation,* and *waste*. Fear of accidents is the most straightforward to deal with.

Today, mankind has developed technologies that are mature enough to make nuclear power safe enough. Since the 1960s, in fact, Western technology has been in place to contain the accident that occurred at Chernobyl.

Fastening not on low risks but on major consequences, Greens can always imagine a catastrophic scenario. An unlikely chain of events is always possible. But as the politics of precaution reveal (see Chapter 3), a Doomsday can be dreamt up for any technology.

What matters is not what is *possible*, but what is significantly *probable*. Here, the historical record *supports* the idea that contemporary nuclear power is safe. The most notorious accident in the West, Three Mile Island, killed nobody – with technologies that are now more than 30 years out of date.

Newly built nuclear power stations will be even safer than those in operation today, if only because materials and electronics are better. Components are more standardised, and thus more reliable. Designs have improved, and incorporate new safety devices.

'Passive safety' features shut down reactors by relying on natural forces such as gravity and thermal expansion, making them more reliable. The name Westinghouse gives the generation of reactors it is building in China is AP1000 standing for 'Advanced Passive' reactor. This is the company's flagship new model, intended to lead a nuclear revival. Passive safety principles cover the isolation of its core and its cooling mechanisms. Westinghouse claims the following advances: [141]

The AP 1000: claimed percentage reductions on previous designs [142]

	reductions on previous designs, per cent
Number of safety-related valves	50
Amount of safety-related piping	80
Amount of control cable	85
Number of pumps	35
Volume of earthquake-proofed building	45

Reactor safety, however, is not just a matter of fewer parts to go wrong and better technology. Without good leadership, as well as good staff motivation, training, and professionalism, high levels of safety will never be achieved. Management competence and the pursuit of shareholder value are questions for every industry, and the nuclear sector should certainly be held to account on these issues. But poor safety practices don't automatically adhere to the nucleus.

Nevertheless, the nuclear industry feels itself on the back foot here as everywhere else. Again and again, it emphasises the need for a 'safety culture' in nuclear plants and – needless to say – among nuclear workers. The US Nuclear Regulatory Commission, no less, wants 'a work environment where management and employees are dedicated to putting safety first'. [143]

Yet in the end safety cannot really come first. After all, if the nuclear industry's overarching objective really did become safety, nuclear power stations would have to be shut down – just in case.

Safety can never be an afterthought. But neither can it become a substitute for the central purpose of an organisation or a society. In the case of nuclear power, the central purpose ought to be *to generate electricity*.

That Westinghouse brands its new breed of reactors around their safety features suggests that it has its priorities wrong. To be satisfied with building any reactor so long as it's safe shows a poor imagination. *What the world needs now are Advanced Productivity reactors.*

Of course, reactors need to be safe. But Advanced Productivity is the kind of inspiring mission that can ground discipline, teamwork, insight and a properly watchful attitude toward safety.

Safety demands more than simply following rules and ticking boxes. It requires a workforce with high morale. Ultimately, that will depend on the support nuclear power finds in society. When friends, neighbours and the media regard reactors as beneficial, commitment to safety on site will be that much stronger.

The Devil's Trinity today (2): unbridled proliferation

In the Carter years, fears of nuclear proliferation put the brakes on nuclear energy, both inside and outside the US. With the end of the Cold War, those fears shifted from the nuclear arms race of the superpowers to the 'asymmetric warfare' – nuclear included – waged by rogue states, terrorist networks and aberrant individuals. As Secretary of State James Baker III put it, proliferation of one sort or another was going to be 'perhaps the greatest security challenge of the 1990s'. [144] This view was bound to have consequences for nuclear power.

The idea of *rogue individuals setting off nuclear devices* goes back, as we have seen with Fred Cook, to 1961. Then, in 1985, a week or so before he approved the sale of arms through Israel to Iran, President Ronald Reagan famously cast the latter as the main example of *outlaw states* run by 'the strangest collection of misfits, looney tunes and squalid criminals since the advent of the Third Reich'. [145] Finally the end of the Cold War allowed a new synthesis of nuclear roguishness to be imagined and feared. People were told that with nuclear proliferation, the bridle was now off.

By 1989, the end of political certainties, burgeoning globalisation and the rise of the Internet seemed to make for an unpredictable world. In 1990, America's Arms Export Control Act targeted states aiding or abetting 'an individual or groups in acquiring unguarded nuclear material'. [146] Individual Russian scientists, disgruntled with Boris Yeltsin's rule, were going to leave the country with bags of nuclear materials for sale on a world black market. Superguns, designed by Canadian rogue scientist Gerald V Bull, were going to give Saddam Hussein's state nuclear weapons with a range of 1000km. [147]

In January 1994, in a speech to European politicians, Clinton formalised the doctrine of rogue states. [148] In the aftermath of 9/11, George W Bush sought to connect Al Qaeda, a rogue network, to a rogue state – Iraq. Notoriously, it was argued that in Saddam Hussein's rogue state, there were going to be weapons of mass destruction. Some of them, too, were going to be nuclear, being based on 'yellowcake' uranium ore drawn from Niger.

Everyone now knows that fears of rogue elements have proven unfounded. But while conspiracy theorists still wax indignant over the dodgy intelligence dossiers that led up to the Iraq war, something seems to have escaped them. When Bush and Blair talked themselves into believing their worst-case scenarios about Iraq, a culture *obsessed by nuclear proliferation ensured that they could also talk quite a lot of other people into their war.*

Fears of nuclear proliferation are quite widely embedded nowadays. These fears impede nuclear technology, which the world urgently needs to put to work.

The US is now keen to see nuclear energy taken up around the world, but it remains fearful of proliferation. Its Global Nuclear Energy Partnership (GNEP) has a convoluted scheme to give developing nations nuclear power with no 'user serviceable parts', so that they will have no access to the nuclear fuel that might be reprocessed into weapons. [149] GNEP's statement of principles, signed in September 2007, says that it aims to 'take advantage of the best available fuel cycle approaches for the efficient and responsible use of energy and natural resources'. [150]

Those are appropriate goals for international co-operation.

But a whole four out of the GNEP's seven principles are concerned with limiting proliferation. These include: enhancing IAEA safeguards, creating a viable alternative to the acquisition of sensitive fuel cycle technologies, proliferation-resistant reactors appropriate for developing countries, and recycling fuel in a proliferation-resistant manner.

Such schemes make it difficult to get the most out of nuclear power. They also perpetuate a world divided between nuclear haves and have-nots. What the developing world needs is not proliferation-resistant reactors, but the best kind possible.

GNEP is supposedly dedicated to the development of nuclear energy. Yet though its main bias is in fact to *limit* the spread of nuclear technologies, even this has not been enough for its critics. In June 2008, the US Congress abolished funding for the GNEP on the grounds that the risks of nuclear proliferation made it a bad idea. [151]

In fact any truly developed country wanting nuclear weapons cannot easily be stopped from getting them. As three experts in international relations and diplomacy write, more than 40 countries today have the wherewithal to make fuel for peaceful nuclear power, 'but this easily can be modified to make material for nuclear weapon'. [152]

Whether nuclear weapons spread, and whether that's a problem, are political, not technical issues. But a foreign policy built on denying most of the world access to modern technology is immoral and unsustainable.

The pattern of nuclear tensions in the world began, as we saw, with the Manhattan Project. Tensions certainly exist today. It's also true that if North Korea or Iran acquired a nuclear weapon, it would change the balance of power with the West. But it isn't true that nuclear weapons are the *source* of all these tensions.

Neither is it true that states such as Iran or North Korea

are so irrational as often assumed. Just as the interests of their rulers differ from those of their subjects, so do their interests differ from those of the West. But there are few grounds for believing that they have a strong motive to begin a nuclear war.

Efforts to stop proliferation in developing countries cannot be compatible with real economic advance there. It isn't just uranium that's required to make a bomb: heavyweight computers, sophisticated electronics and advanced materials are needed too, as well as a strong infrastructure. Therefore efforts to stop proliferation can know few limits. The attempt to prevent Iraq obtaining nuclear weapons involved controversy over aluminium tubing. And under UN resolution 661, the export of pencils to Iraq was limited to certain quantities. [153] Perhaps the graphite in them might be used in a reactor.

The attempt to stop nuclear proliferation through restricting access to technology can only end in large parts of the world not developing to the extent that they should.

The world will have to solve the problems of international relations. But it can't just wait in hope for the benefits of civilian nuclear energy.

When more energy helps speed developing countries out of poverty, the world just might become a more peaceful place.

The Devil's trinity today (3): waste, its disposal, and the problem of future generations

For many people, the killer argument against nuclear energy remains *waste*. Governments have dithered over approving new reactors. But that is as nothing compared to their record on nuclear waste.

In the US, studies for a waste repository at Yucca Mountain, Nevada, began in 1978. When, in June 2008, the Department of Energy submitted a 10,000 page, 12-volume application for a licence, the Nuclear Regulatory Commission responded with the view that it anticipated completing a safety analysis and giving an answer sometime before... 2013. [154]

Why the delay? In 1991 social psychologists Paul Slovic,

James Flynn and Mark Layman published the results of four surveys of Americans – 3,334 people in total. In each survey, respondents were asked freely to associate words with the idea of a nuclear waste repository. Slovic and his colleagues reported that the responses revealed an aversion so strong, it could not justly be labelled a simple dislike. [155]

Perhaps that's not a surprise. But as good social scientists, Slovic and his colleagues took nothing at face value:

> 'One might expect to find associations to energy and its benefits – electricity, light, heat, employment, health, progress, the good life – scattered among the images. Almost none were observed.'

People were asked about a nuclear waste repository rather than waste itself. As a result, the researchers expected, 'following the predominant view of experts in the this field, to find a substantial number of repository images reflecting the qualities "necessary" and "safe"'. Yet as it turned out, most respondents did not form such images.

Slovic and his colleagues offered two explanations for what they had found:

1. The 'sprouting' of nuclear power in the aftermath of Hiroshima
2. Public fears and opposition to nuclear waste disposal plans could be seen as a crisis of confidence.

The team described the latter as 'a profound breakdown of trust in the scientific, governmental and industrial managers of nuclear technologies.'

As this chapter has shown, both points are important. They confirm that history, more than physics, chemistry or biology, is what makes nuclear power so much the object of hysteria.

We add two further points. First, any *and all* waste today is seen to be a big problem. Second, the long timescales associated with nuclear waste conjure questions about the long-term future.

How present generations relate to the future is a question that society no longer has easy answers for. Without a sense of progress, that human beings can make the world a better place for their descendants, people can slip all too easily into focusing on the dark side of the future.

Here the worst-case scenario thinking of the Precautionary Principle finds a grander canvas on which to paint. Projected thousands of years into the future, there is even less of a sense of reality to constrain fantasies of disaster.

Let's get precise about the longevity of radiation

Spent nuclear fuel contains a mixture of components – some of which only lose their radioactivity after hundreds of thousands of years. These are timescales beyond everyday imagining; yet it's worth remembering that they describe how long it takes for radioactivity to decay completely, not how long it takes to become tractable. In fact, spent fuel is actually less radioactive than uranium ore after 'just' 10,000 years. [156]

Cadmium, mercury and lead, by contrast, have chemical toxicities that persist forever. Compared with nuclear waste, too, these naturally occurring elements are much more prevalent on the Earth:

Longevities and quantities of some toxic substances

Substance	half-life, years [157]	annual production, tonnes [158]
Cadmium	infinite	19,900
Mercury	infinite	1,500
Lead	infinite	3,500,500
Uranium 235	703,800,000	41,279
Plutonium 239	24,110	60

Contrary to the expectations that follow from the Precautionary Principle, mankind is not in the process of continually discovering new problems with radioactive waste. Its hazards have been well understood for decades. Indeed, what has happened over the years is that while technological advances have made nuclear waste easier to manage, there has been a decay of rational discourse about it.

Actually, it is human fears that have taken nuclear waste, a resource that can be reprocessed into new fuel, and transformed into a problem for which there is no solution.

Fusion deserves a higher level of commitment

At present, nuclear energy is based on fission: the breaking up of large atomic nuclei into smaller parts. An alternative prospect is fusion: the joining together of light nuclei such as hydrogen, helium and lithium. This process is what powers the sun and the stars.

For 50 years, fusion has held promise as a theoretical possibility that is potentially cleaner and capable of producing more fuel than conventional nuclear energy. But the practical engineering of fusion has proved a big challenge.

Fusion energy has been released on Earth in hydrogen bombs. But for energy generation, it needs a more controlled release. So far, efforts have been focused on containing the nuclear reaction in a special 'magnetic bottle', which is suspended away from the walls of the reaction chamber. This is the approach taken by the ITER project, a $9.3 billion joint international R&D effort in which the EU, Japan, China, India, Korea, Russia and the US are partners.

Other strategies to controlled fusion include inertial confinement, an approach taken by the European-led HiPER project, which uses powerful laser beams to confine the fuel. While these methods have proven to be less popular than magnetic confinement, they may eventually provide the solution.

That said, the difficulties of controlling the release of energy, such as designing materials that can survive the harsh conditions of the reaction chamber, remain formidable.

Another problem is to do with funding. The pressure on science funding is to show that research will have an immediate payoff. For fusion scientists, that can mean overselling the practical benefits of their research over long time scales that are necessarily speculative.

Given these facts, it is unlikely that fusion will make a real contribution to energy before the middle of the 21st century. A better approach would be for society to have more confidence in funding basic research. If that can be done right, *Energise!* believes that, after 2050, the fundamental advances will be in place to make radical energy technologies such as fusion a reality.

Summing up this chapter

The physics of nuclear power give every reason to favour it as a key component of energy supply in the future. However, politicians and officials in the West, and especially in Britain, are broadly unable to muster strong arguments for nuclear power. They neglect how reactors are the flat-out, low fuel cost energy workhorses of today and tomorrow, ready and able to meet the world's demand for energy right now. Instead, they promote nuclear power in terms of it not generating CO_2, or as an insurance policy against turbulence in the East.

Renewables will always beat nuclear into submission in that kind of contest.

Barack Obama's policy on nuclear energy is instructive here. For him the merit of nuclear reactors is modest: since they account for more than 70 per cent of America's non-carbon generated electricity, eliminating them would make it unlikely that he could meet his goals around climate. But *expanding* US nuclear power is, for Obama, another matter. Following the precedent of Carter, Obama says he will make 'safeguarding nuclear material both abroad and in the US a top anti-terrorism priority'. That must be fixed before another reactor is built in America. An alternative to Yucca Mountain must also be found, because Obama rejects it as a site for nuclear waste. Meanwhile, waste stored at current reactor sites should be contained with the best technology available. [159]

Sensible prudence coupled with traditional American commitment to high technology? Perhaps. Yet it seems that Obama will make the difficulty of long-term storage of nuclear waste reason enough to abandon plans for new reactors for the foreseeable future. Of nuclear waste, Obama has said in an interview that

'... if we could figure out how to store it safely, then I think most of us would say "That might be a pretty good deal".' [160]

Yet here he wrongly blurs the distinction between long-term storage, which will be taxing but not intractable, and the shorter-term variety, which has long been sorted out.

Just before expressing his hopes for nuclear, Obama's fears and uncertainties ran away with him. In the same interview, he said:

'Right now we don't know how to store nuclear waste wisely, and we don't know how to deal with the safety issues that remain, and so it's wildly expensive to pursue nuclear energy.' [161]

But right now it *isn't* wildly expensive to pursue nuclear energy: if nuclear is so expensive, why is the East so keen on it?

Nuclear power started off with the best of hopes. But in the 1970s it received a blow that, in the process, also hit the whole reputation of science. Looking back on the generation of British who had gone to Oxford, Cambridge or the LSE between 1919 and 1951, the brilliant English academic Noel Annan wrote that, in the 1970s and around the issue of science, its members

> '… heard the rumble of discontent. They believed scientists improved the quality of life, but their successors were at first sceptical and then hostile…. Then in the 1970s ominous signs appeared. Those in CND who wanted to ban the bomb now wanted to dismantle all nuclear power stations. Physics was no longer a neutral subject and physicists could no longer plead that they were the dispassionate observers of a power whose use for good or evil depended on politicians.'

'For the left,' Annan continued, 'science became an aspect of American imperialism.' [162]

The rancour around nuclear power and science took a long time to mature; yet it's possible to see the very first beginnings of it around the attacks on Hiroshima and Nagasaki. Why? Because use of atomic weapons didn't just usher in the Old Scientism, but also contained within itself the possibility of a counterattack against both the military and the civilian use of nuclear physics.

The counterattack emerged within Western societies and, indeed, within Western elites.

With radiation and nuclear weapons tests together forming the logical and historical link between anti-war and anti-reactor feeling, the world has for decades conferred a mystical and Faustian significance on the atomic nucleus. As a result, the power of that nucleus has been politically transmuted into a bizarre and seemingly autonomous force, stripped of the social relations that surround it. The nucleus is held to be intrinsically dangerous. Though the word 'proliferation' is usually focused on the East, its repeated deployment also suggests that the military use of the nucleus is an unstoppable reaction all its own – something not subject to the whims of mankind.

During the Cold War, fear of a nuclear strike embraced the schoolroom and the fall-out shelter, and sometimes took the form of dissent about things nuclear. However, since the end of the Cold War, fear of global warming has come to embrace still more spheres of human life: the home, the car and the plane. Yet importantly, the anti-nuclear dissent of the past has proved to be a preface for today's conventional wisdom about climate change. The trump card of anti-nuclear activists is the doctrine that future generations will hate the current one for imposing upon them irreversible nuclear waste. And the trump card of climate alarmists is similar: the shocking selfishness of human beings today should not be allowed to impose an irreversibly hotter planet on future generations.

Always dogged by dissent, the Old Scientism began to develop cracks by 1975. However, it took the end of the Cold War to give dissent the upper hand.

As we saw in Chapter 3, the IPCC was already up and running by this time. Worry about climate change now had inter-governmental support; and with that support, the New Scientism was born. The New Scientism had much more elite backing than the old anti-nuclear dissent.

Today's critics of nuclear power find it hard to put science on a pedestal the way climate obsessives do. Where the two groups overlap is giving a special place to uncertainty. Uncertain, nightmarish leaps in climate find a parallel in the uncertain, nightmarish effects of low-level radiation.

What the New Scientism retains from the New Left is the hatred for science and technology. Instead of faith in science's exhilarating interventions in the natural world, science is summoned as something that must be personally obeyed (climatology), or something that must be feared and, implicitly, wound up (nuclear physics).

Chapter 5 turns to fossil fuels. In it we show that mankind's working up of nature's creations is no more a Faustian bargain than is nuclear power.

It's time to forget Dr Faustus.

Rating nuclear technologies

To understand the table of ratings that follows, it's important to know the basics of nuclear reactor design.

In a reactor, energy is released when the nuclei of uranium atoms break up. The atomic nucleus is made up of two types of particle, protons and neutrons. The splitting of a large nucleus such as uranium results in two fragments, each consisting of about half the original nucleus, together with some excess neutrons.

The extra neutrons are the key to energy generation. They play two roles. First, by colliding with further uranium nuclei, they trigger further splits and sustain a chain reaction. Secondly, they carry off much of the energy.

A reactor's *coolant* prevents it overheating through absorption of the energy from the neutrons and nuclear fragments. The energy captured is then used to generate steam to turn an electric generator.

The main distinctions between different reactor designs revolve around the coolant used. Water and gas are the options used at present, while designs using molten salt and molten metal are under development.

A second important choice of technique with reactors concerns the way in which the speed of neutrons is regulated between collisions. Most commonly, graphite is incorporated in some way as a *moderator* of neutron speeds. An alternative is a reactor that works with neutrons working at higher speeds. It's this that accounts for the word 'fast' in some reactor names.

A final alternative is nuclear fusion.

Between the different kinds of reactors listed in the table below, nuclear energy could be providing 25 per cent of world electricity by 2050. If world electricity quadrupled between 2000 and 2050, then 25 per cent of electricity in 2050 is the same as all the electricity produced in 2000. Our plan calls for about 2000 new power stations, worldwide, by 2050, or almost one a week. Gordon Brown has called for 1000; the IEA, for 1250.

If the UK invested in proportion to the number of its inhabitants, it would build a reactor roughly every two years – that is, about 10 power stations by 2030. The real ambition of our plan, however, lies in the idea that every country should do the same. China would have to build 20 times as many, or 200 by 2030.

At the moment, the Chinese authorities expect to build only 150 new reactors by then; but the country's plans keep getting revised upwards and, if present trends continue, will very likely beat our target. What we are suggesting is that the rest of the developing world can and should soon start building like China.

Rating selected types of reactor

Type of reactor	Period of take-off	Top share of global electricity, 2050	Advantages	Disadvantages
Pressurised Water Reactor / Boiling Water Reactor	1970-1990	5 per cent	Existing technology	Does not handle fuel and waste generation well Does not integrate best of new technology for economy and safety
Gas Cooled Fast Reactor	2025-2045	20 per cent, when combined with other types of reactor	CHP; hydrogen production; industrial applications Burns many fuel types, leaving little waste	Technology less developed than Very High Temperature reactors which have similar application
Lead Cooled Fast Reactor	2020-2040	20 per cent, when combined with other types of reactor	Modular mass production Long periods (10 years+) between refuelling	More suitable for relatively small scale applications (10s-100s MW)
Molten Salt Reactor	2025-2045	20 per cent, when combined with other types of reactor	Efficiently burns fuel without need for extraction and reprocessing	Complex technology Solves a problem that is not critical
Sodium Cooled Fast Reactor	2025-2045	20 per cent, when combined with other types of reactor	Can use waste from other reactors as fuel	Complex technology
Supercritical-Water-Cooled Reactor	2020-2040	20 per cent, when combined with other types of reactor	Builds on best of existing technology Simple and efficient electricity production	Inflexible in applications and scale
Very High Temperature Reactor	2020-2040	20 per cent, when combined with other types of reactor	Can produce heat for CHP, hydrogen production, and water purification	Will be outclassed by flexible fuel handling of gas cooled fast reactors
Fusion Reactor	after 2050	less than 1 per cent	Abundant fuel, low pollution	Serious development required

Generation	Resources required for development	Key research problems	Rating now /10	Rating for the future / 10
3G	Uranium Steel and concrete	None	9	3
4G	Uranium Steel and concrete	High temperature materials Design of fuel elements	In development	9
4G	Uranium Steel and concrete	Molten lead and lead alloys for use as coolant	In development	9
4G	Uranium Steel and concrete	Molten salts which will both carry fuel and act as coolant	In development	7
4G	Uranium Steel and concrete	Improving inherent safety of design	In development	8
4G	Uranium Steel and concrete	Corrosion and radiation resistant materials	In development	9
4G	Uranium Steel and concrete	High temperature materials Waste management	In development	8
One	Lithium, (an abundant mineral), deuterium (from seawater), steel and concrete	Radiation resistant materials, plasma ignition and containment	In development	9

Is nuclear power intrinsically uneconomic?

An enduring charge made against nuclear power is that it has always promised to be cheaper than other kinds of energy, but has never delivered on that promise. For critics this makes it intrinsically uneconomic. In fact, however, the accusation says more about those who level it than their target.

First, given the widespread historical amnesia that afflicts society nowadays, it's worth getting nuclear's development in perspective. Because of its late arrival on the energy scene, nuclear is a young industry – and, in considering its economics, allowances need to be made for this. Since Calder Hall, the civilian nuclear sector has grown up very fast. It's

- less than half as old as the oil sector
- less than a third as old as the natural gas sector
- about a fifth as old as coal.

As large-scale, intricate technologies go, nuclear's growth has occurred at an extremely rapid pace.

Second, a whole variety of reactor models has been experimented with, slowing maturation and thus economic viability. Of course, that's to be expected of any new industrial sector. But there's something else as well: nuclear's military origins not only made it a profoundly national affair, impeding international collaboration for years, but also distorted its development, leading to an above-average number of false starts.

With nuclear, modest economies of scale only really emerged in the 1960s, when PWRs matured as a reliable and reproducible design. Given its 16 per cent share of world electricity supply, nuclear can be said to have gone from nothing to something substantive very quickly. Of course, nuclear industry boosters have made silly claims for it in the past – above all, that it would be 'too cheap to meter'. But that should not blind people to what the industry has, despite everything, achieved.

Third, it's only after examining the ways in which nuclear's opponents systematically exaggerate its risks that a rounded judgment about costs can be made. Irrational opposition leads to spiralling costs – only for Greens to turn round and say 'I told you so'.

Fourth, environmentalists exhibit double standards around state subsidies for nuclear. It's true that over the years and even today, nuclear has enjoyed some significant support from the state. But what kind of energy generation is free from such support? Greens attack nuclear, and also fossil fuels, for the featherbedding they get from the state. Meanwhile, free-marketeers one-sidedly attack renewables on the same grounds. The truth is that the state assists all forms of energy generation. Nuclear is by no means exceptional in this regard.

Finally and most importantly, nuclear power is characterised by technological advance, which has lowered and will continue to lower its costs. This sounds obvious; but like their inspirations Malthus and Marshall (see Chapter 2), nuclear's opponents don't incorporate technological development into their political economy. Alternatively, Greens are hopeful that costs will come down with, say, solar power, but do not extend that hope to nuclear.

Again, double standards are apparent here. Because the very thought of interference with sub-atomic nature, and of dealing with radioactive waste, is anathema to them, Greens don't entertain the idea that more sophisticated nuclear technology brings lower electricity prices. On the contrary: to improve nuclear is, for its enemies at least, a contradiction in terms.

Actually nuclear has improved, despite the fact that opportunities for reaping economies of scale have so far proved rather limited. With fewer than 500 reactors working in the world today, making reactors in large batch runs, with modular components produced in still greater volumes, remains a task for the future.

When they come to the economics of nuclear power, the

Greens' biggest complaint is about the costs of decommissioning. In addition to wastes produced during operation, the power station itself must be taken apart at the end of its life and radioactive components disposed of. All the steel structures that contained the fuel and coolant are radioactive waste.

The arguments that we have made above apply.

It is hard to get an accurate idea of just how much decommissioning will really cost. In July 2008, newspapers reported an internal audit claiming that the financial affairs of the UK Nuclear Decommissioning Authority were subject to 'inherent risks', that budgetary problems were plagued by 'misunderstandings, unminuted meetings and lack of sufficiently trained staff'. Over 12 months, cost estimates for dealing with Britain's waste had risen by £10bn to £83bn. [163]

With nuclear, today's precautionary climate is putting an upward pressure on costs of waste disposal. Inflationary problems, however, knows no upper limit. These kinds of problems, however, have long been known to afflict large government projects – regardless of whether they involve radiation.

There are private sector alternatives to direct government funding, such as prepayment, a sinking fund, or the purchase of insurance. But since the Crash of 2008, none of these is guaranteed to operate smoothly.

Given the quantities of electricity generated, the costs of decommissioning are not so large. Managed properly, they should probably run to a few hundred million pounds per reactor. That represents a fraction of a penny per kilowatt hour generated.

There are other ways to bring down costs: economies of scale, reactors with longer lives, and clever recycling of contaminated steel from old to new power stations.

Let's get precise about the size of nuclear waste

Because uranium has such a great energy density, both the volumes of fuel going into reactors and the volumes of waste coming out are small. So despite the fierceness of its radioactivity, the limited volume of nuclear waste makes it a modest problem for society to solve. The physical scale of nuclear waste is nothing compared to, say, the Channel Tunnel.

Nuclear waste management is certainly a complex and particularly dangerous business, demanding professionalism, teamwork and calm. But that only makes it like many other important endeavors in life.

The UK has a historic legacy of nuclear waste that, in volume terms, is estimated to total 476,900m³. [164] But how much is that? Take the cube root of 476,900m³, and it's 78 metres. In other words, all the existing nuclear waste in the UK, generated over more than 50 years, occupies about the same envelope as 27 floors of Canary Wharf Tower, east London.

And new waste promises to be pretty modest too, for modern nuclear plants don't make much. Though it has since insisted that it only deals with existing waste, the UK's official Committee on Radioactive Waste Management has suggested that if the country's current level of nuclear capacity were replaced with new-build, the volume of existing waste stocks would increase by about 10 per cent. [165] In that case, the UK would have to add 36 x 36 x 36m of nuclear waste – an envelope equivalent to fewer than three more floors of Wharf Tower.

Furthermore, more than 95 per cent of the radioactivity in existing waste is concentrated in high-level waste, spent fuel and plutonium. These components make up just 15,900m³, or 3.3 per cent of total packaged waste volume. [166] Altogether, that's less than one floor of Canary Wharf Tower.

To store such a modest amount of nuclear waste is not beyond British engineers. The alternative is to concede that, in the 21st century, handling these amounts of waste is beyond the wit of man.

Taking a position on radiation standards

For practical regulatory purposes, the 'linear no threshold' model, makes sense. However, a look at radiation standards shows that, under the influence of different politics, the same science of radiation can result in different standards.

The first federal radiation standards in the US were based on the 1956 work of the National Academy of Sciences' committee on the Biological Effects of Atomic Radiation (BEAR). [167] The standards began to bring under control the very high exposures produced by the rush to create a nuclear weapons industry.

But their distinguishing feature was that the standard for the maximum exposure for workers was set at 50 mSv, 10 times higher than the maximum for the general public. The nuclear industry and the AEC complained that without this measure, standards would be impractical.

Our view is that the 1956 maximum for the public is broadly right. It is a little above, but nevertheless broadly comparable with natural background levels of radiation. Unlike, say, smoking, the effects of natural background radiation are so small, they enough that they cannot be disentangled from the multitude of complex processes that lead to cancer. The same will be true of an increment from artificial radiation. This is a level of risk that most people are happy to take in return for the benefits of a technological society.

The maximum public exposure of 10 mSV is in any case only relevant to the few individuals who, for whatever reason, accidentally come into close contact with nuclear wastes. Most people living within 50 miles of nuclear power station, for example, will receive more like 0.0001 mSv per year.

There seems no reason, however, why levels for workers should be higher than those for the general public. It's possible to imagine emergencies in which workers might be exposed to high levels. But there's no reason why routine work in the nuclear industry should be more dangerous than any other occupations.

In 1977 the International Committee on Radiological Protection (ICRP) helped drive a major revision of radiation standards. In practice, the maxima for workers were lowered to match maxima for the public, while allowing for exceptional circumstances.

That seems to us appropriate. But by 1977, appropriate caution about radiation started to be accompanied by over-caution. In a seminal if controversial paper, three researchers came up with new evidence of cancer among workers at Hanford. [168] Alongside the new standards, therefore, the ICRP developed the philosophy that exposure should be As Low As Reasonably Achievable, or ALARA.

Sensible? Perhaps. But by omitting any mention of a limit below which there is no concern, ALARA opened the door to regulating lower and lower levels of radiation. ALARA could always be that little bit lower, even if the quantities of radiation dealt with are far lower than the natural levels that people are exposed to in everyday life.

Just as some statements of the Precautionary Principle specify that measures should be 'reasonable', so does ALARA. But that does not prevent it from being a highly dubious piece of regulation.

Toward a New Carbon Infrastructure

When environmental campaigners attack carbon, they use the word as a synonym for fossil fuels.

They argue that:

1. Coal-fired power plants generate more CO_2 per unit of energy than any other source of energy. Such plants promise to reinforce China's status as the world's top emitter of GHGs. Coal supplies may be around for two centuries or more, but it's certain that *burning coal accelerates global warming*. Burning coal is a danger to the planet, so we should leave all of the world's coal in the ground, right now.

2. Dependence on natural gas leaves Europe hostage to unstable regimes, especially in Russia and the Middle East. Gas supplies may be around for quite a few decades, but already *gas is a threat to national energy security.*

3. Like gas, oil is also threat to energy security – and not just in Europe, but also in the US. On top of that, oil is the fossil fuel that's most rapidly being depleted: supplies will only be around for a very few decades. Thus the world faces, or has passed, a peak of supply. One way or another, *humanity faces peak oil.*

Environmentalism perceives fossil fuels as an addiction (see Chapter 2). It especially condemns the repeated use of oil for personal transport in cars and planes.

In fact, fossil fuels are the largest and most versatile energy source that humanity has ever mastered. And even better will be their more broadly based successors, the carbon-based fuels that will be derived from what we call the *New Carbon Infrastructure.*

Sticking up for carbon, seeing CO$_2$ as more than just a problem

For climate change activists, fossil fuels prompt a simple worldview: *cut carbon out*. But the reality will be different. The world will never move into a 'low carbon' economy. Instead, it will one day control CO$_2$, put it to work, and, through a variety of industrial recycling techniques, organise a New Carbon Infrastructure.

With a New Carbon Infrastructure, the versatile chemistry of *carbon*, as distinct from *CO$_2$*, can be made to generate new transport fuels (and, as it happens, new materials). The world will not have entered a New Carbon Economy – but the quantity and quality of carbon-based fuels will be unrestrained by the Earth's fossilised endowments.

Fuels that come from fossils are not the only source of carbon-based fuels. Biofuels, which are based on agriculture and are now the subject of an outcry, are in fact the modern era's first mass-scale transport fuel that does not rely on fossils. Far from lamenting society's alleged addiction to fossil fuels, the energised citizen will proclaim that carbon-based fuels will come from both old and new sources. They will come from:

1. **The fossilised remains of living organisms**, which, many millions of years ago, through the process known as photosynthesis, used sunlight to pull carbon atoms out of atmospheric CO$_2$, string them together with hydrogen extracted from water, and so grow and create hydrocarbons. These fossils have stored energy for tens of millions of years in the chemical bonds among the carbon and hydrogen atoms that make up coal, natural gas and oil. [1]

2. **Crops, plants and algae** – living organisms which perform photosynthesis to grow, and can be harvested and processed into new, artificial hydrocarbons.

3. **Biotechnology**, in the shape of the genetic modification

of crops, plants, algae and other living organisms. Biotechnology will speed the growth of these life forms and speed their absorption of CO_2. It will also help mankind make artificial fuels that are easy to process, have high energy densities, are compatible with existing infrastructure, and are convenient to employ.

4. **The artificial synthesis of hydrocarbons**, using solar or other forms of energy (nuclear, geothermal), to make new hydrocarbon fuels. Artificial synthesis is still in the laboratory, but in principle it means playing the same trick as photosynthesis, only without relying on living organisms.

Like it or not, the burning of coal will remain an important source of electricity. For a long time to come, residential, commercial and industrial heat will be supplied by gas. Eventually, however, carbon-based fuels will find their majority application in personal and freight transport. Indeed, the high energy density of carbon-based fuels gives them great prospects in transport – prospects which ensure that a future of purely electric forms of transport is, with the exception of trains, somewhat remote.

The idea of generating more energy through human thoughtfulness and ingenuity applies to carbon-based fuels as much as to any other. And if people can be energised by that idea, they'll need to

*   stick up for carbon, a wonder element
*   present CO_2 as an opportunity, and not 110 per cent a problem.

CO_2 is a problem. But by managing it carefully, the world need not deny itself the remaining benefits of fossil fuels.

The world has a lot to look forward to with carbon-based fuels. In the future, old and new hydrocarbons will power a whole lot more human and freight transport, and will especially assist motorisation and the growth of air travel in the East.

The West frets about China sourcing fossil fuels from Africa – fuels with which China will generate electricity, mobility and CO_2. More broadly, the West frets about fossil fuels and East-West relations. It worries that Asian demand will

*   revive high energy prices once the recession has lifted
*   fast deplete the Earth's remaining stocks of oil.

But against these rather exaggerated dangers, unleashing the human energy and the mobility of billions in the East is a prize worth aspiring to.

This chapter looks first at coal. Then it moves to gas, before turning to the biggest source of anxiety: oil.

Coal: the most maligned fuel

Coal is disparaged as the dirtiest of fossil fuels. In Britain, the Institute of Public Policy Research, a New Labour think-tank, believes that the building of new coal-fired plants in Europe should, at least temporarily, be limited. It points out that:

*   Denmark has banned new coal plants since 1990
*   Canada will insist that such plants be equipped with CCS from 2018
*   California's emissions performance standards now dictate that any new coal-fired plants are built either with CCS or with CHP. [2]

In Washington, DC, another think-tank, the Center for Global Development (CGD), takes a similar view. CGD senior fellow David Wheeler describes the Tata Ultra Mega project, a 4 GW coal-fired power plant at Mundra, Gujarat, India, as 'obsolete, unnecessary, ultra-dangerous for the planet'. [3] And in Australia, Professor Tim Flannery, a best-selling author on climate change, hates coal. He's outraged at the 'irresponsibility' of those who burn it to supply his country with electricity. [4]

Before the Crash of 2008, three of Wall Street's largest

investment banks – Citigroup, JP Morgan and Morgan Stanley – announced standards that would make it harder for companies to receive financing to build new coal-fired in the US. The banks said that a government-regulated cap on emissions was inevitable, that such a cap would greatly add to the cost of electricity generation, and that, as banks, they would encourage energy efficiency and renewables before giving their support to new coal-fired power plants. [5]

Given all the innovations that were made in investment banking before the Crash, it's striking that Citigroup, JP Morgan and Morgan Stanley could not bring themselves to support technological innovation in and around coal-fired plants. These institutions sought measures that were more to do with finance and markets than with useful energy infrastructure. While they gave the usual support to energy efficiency and renewables, the three banks did not have it in them to mount an aggressive plan on CCS.

Wall Street banks put themselves in the vanguard of the fight against coal in the US. Yet all over the world, and above all in Asia, coal, with or without CCS, will continue to power the world well into the 21st century.

No coal, no future

Once, coal brought power to the steam engines of the industrial revolution. It provided warmth, too, in domestic heating. Today, it powers electricity and is more and more industrial, rather than residential, in application. In 2005 there were just over 700 billion tonnes of reserves of hard coal and lignite in the Earth. North America had about 200bn, Russia and its environs had 150bn, India and China 75bn each, and South Africa, 40bn. [6] The contribution of coal to power generation in different countries reflects its disposition:

Estimated proportion of electricity generated by coal fired plants, selected countries, 2006, per cent [7]

Country	
Poland	93
S Africa	93
Australia	80
China	78
Israel	71
Kazakhstan	70
India, Morocco	69
Czech Rep	59
Greece	58
USA	50
Germany	47

Coal can certainly be dirty. Much – although not all – of the smog that blights Chinese industrial cities originates in coal burning. Just like the London smog that was ended by the 1956 Clean Air Act, coal in China causes problems when burnt in small boilers, or for domestic heating. The solution is a shift to gas and electricity. [8]

Cheap and abundant, more and more of the world's coal will tomorrow be used to generate that electricity. Coal will also be burnt in new ways (see panel below).

Burning coal better

Coal pollution can simply be cleaned up, without an increase in efficiency. On the other hand, when coal is burnt more efficiently than in the past, it burns cleaner: less coal is used to produce the same amount of energy. Historically, four broad techniques have been applied around these two approaches.

In the West, **filters and scrubbers** have had no effect on either efficiency or CO_2, but have cut the rest of the pollution contained in the flue gas exiting from coal-fired power stations. Filters are used to remove small particles from that gas – including those containing heavy metals such as lead. Scrubbers remove the sulphur dioxide that would otherwise cause acid rain: typically, limewater sprays are used to dissolve it. Scrubbers have also been developed to remove the mercury from coal. As the East develops, it will fit more filters and scrubbers.

In **pulverisation**, coal for use in power stations is ground into particles smaller than a millimetre in diametre. That allows for fast, efficient combustion. Pulverisation is now a standard technique for new coal power stations.

To get the most out of pulverisation, combustion burns the particles while they are suspended in a jet of air. This technique has yet to become standard.

Supercritical technology brings steam to turbines at very high pressures and temperatures; ultra supercritical refers to still higher temperatures and pressures. In these zones, the distinction between liquids and gases vanishes, and heat energy is turned into electricity very efficiently.

Britain's first supercritical plant, at Kingsnorth, Kent, remains a controversial proposal (see below). But in China, more than 60 ultra-supercritical generators are already under construction. The government's chief planning agency, the National Development and Reform Commission (NDRC), has said that all future coal-fired units rated above 0.6 GW must meet ultra-supercritical standards. [9] New steels are a key technology enabling boilers and turbines to cope with supercritical steam.

Integrated Gas Combined Cycle power stations will take coal and convert it into carbon monoxide gas for combustion. That allows for turbine systems to be more efficient. Waste heat from IGCC stations can easily be fed into combined heat and power. IGCC could also be integrated with coal-to-liquids technology or hydrogen production.

IGCC plants could produce a variety of fuels and chemical products alongside heat and electricity, pointing the way to the New Carbon Infrastructure. Mark Jaccard, environment and resource management professor at Simon Fraser University, Vancouver, believes that multipurpose IGCC plants could not only be financially viable, but also exciting for engineers to design. [10] As yet, however, they remain on the drawing board.

Representative figures for the performance of advanced coal technologies [11]

	Efficiency, per cent	Consumption of coal per hour, kg, for an output of 500MWe
Subcritical pulverised	34.3	208,000
Supercritical pulverised	38.5	184,894
Ultra-supercritical pulverised	43.3	164,000
Subcritical fluidised	34.8	297,000
IGCC	38.4	185,376
Average of existing US power stations	33	

As coal's non-GHG pollutants are cleaned up, so the CO2 that coal generates will become a larger issue. It's true that CCS can remove CO2 from the waste stream of power stations and store it away from the atmosphere; but, like IGCC, CCS is still at a primitive stage.

Investment in both technologies is urgently needed if a new round of coal-fired plants is not to face opposition. But there is hope, all the same. At Imperial College, London, Jon Gibbins and Hannah Chalmers believe that there's a need for two learning cycles before full-scale adoption of CCS. They envisage a first tranche of demonstration projects quickly implementing a range of concepts at capacities of between 100 and 500 MW. A second, larger tranche would consist of semi-commercial projects in which industry pursues those designs most suitable to its needs. Working with a timescale of four years for construction, and longer for the learning of operational lessons, Gibbins and Chalmers argue that, with full political and economic backing, *CCS could become standard* – at least in the developed world – by the 2020s. [12]

Billions need coal (1): problems beyond climate

When Greens worry about coal, they worry most of all about China. They are right when they point to the dramatic rise in coal burning in China. To build just one coal-fired power plant every week, let alone the two more often cited, adds up over a year to the equivalent of the whole of the UK's generating capacity. [13]

There is, however, something of a rush to judgment about Asia and coal.

Separately from the issue of CO_2 emissions around the burning of coal, Joseph McConnell and Ross Edwards at the Desert Research Institute (DRI), Reno, Nevada, have traced the presence of the toxic heavy metals thallium, cadmium and lead in the Arctic. Analysing an ice core extracted in Greenland, the two scientists found that, contrary to expectations, the Atlantic sector of the Arctic revealed markedly *decreased* coal burning emissions in North America and Europe during the second half of the 20th century. This was a result both of improvements in combustion technology in the West, and of a shift from coal to oil and gas as the primary fuel source there.

In their conclusions, McConnell and Edwards made an important flourish. They hypothesised that thallium, cadmium and lead are currently increasing in the Pacific sector of the Arctic, because of 'pollution from rapidly growing Asian economies predominantly fuelled by coal combustion'. [14]

Now, it may be that Asia is set to repeat the ill effects of burning coal that were once seen in Europe and North America. China's rapid catch-up of the West, after all, has sharply raised levels of pollution. But the record confirms that Europe and North America have been able to cut coal-related pollution – and not just by burning less of the stuff.

The very speed of China's advance also means that it's unlikely to go on building or operating the same old coal-fired plants – complete with heavy metals emissions that are headed for the Arctic – for decade after decade. Sooner or later, countries like China and India will begin to cut, rather than increase, the pollution they generate through the use of coal.

China and India need not endure the protracted industrialisation that Western countries had to go through more than a century ago. By historical standards, the East, now fast into coal, is likely to be fast out of it.

Billions need coal (2): the effect on climate

In 2007, the top climate change official at the UK Foreign Office, John Ashton, came out into the open with the self-seeking argument that Greens have long made about China and the world's climate. Rich nations, he said, had to *set an example* of low-carbon development for China to follow. [15]

Such a statement evokes the fake generosity and moral rectitude of an imperious parent in relation to an errant child, and is no doubt apprehended as such in China. But in fact the patronising tone is entirely out of place. If anything, it's the Chinese who are setting an example to the world.

It's China that has pushed ahead with new coal technology. A recent retrofit of the 1 GW Gaobeidan power station, which alone provides a tenth of Beijing's power and a third of its hot-water heat, has begun to strip out a small fraction – less than one per cent – of the CO_2 in the station's exhaust gas. [16]

That isn't much. The process only involves carbon capture, and doesn't run to the storage part of CCS. But before critics thumb their noses, there are three points to note:

1. A reduction of one per cent in CO_2 emitted is infinitely more than what's happening around UK coal-fired power plants
2. With CCS technology in its infancy, it's significant that a first step toward it seems to be happening in the East, rather than in the West
3. From the time of announcing the retrofit, China took only nine months to commission, design, build and complete it.

For Greens, tablets of stone from the Science of Climate Change suggest that the growth of Eastern economies could, now the Bush years are over, be the world's most dangerous example of *anti-social behaviour*. Of course, to cover their backs, British Greens like to implicate the West in China's misdeeds, pointing to the factories it has set up there. Inadvertently mixing biochemistry, ethnicity and carbon footprints, Greenpeace UK director John Sauven likewise insists that:

> 'The average Chinese emits just 3.5 tonnes of CO_2 per year, whereas Britons emit nearly 10 tonnes and Americans 20 tonnes.' [17]

Meanwhile the WWF urges China to pay for the negative externalities caused by burning coal through an energy tax – imposed by the authorities in China, presumably, and certainly 'applicable to all consumers of coal'. [18]

In effect, this is a call for millions of Chinese to pay more for the privilege of staying warm in the winter. That's rich coming from the panda-friendly WWF, a $134m organisation headquartered in prosperous Gland, Switzerland. [19]

British Greens mobilise against coal

In September 2008, a British jury listened to evidence on the gravity of global warming. NASA's James Hansen, who had made a special visit from the US to Maidstone, Kent, addressed the court. After that, the jury acquitted six climate change campaigners.

All smiles: the Kingsnorth Six outside
Maidstone Crown Court, September 2008.
Picture credit: Rose/Greenpeace

The Kingsnorth Six had been arrested after taking action against the plans of E.on to build a new coal-fired power station next to the company's existing plant at Kingsnorth. That the design of the station would allow it to accommodate CCS in the future had not dissuaded the protesters. In their protest climb of a 200m high company chimney, they had caused £30,000 of criminal damage; but the jury decided by a 10:2 majority verdict that there had been no criminal intent.

Environmentalists savoured the moment. In their account the common people, through the courts, had registered what could prove a turning point both for Green activists and for choice of technique in UK energy supply. As the environment correspondent of the London *Guardian* wrote:

> 'In the past decade, prosecutions of protesters against GM crops, incinerators, new roads and nuclear, chemical and arms trade companies have all collapsed after defendants argued that they had acted according to their consciences and that they were trying to prevent a greater crime. Greenpeace itself has a four-nil record against the crown using the same defence and was widely known to be seeking a jury trial to present complex arguments about coal and climate change.' [20]

'The public,' said one solicitor specialising in environmental matters, 'is increasingly speaking through the courts.' He went on:

> 'Politicians and companies have not understood that most people now understand the issues. There's a feeling that government and the authorities have not been paying sufficient heed, and that the courts are righting the balance.' [21]

In a truly Dickensian manner, 'mobilisation' against coal in the UK takes place not on the streets, but through the courts. It is a mobilisation not just against global warming, but also against more energy. [22]

In this, the Kingsnorth affair typifies the democratic rhetoric and authoritarian content of environmentalist hostility to coal-fired power generation.

Coal's defenders favour autarchy, and are complacent

Those who defend coal in the US do so, predictably, on the grounds of energy security. The American Coalition for Clean Coal Electricity (ACCCE) proclaims:

- indigenous supplies of coal in the US mean that 'we do not have to rely on foreign imports from politically volatile parts of the world'
- Coal-to-Liquid (CTL) technologies, in which coal is converted to high energy density fuels for transport and other applications, can form a substitute for foreign oil, if the latter is priced at $54 or higher – 'based on a US Department of Energy formula'. [23]

Arthur Scargill goes even further than the ACCCE in his desire for coal-based autarchy in energy. He believes that, if 250m tonnes of indigenous deep-mine clean coal were produced each year, Britain could extract from it 'all the electricity, oil, gas and petrochemicals that our people need'. [24] Here, by implication, CTL can rid Britain of all dependence on foreign gas and oil.

These visions of a coal-laden, oil-free future are ridiculous. America and Britain cannot go back to the 19th century. The oil industry and the globalisation of oil supplies are facts of life.

The ACCCE claims that the US coal industry has spent $50bn reducing emissions per billion kilowatt-hours by 74 per cent. That's not too bad; but the advance must be measured over all of 30 years – from 1970 to 2000. [25] It represents relatively slow progress.

Equally, it's all very well for the ACCCE to say that several billion dollars are being spent on more than 300 research projects into clean coal around the US, 'each one breaking new ground and helping pave the way for an energy independent future'. A more pertinent opinion is probably that of James F Roberts, chairman and CEO of Foundation Coal Corporation, as expressed to a Senate summit on energy on 13 September 2008. Roberts said that the $3.5 billion spent on CCS in the US over the previous decade was 'wholly inadequate'. [26]

Roberts was right. In 2007, US coal mining had an annual turnover of more than $25bn. [27] Coal-fired power stations in the US bought most of the coal produced – but these purchases of supplies formed only part of their general turnover. From 1998 to 2007 in the US, it's clear that R&D spending on CCS, performed by electricity generators reliant on coal, equated to less than one per cent of electricity sales. More importantly, actual progress with CCS, as we shall see shortly, has been slow.

Roberts added a further barb to his indictment. From 2003 to 2008, in just five years, the Bush administration had gone from planning to equip a 275 MW coal-fired power station with CCS, at the cost of $1.8bn, through to cancelling the project. This was despite the significant fact that FutureGen, as the scheme was called, enjoyed some financial support from China, India, Australia, South Korea, and Japan. And what did Roberts have to say about this? He called the US Department of Energy's decision to withdraw funding for FutureGen 'incomprehensible'. [28]

Barack Obama's election policy on energy committed his administration to developing, through public-private partnerships, five commercial plants with CCS. But delay with these could well occur. Certainly, the US coal sector has been enormously unsuccessful in moving both itself and Washington toward clean coal. It took the financial crisis of 2008 for US legislators to agree to put less than $3bn into clean coal and CCS. [29]

It's a similar story in CTL technology. As the ACCCE points out, Germany and South Africa have been gasifying coal and turning it into low-sulphur diesel and jet fuel for decades. But where's the progress in the US?

The quantity of fuels made from CTL methods will never be enough to drive more than a fraction of transport in future. But, like clean coal, CTL has suffered from neglect.

The backers of these technologies seem to be as complacent and as sleepwalking as their detractors.

CTL: converting coal to liquids

CTL technology was initially developed by the pariah economies of pre-war Germany and post-war South Africa. Both countries had access to coal, but not to oil. Then, in the 1980s, the US government funded the now-forgotten Synthetic Fuels Corporation, in a bid to push America toward energy independence.

With the advance of science, chemical engineers gradually became more practised in transforming carbon from solid to liquid forms. Still, despite more than 50 years of work in the area, the *Guardian* newspaper helpfully opened a recent story on CTL with the sober remark: 'Energy companies are planning to revive a polluting technology developed by the Nazis to replace dwindling supplies of oil with synthetic fuels derived from coal'. [30]

A first route to liquefaction begins by converting coal to carbon monoxide gas and then to liquids. Known as the Fischer-Tropsch process, this technology is also used in IGCC. A second, more recent technique is direct liquefaction. Here coal is dissolved in a solvent in the presence of hydrogen.

CTL technologies are expensive. However, should demand for oil outstrip supply, they could form a useful backstop.

Since the 1970s, South Africa's Sasol has pioneered industrial-scale CTL facilities. In China, the Shenhua Group has more recently launched a direct CTL plant in Inner Mongolia, and, in partnership with Sasol, has commissioned a feasibility study for an indirect coal liquefaction plant in Ningxia Hui, a much smaller autonomous region south of Inner Mongolia. [31]

The Inner Mongolia plant produces some 20,000 barrels

of oil a day. However, this is very little compared with an all-China consumption figure in excess of seven million barrels. What's more, China's NDRC has put further CTL projects on hold, describing them as 'a technology-, talent- and capital-intensive project at an experimental stage with high business risks'. [32]

These developments confirm that CTL will likely remain a niche technology. It will never be the mainstay of oil production.

Coal and CCS

Given that CO_2 causes problems in the atmosphere, why not capture it and store it away from the atmosphere?

One reason is the vast amounts of CO_2 a power station generates. Pollutants such as sulphur dioxide form only a small part of flue gases, and are therefore relatively easy to remove. A 500 MW coal power station, however, produces about three Mts of CO_2 a year. In total, the CO_2 emissions made by US coal-fired plants amount to 1.5 Gt per year – three times by weight, or a third by volume, of the natural gas annually transported through America's gas pipelines. [33]

Applying CCS to even a fraction of CO_2 emissions will be a major effort. Furthermore, unlike nuclear power, renewables or even measures improving energy efficiency, CCS can only clean things up: it can never add to the generation of electricity.

However, as the world first increases consumption of coal, and then makes a transition away from it, CCS is both technically feasible and likely to figure prominently in coming decades.

The expense of CCS derives from both the capture and the storage of CO_2. Driving CCS itself requires a ots of energy. To bury their CO_2, both IGCC coal plants and also gas power stations will require between 10 and 25 per cent more energy than they would need without CCS. As for CCS on conventional coal-fired electricity plants, it will requite as much as 40 per cent more energy.

For conventional coal power, *capture* happens by absorption of CO_2 from flue gases using ammonia, or related chemicals known as amines. But there is potential to improve on ammonia. Omar M Yaghi, professor of chemistry at the University of California in Los Angeles, has created exotic substances composed of molecular-scale tunnels that can trap CO_2. [34]

After the CO_2 has been captured from waste, it's released in pure form for *storage*. Today, the kind of storage that's most straightforward and developed involves pumping the captured CO_2 underground. In a brilliant address to scientists in London in 2007, Peter Styles, director of the Research Institute for the Environment, Physical Science and Applied Mathematics at the University of Keele, pointed out that while the world emits 25 Gt of CO_2 each year, the geological capacity exists to store no fewer than 11,000 Gt. [35]

As Styles noted, CO_2 can be compressed and pumped down into a variety of geological formations, including oil and gas fields that have largely been used up, coal beds that are too difficult to mine, or aquifers – underground rocks which are porous and saturated with water.

Oil and gas companies already have experience with this technology. CO_2 is already pumped into oil reservoirs – not primarily for environmental reasons, but rather to increase extraction by forcing more oil out. A similar technique could be used with coal seams that have not been mined. When CO_2 is pumped into coal, the process releases methane, which can be used as fuel. Meanwhile the CO_2 is retained underground.

To make a difference to climate change, the CO_2 stored should stay put for many centuries. For safety, there should be no large sudden leaks that might poison the local population. If stored in an aquifer, the CO_2 should not dissolve toxic metals, only to flow upward to contaminate surface ground.

To meet these requirements, CCS will need more research. However, given that geological formations have trapped hydrocarbons such as natural gas for hundreds of millions of years, finding suitable sites that can safely contain CO_2 for millions of years shouldn't be too difficult.

If the overall process offered by CCS makes for cleaner

power stations, people shouldn't be afraid of the extra energy inputs that process demands. What matters is the ability to make energy cheaply – so that mankind can afford to spend more energy on cleaning up the energy it generates.

In Britain, CCS has suffered from the same stop-go bungles as the FutureGen project in the US. [36] But how far away is the general application of CCS? A straw poll at a Royal Society workshop on CCS, held in London 2007, found that the majority of those present believed that non-commercial CCS could be demonstrated at large scale by 2014, with commercial operation achieved between 2016 and 2020.

As a report on the workshop later observed, these 'challenging' targets

> '… can only be achieved if the urgency expressed by some spreads to become a demand of the majority. Experienced practitioners warned that the remaining technical, economic, policy, regulatory and legal issues must be addressed simultaneously and with determination.' [37]

That's the right spirit. But at the moment capture and storage are being developed separately, and full CCS is being applied at only a few locations around the world, such as Statoil's Sleipner West gas field in the North Sea.

To make CCS real will take much more investment.

Summing up on coal

The surprising thing about coal is how dependent the world still is on it. Coal remains the leading source of electricity in Europe, and has a prominent role in most parts of the world. In the US, concerns about GHGs forced the delay or cancellation of more than 50 proposed coal-fired power plants in the 12 months to March 2008. Yet at the same time, US exports of coal were, before the Crash of 2008 at least, tipped to reach $3.75bn. [38]

The world has been burning *more* coal, and the proportion of the world's coal burnt by China has risen from 24 per cent in

1990, to 29 per cent in 2000 and an astonishing 41 per cent in 2007. [39] Indeed, these figures look set to rise higher still, given that China will continue to burn more and more coal for at least the next 20 years.

The immediate prospects for coal, then, are largely Chinese. So it's in China that clean coal technologies will have their greatest opportunity.

In a further 20 years' time, three developments are likely. First, the most rapid phase of expansion in Chinese demand for energy will be over. Second, China will be rich enough to invest in cleaner forms of energy. Third, renewable energies will be cheaper and more effective. For all those reasons, Chinese emissions of CO_2 from burning coal are likely to peak in the 2030s.

Other parts of the developing world are likely to follow a similar path. Economic growth in India, and perhaps Africa, could be too fast for nuclear power and cheap renewables to meet burgeoning demand for energy.

In 2007 India burnt less coal than China did in 1971. [40] However, India has now embarked upon a rapid expansion in coal burning. To the extent that there's an economic take-off in parts of Africa, an increase in the use of coal can be expected there, too.

In developing countries, coal's low cost will tell in its favour, For all its faults, no other method is available to lift hundreds of millions out of poverty.

With the real growth in coal use outside China still to come, China's advances in clean and efficient coal technology promise to have international application. If China can bring CCS in line by 2020, it will be in a strong position to lead the subsequent cleanup of those parts of the world that are still industrialising.

In the West, it's likely that the generation of coal power stations that's to be built over the next decade may be the last. As a result, CCS may well prove a purely stopgap measure, since after 2030, electricity generation with fuels other than coal will take over from coal-fired plants.

Still, coal does not deserve the Kingsnorth protesters. Like many things in life, it is not ideal; but also like many things, it can and should be changed. To wish coal away is to imagine that America, China, India, Germany, Poland and South Africa can do without it. They cannot.

By making a special issue of coal sootiness, those who vilify it reveal their own snootiness. Determined to set an example to the world, the EU has declared CO_2 emissions targets that already pose particular difficulties for its new, coal-dependent members in Eastern Europe – beginning with Poland. Isn't it likely that the EU will adopt a still more supercilious policy further East, in relation to China and India?

Critics of coal quickly lose sight of its useful aspects. They likewise neglect the human circumstances, and the jobs, that accompany the mining and burning of coal. More than 50 years ago, George Orwell wrote eloquently about that side. [41]

Coal is what we make it. With a combination of clean technologies, including CCS, it can be not just an inescapable source of electricity for years to come, but also one that will have less and less of an effect on the Earth's atmosphere.

The Green campaign against coal is a negative one, focused on fear that runaway coal burning will destroy the world. Greens would do better to ask why people burn so much coal. The world will move away from coal, but will do so over decades rather than overnight. Shutting down the combustion of coal would be shutting down the lives of billions.

History shows that burning coal better makes a big difference to pollution. Today's coal technology is better than that of the past, but still crude compared with what's to come. Coal will be used in new ways as part of the New Carbon Infrastructure. Coal burnt underground or in IGCC power stations will provide not just energy, but also hydrogen that – as we discuss below – will be combined with other forms of carbon to make new fuels.

In the New Carbon Infrastructure, too, even the CO_2 that's sequestered underground through CCS may be put to work one day to make new fuel.

Not much natural about natural gas

Natural gas, or methane, is the least contentious of the fossil fuels, primarily because it burns cleaner than coal or oil, and produces less CO_2 per unit of energy. Among OECD countries, natural gas provides an increasing share of energy for heating, cooking and electricity generation, accounting for 22.6 per cent of all primary energy in 2007. [42]

Gas is much sought after. Until the Crash of 2008, prices rose alongside those of coal and oil. Supply was tight, and demand buoyant – especially from Asia, and everywhere from *electricity generation*. Driving barely seven per cent of the EU's electricity in 1990, gas is likely to account for 25 per cent of it in 2010. For Italy and the UK in 2020, the figure is forecast to be 60 per cent. By 2010, gas-fired electricity in the EU is set to surpass that made by nuclear power stations, standing second only to power derived from coal. [43]

Natural gas doesn't escape Green criticism. While its GHG emissions are lower than those associated with coal and oil, they are significant: even now there's work to be done to cut them, by developing commercial applications for the 150 billion cubic metres (bcm) of the world's gas that's still flared, or simply burnt off at the top of stacks. Some environmentalists also predict *peak gas*. Above all, both officials and Greens worry that Europe's rising dependence on natural gas, and on *imports* of gas, makes it vulnerable to threats and actions by Russia, which has the world's largest reserves, and already exports enough gas to Europe to account for 24 per cent of its consumption. [44]

In the EU, aged nuclear- and coal-fired power stations are scheduled to close. The EU's targets for renewable energy may not be met. Thus the IEA reckons that EU imports of gas will rise from 320 bcm in 2004 to 540 bcm in 2020, by which time they will take no less than 77 per cent of electricity supply. 'Gas and electricity security,' the IEA observes, 'will become increasingly intertwined.' [45]

Gas is therefore a sensitive issue and, after the dispute between Georgia and Russia in August 2008, sensitivity to Russia's actions – especially with pipelines – is acute.

Europe's feeling that *gas from somewhere else is a threat to national security* is especially pronounced in Britain. There, by 2020, imports of gas are officially predicted to rise to about 80 per cent of the total gas used. Predictably enough, the main way in which the government seeks to reduce this 'dependence' to 60 per cent is

> '… directly by reducing demand for gas i.e. in heating our homes; but also indirectly by reducing demand for electricity so reducing the need for new gas-fired power stations.' [46]

Here, gas security is held up as a national call for energy conservation.

In 2016/17, UK import capacity from Norway could represent some 16 per cent of peak supply capacity. In the same year, gas coming from the Continent could represent 14 per cent of peak supply capacity, with Russia merely one among a range of nations sitting behind these imports. The British government insists, then, that the UK 'is not significantly dependent on any single country supplying to the EU market'. [47] At the same time, it worries that Britain will be

> '… more exposed to the risk and impact of any overseas disruptions to energy supplies as supply routes become longer and across more countries.' [48]

The fear is, then, that dependency on gas is at the heart of Britain's general lack of independence in energy. In the view of Jill Kirby, an adviser to the Conservative Party and director of the Centre for Policy Studies, the UK's a soon-to-be-yawning 'energy gap' will be compounded by its dependence on gas bought from Europe, Russia and the Middle East. [49]

As much as coal in climate change negotiations and American revulsion against dependence on foreign oil, dependence on foreign natural gas is posed as a problem for

Europe – Britain included. Yet in fact the story of gas effectively illustrates how innovation can create new sources of energy supply. Because of technological developments in its transmission and distribution, there's now a nascent global market for natural gas, not just a collection of large regional markets. While geological dispositions of natural gas obviously remain important, there's less and less that's natural about the production and delivery of natural gas.

A shifting dash for gas won't readily exhaust reserves of the stuff

Natural gas has been found alongside oil since drilling began in the 19th century. Yet until relatively recently, gas was regarded as a waste product. The problem was that natural gas was hard to transport. As a result, it was often simply flared or pumped back into the ground.

It was only after the Second World War that advances in the metallurgy of steel made large-scale natural gas pipelines a practical proposition. Even so, gas is harder to pump through pipes than oil. The construction of very long pipelines over thousands or miles, like the construction of deep-sea pipelines, remains difficult.

After the oil shocks of the 1970s, increased interest in natural gas led to the development of liquefied natural gas (LNG). Here natural gas is cooled until, at a temperature of –160° C, it liquefies, allowing it to be transported by specialised container ships. Today the top LNG exporters are Indonesia, Malaysia and Qatar. Liquefaction was first applied at Arzew, Algeria, in 1964. The liquid gas was exported to Europe. Then, in October 1969, another milestone was passed when a shipment of LNG left Alaska for Japan. [50]

By the 1980s, a second technological advance came to gas in the shape of combined cycle gas turbines (CCGT). Here, gas is mixed with air and burnt in a turbine similar to a jet engine. Rotation of the turbine turns a generator. The trick played with CCGT is that the hot gas that forms an exhaust for this first process is then used to heat water to steam, which is used to turn a second, steam turbine. That increases the efficiency of the whole process from about 30 per cent to more than 50 per cent, almost doubling the amount of electricity produced from a given amount of gas.

CCGT technology was not just efficient; it was well suited to relatively small scale, incremental increases in electricity production – increases from one to a few hundred megawatts. With liberalisation of the energy industry in the 1980s, small independent producers flocked to join the market, serviced by businesses such as Enron that sold them the gas. Between 1987 and 1993, the share of new world electricity generating capacity taken by CCGT rose from 10 more than 35 per cent, with orders filled by companies such as GE, Siemens, ABB and Westinghouse. [51]

Across the world there was a 'dash for gas' – above all in Britain, where drilling under the North Sea added to its attractions. But today the dash for gas has slowed up a little in the West, as European electricity generators replace old gas-fired plant, diversify their sources of energy, and concentrate on vertical integration – following the path of the Franco-Belgian group Suez, which sources, transports and makes electricity from gas. [52]

Demand for gas has shifted to Asia, the Middle East and Russia. Perhaps the Crash of 2008 will moderate that demand; but, well before it, there were Green warnings about peak gas. In 2004, when the noise about peak oil was beginning to become audible, Julian Darley, founder of the Post Carbon Institute, claimed that it was possible that the world could have to suffer 'a gas peak-plateau somewhere between 2010 and 2025'. [53]

Since then, peak gas has become the less media-worthy cousin of peak oil.

In fact, there's no immediate geological limit to the availability of gas. Competition between Russia and Norway to open up the Arctic Circle, for instance, is a sign that the market for gas is expanding, not that gas is running out. The US Geological

Survey's 2008 Circum-Arctic Resource Appraisal suggests that the Arctic Circle, which encompasses six per cent of the Earth's surface, has gas reserves of 1.67 trillion cubic feet. [54] That's equivalent to a whopping 47 thousand bcm, or more than 100 times the EU's current annual imports of gas.

Take the Norwegian Snøhvit (Snow White) gas field in the Barents Sea. It began production in September 2007. It contains 193 bcm of gas, with production expected to extend until 2035. Statoil discovered the field in 1984, but development only began in earnest in 2002. Production takes place 145km offshore, but no fixed or floating offshore structures are needed. Instead, wells on the sea floor at depths from 250 to 345 metres pipe gas to an onshore facility where it's liquefied before being taken abroad by oil tankers. About 700,000 tonnes of CO_2 extracted with the gas is pumped back underground each year. [55]

In the longer term, there are vast amounts of what are termed 'unconventional' sources of natural gas, including gas

- absorbed in coal beds
- trapped at high pressure in impermeable rocks
- trapped at high pressure and dissolved in underground aquifers
- in the form of so-called methane clathrates – ice which, found under or on the ocean floor, traps methane within its crystal structure.

With these sources, gas is (or will be) hard to extract. But in the US, gas extraction from sedimentary rock, or shale, in the middle of the continent and the Appalachian Basin, is on the up. It has risen rapidly in the past six years, and great potential is seen in the Rocky Mountains. [56]

Gas as poor European energy security

For followers of the Precautionary Principle, both gas pipelines and LNG are a bad idea. A new pipeline being laid under the Baltic Sea to distribute Russian gas across Europe drew the objection from Greenpeace that it involved 'huge risks'. Greenpeace said: 'We do not know what will happen when the seabed is disturbed'. [57]

LNG suffers from even more vigorous protests. Across both Europe and America, wrangles over planning have delayed the development of LNG. [58] Typically enough, LNG terminals for offloading the gas are seen as potential targets for terrorists. [59]

In the case of Russia, tactics with customers have been harsh, as when Gazprom cut off supplies to the Ukraine in 2005-2006. [60] Yet Europe's underinvestment in the infrastructure of gas has had much more serious consequences than Russian 'blackmail'. In the UK in 2006, a cold snap led to a quadrupling of prices and forced cutbacks for industrial users. Yet that was down not to Vladimir Putin, but to the closure of the UK's main facility for storing gas. [61]

Even where problems go beyond those of capacity, expanding supply can only ease tensions. It's hardly surprising that British officialdom fears being caught out by foreign gas producers. Compared with the levels of 2005/6, more than 5.5bcm of new gas storage is under construction, planned, or proposed – so doubling UK gas storage capacity by 2005. However, as the government concedes, this expansion of storage is only going to happen 'if projects are not unduly delayed by planning, technical, or other factors'. [62]

Politics in Britain today makes that a big 'if'. The serious investment in gas infrastructure that's needed may well be further delayed by the Crash of 2008. Yet for Britain and the rest of Europe, there is simply no alternative to turning abroad for gas supplies.

The globalisation of gas

After an offshore earthquake in July 2007, Japan shut down its seven nuclear reactors at Kashiwazaki-Kariwa and kept them shut. To make up for its lost nuclear power, the country Japan proved ready to pay high prices for LNG cargoes, In turn, Japanese demand for LNG has affected gas prices in the UK. [63]

This globalisation of the market for gas needs getting in perspective. In 2006, about 4.8 bcm, or six per cent of the Atlantic region's production of LNG, was diverted to Asian markets. The following year, similar diversions reached 12.5 bcm. However, although LNG afloat can be promptly diverted to different global buyers in a way that gas in pipelines cannot, LNG represents only seven per cent of global gas sales, compared to 15 per cent for seaborne coal as a proportion of global coal sales, and 48 per cent for seaborne oil. [64]

The globalisation of gas is in its infancy. Yet it is inescapable. As Ed Crooks, energy editor of the *Financial Times*, has laconically observed:

> 'As is widely appreciated in continental Europe, the EU cannot simply cut itself off from Russian gas, or even reduce demand… Geography and economics dictate that the EU is dependent on Russia for gas, whether politicians like it or not.' [65]

Given these facts of life, why does gas security so obsess Western elites? Why, for instance, is *Petroleum Economist* magazine able to sell simple maps of existing East-West gas pipelines in Eurasia for as much as £295 each? [66]

The answer is that pipelines control the gaze of European officials not because of their intrinsic qualities, or those of oil, but on account of a different reason altogether: Europe's historically fractious political relationship with Russia. The right-wing Heritage Foundation, Washington, argues that it's through gas that Russia has been 'consolidating its grip on the economic lifeblood of Europe'. [67] But as the hackneyed idea of gas as economic lifeblood suggests, methane cannot easily be held culpable for poor East-West relations.

Russia is assertive toward Europe in gas because that's one of the few levers it has. As a US State Department official notes, abbreviating the North Atlantic Treaty Organisation, Russia has been 'having to swallow everything the West wanted to do, like expand NATO and put missile defence in Poland'. [68] Russia needs to diversify its customer base for gas, and signed a deal with China in 2006 to do just that. Feeling competition from Nabucco, a 3,300km pipeline through which the EU hopes to bring gas from the Caspian region to Austria, Russia plans to bring gas to Europe through new pipelines.

Given the impact of the Crash of 2008 on the Kremlin, however, Russia will be in little position really to dictate terms to Europe on gas. Norway and Algeria will anyway add new pipelines to rival Russia's.

To harp on about Europe's poor security in gas is to turn the clock back. The globalisation of gas may be slowed by protectionism, but is broadly unstoppable. Three aspects of the development of LNG clearly show this.

First, LNG projects in Sakhalin Island off Russia's Pacific coast, in Yemen and in Tangguh, Indonesia, only obtained approval because each could promise customers in both Eastern and Western markets. Second, it's estimated that LNG will make up about 20 per cent of OECD gas supply as soon as 2010, and 14-16 per cent of global gas demand by 2015. [69] Third, the global average shipping distance for LNG has increased. In 2000, it was 5700km; in 2006, it was 6400km, and in 2007 it was 6700 km. It could be more than 8000km in 2010. [70]

In South Korea, Samsung is building Q-Max tankers, each of which will carry more than a quarter of a million cubic metres of LNG, twice as much as standard tankers today. [71]

Today gas is less and less a regional business. More and more, it's an international one.

Summing up on gas

Demand for gas will continue to rise in both the developed and the developing worlds. In the past 20 years, gas has established itself as a significant component of energy supply, replacing coal and especially oil in electricity production. In the next 20 years gas will hold, if not increase, its share of electricity generation.

In the longer term, as renewables and nuclear play a

285 TOWARD A NEW CARBON INFRASTRUCTURE

greater role in the making of electricity, the major application for gas will prove to be heating. Burning gas directly for heat is more efficient than burning gas for electricity and then using electricity for heating. Just as importantly, many people will prefer to cook on flames, and find gas heating more convenient than electric.

Gas will retain its role for heat in the developed world, and will also be applied in the developing world, where many millions still rely on wood, dung or coal fires.

Today's big supplies of gas will be supplemented by new unconventional sources, and sources that lie in places more distant than those that have yielded gas thus far.

Gas will also become part of the New Carbon Infrastructure. As we discuss below, it will be used to upgrade oil sands into liquid fuels. Gas will be converted into methanol for use in fuel, or as an alternative to liquefaction. There will also be new sources of gas – from coal, and from biomass.

Altogether, gas has a great future. Nevertheless, fears about it run deep in Europe. Gazprom has an agreement with Equatorial Guinea to develop an LNG plant; also, with the Nigerian state National Petroleum Corporation, with whom it signed an agreement in September 2008 to work toward a joint venture to produce and transport gas. Finally, Gazprom's chairman, Alexei Miller, hopes to hold three or four meetings a year of the world's leading gas producing states: Russia, Iran and Qatar. [72]

These developments fill European leaders with consternation. It's important to note, however, that neuroses about Russian and gas pipelines, though longstanding, have a very contemporary twist. They're to do with the general apprehensions of a new century. They relate to that sensitivity to global disruptions which was so heightened by 9/11.

In 1997, the experienced American geo-strategist Zbigniew Brzezinski, discussing what he called the Eurasian 'chessboard', devoted special attention to what he called 'The Global Zone of Percolating Violence', an ovoid region stretching from Egypt, Sudan, Yemen and Saudi Arabia in the south west through to Kazakhstan in the north east. But what was it that worried Brzezinski in those days? Was it gas pipelines – or oil supplies, for that matter? No, what worried him was the fact that about 25 states were in his Zone, that they were ethnically and religiously heterogeneous, that they were politically unstable, and that some of them were in the process of acquiring nuclear weapons. Neither gas nor oil figured anywhere in Brzezinski's 200-page book. [73]

Today, by contrast, the chessboard that's discussed in Eastern Europe, the Caucasus, the Caspian Sea and Central Asia is all about gas and oil.

When Russia cut off gas to Ukraine, the effects in Europe lasted only days. Nevertheless, the incident was taken as a signal of European vulnerability. It's rarely remembered that, with the Ukraine, Russia's ambitions were mainly about ending Soviet-era subsidies for gas, and instead selling it at market prices. They had little to do with the Kremlin enforcing regional- or super power status. In fact, Russia and the Ukraine quickly negotiated a new contract. That Europe felt the need to turn this relatively mundane incident into a melodrama gives a strong sign of the worry that surrounds gas today.

Gas and the grids it moves around have come to a prominence that cannot just be put down to economic factors. For all the insecurity associated with Russian gas in Europe, the bigger insecurity is about the times, the EU's poor leadership, and about what it is to be a European.

Nothing fascinates catastrophists like oil

Unlike nuclear and renewable energy, oil, the largest of the three industries, has, since the 1960s, boasted a serious semi-popular literature. [74] And although, as we shall see, the data on oil is widely held to be suspect, it's certainly voluminous. [75] Yet in 2004-5 there was an outpouring of new, more critical analysis. Even though the categories overlap, three broad groups of authors emerged:

1. Industry insiders who had gone over to apocalyptic – if not apoplectic – visions of the future [76]
2. Professional journalists who had belatedly discovered that big Western oil companies are guilty of nefarious practices, particularly in the Middle East and elsewhere [77]
3. Academics who had become concerned about the advent of peak oil. [78]

Although the oil industry is very large and influential, it is the critics of that industry, rather than its advocates, who today dominate publishing on it. There are no confident defences of oil to be found in today's bookstores.

Critics of oil also have some purchase on the cinema nowadays. Paul Thomas Anderson's *There Will Be Blood* (2007) told the story of a monstrous oilman. A year before it, *A Crude Awakening: The Oil Crash* (2006), an 85-minute Swiss colour documentary and winner of many awards, featured the survivalist lawyer Matthew Savinar, kitted out in camouflage fatigues, with a massive supply of bottled water behind him. It also featured just about every prominent and respectable opponent of oil, in a compilation of prejudices about it. [79]

For a black viscous fluid, oil is now surrounded by a remarkable sociology. So fascinating and supposedly imminent is the prospect of peak oil, in fact, that critics under its spell have extended the concept to predict the demise of coal within little more than half a century. [80] Indeed Richard Heinberg, whom we met in Chapter 2, has covered all the bases – uranium included – in his deliriously titled collection of essays, *Peak Everything* (2007). [81]

What has brought about this remarkable sociology?

Standard Oil vs the little guy

Restricted consumption within the US, the English socialist JA Hobson said in 1902, accounted for big American capitalists in oil and other commodities engaging in investments abroad, and

engaging, too, in the policy of imperialism (again, see Chapter 2). But there is another historic factor in contemporary narratives about oil, and it too relates to Hobson's era.

Founded in 1870, Standard Oil became within 10 years the largest refiner in a cartel that controlled nearly 90 per cent of US refining capacity. In 1879, after lobbying by small oil producers, a Pennsylvania Grand Jury indicted John D Rockefeller and several of his associates on charges of criminal conspiracy. Recovering from that successfully, Standard Oil became in 1882 the first US organisation to form a *trust* – a unified company, run from a central office as a legal entity, by a board of trustees in whose hands were placed the stock of all the companies owned or partly owned by Standard Oil. As Daniel Yergin remarks, the legal concept of the trust developed by Standard Oil was a response to the judicial and political attacks made on the company in the late 1870s and early 1880s. [82]

By 1888, John D Rockefeller's life work had become a vertically integrated multinational corporation. Over the next decade, America met with strikes, financial panic, economic depression and war. Yet Standard Oil still managed to generate a special kind of disgust.

Before and after the turn of the century, *McClure's*, the mass-circulation magazine that began the movement known as muckraking journalism, paid $4000 a piece for a famous series of articles critical of Standard Oil written by the respectable Pennsylvanian biographer Ida Tarbell. [83] From then on, mass American suspicion of trusts focused strongly on Standard Oil. In particular, distaste for the company reached a climax in 1911, when the US Supreme Court broke the company up. The Court's judgment, indeed, was hailed as a victory for the anti-trust cause.

Rapacious rascal: Standard Oil's John D Rockefeller, around 1885. Through assiduous legal and organisational changes, Rockefeller inadvertently made oil come to symbolise everything that was unacceptable about American capitalism

During the Gilded Age and before the First World War, then, the broad American movements known as Populism and Progressivism viewed the Rockefeller dynasty and Standard Oil as the pivot of rapacious, unjust monopoly power. As Alabama University professor Tony Freyer notes in his fine study of the regulation of big business in the UK and US, by 1888 the clout wielded by trusts was regarded as a threat to the citizenry by Democrat and Republican alike. [84]

Interestingly, America's bipartisan regulation of Standard Oil and the trusts in defence of 'the little guy' was not the path preferred by socialists in Britain. While they eventually favoured taxes on excess corporate profits and controls on product prices, men like Hobson, Sidney Webb and, later, Ernest Bevin, saw 'combinations' as efficient, inevitable, and a step toward state ownership. [85]

These things are worth remembering now, even though US anti-trust regulation in the early part of the 20th century was very different from the fearful and pervasive 'consumer protection' that informs the state's attitude to energy companies today. The record suggests that antipathy to Big Oil, while partly grounded in worries about its effects on foreign policy, begins also from American *reaction* against *trusts* and *large companies*. [86] In both cases, the anger aroused had little to do with oil as a commodity.

The anti-trust reaction against oil was broadly conservative, too. As the historian Richard Hofstadter has pointed out, the chief themes of the muckraking approach that fanned it were:

1. Evil-doing by respectable people is the real character of American life – *corruption is everywhere*
2. The mischief demands better *laws*
3. Everyone must take *personal responsibility* for dealing with the mischief.

For Hofstadter, the sense of universal personal responsibility summoned up by the Progressive mind was rural in mythology,

Protestant in outlook, and accompanied by copious amounts of personal *guilt* about the state of the world. [87]

Altogether, then, oil has had a special and negative significance for every American given to impulsive anger and lazy thought about capitalism. This is a historical factor in today's opposition to oil – important, but quite distinct from modern-day causes of loathing.

After 9/11: Saudi conspiracy, deep doo-doo security, uncertain reserves – but a certain peak

There are three salient features of the new critique of oil, as distinct from the old one.

First, in the wake of 9/11 and the second Iraq war, conspiracy theories swirled around oil, George W Bush and the Saudis.

In 2004, *Fahrenheit 9/11*, a film by the mudslinging director Michael Moore, became a particularly popular version of these theories. Moore's montage made a multitude of vague allusions. Foremost among them, however, was a penetrating idea: that Bush's personal involvement in the Texas oil industry relied on money from the Saudis and from the Bin Laden family.

The cognoscenti gave *Fahrenheit 9/11* a Palme d'Or at the Cannes film festival, as well as a 15-minute standing ovation. The film went on to gross more than $250m around the world. Yet the links Moore adduced between oil, US foreign policy and the Saudis are little help in any attempt to understand the political and economic impact of oil. And that deficiency is shared by all the other accounts of the Iraq war that see oil and its agents simply as an endless Machiavellian intrigue.

The man who inspired *Fahrenheit 9/11* was the distinguished East Coast journalist Craig Unger. In 2004, Unger published *House of Bush, House of Saud: the Hidden Relationship Between the World's Two Most Powerful Dynasties*. [88] In it he traced a total of $1.4bn sent by the House of Saud to individuals and entities tied to the Bush family, beginning way back. In his concluding chapter, Unger wrote:

'Even if the president were somehow immune to the fact that in large measure he owed both his personal and political fortunes to the Saudis, it would be astonishing if he did not fall prey to a kind of groupthink as to who they really were... Never before has an American president been so closely tied to a foreign power that harbours and supports our country's mortal enemies.' [89]

Here is a literal, economic and psychological panorama of a man whose oil interests have led him to sup with the Devil. Similarly, Unger holds that the post-war relationship between the US and Saudi Arabia 'was a coarse weave of money, power, and trust.' He concludes that the 'real story' behind Bush's War on Terror is 'full of startling paradoxes and subtle nuances'. [90]

This portrait of Texan and Saudi oil is painted with a broad brush. Stephen Gaghan's movie *Syriana* (2005) also sums up the contemporary mood of unfocused oilophobia. In *Syriana,* the viewer is disorientated, as crucial details of the film are left unclear. *Syriana* concludes with American militarists in dark suits taking out Prince Nazir, a top liberal Arab opponent of oil dependency – remotely, from a darkened room with screens back in the US. The only point that is unambiguous is that collusion between the US government, oil companies and corrupt Arab leaders is leading to recruitment of terrorists among the desperate poor, somewhere 'over there'.

The second aspect of the new critique of oil is to do with its dangers to the environment and to international relations. We saw in Chapter 4 how nuclear power lost its purely economic status in the late 1970s, when it began to be apprehended primarily as a risk. With oil, this process only really accelerated much later – after climate change had entered the public eye, and after 9/11

Misleading movies: film director Michael Moore. *Fahrenheit 9/11*, his indiscriminate collection of conspiracy theories about 9/11, Bush and the Saudis, helped give oil a new toxicity among the American public

had underlined the instability and general peculiarities of Saudi Arabia. Since 2001, however, establishment figures have come to share with oil's critics both a visceral revulsion against oil as a dirty man-made substance, and a geopolitical approach to it that talks up energy security, but boils down to doom and more doom.

According to Juan Pérez Alfonzo – a co-founder of OPEC, no less – oil is the 'excrement of the devil'. [91] On the other side, Senator Richard Lugar, a Republican, claims that addiction to oil represents a six-pronged threat. It encompasses:

1. Vulnerability to natural disaster, war and terror
2. Competition with rising nations such as China and India
3. Being held hostage by nations such as Russia, Iran and Venezuela
4. Corruption and funding of authoritarian regimes
5. Climate change
6. Impoverishment of the developing world through rising energy prices. [92]

On both sides of the indictment, oil is invested with a spectacular ability to wreak havoc.

Third, there is deep uncertainty about the state of oil reserves, particularly in the Middle East. At the same time, uncertainty about oil reserves coincides with a deep sense of certainty that a transition past a peak of finite and increasingly scarce oil supplies is underway.

In climate matters and in nuclear affairs, there is a pervasive sense that outcomes can never be certain (see Chapters 3 and 4). The same applies to oil. For Matthew Simmons, perhaps the most persuasive peak oiler, 'the concept of "fact" becomes problematic when the Saudi Arabian oil industry is concerned'. [93] Looking at Saudi Arabia's largest oilfield, Ghawar, and the state-owned oil giant Saudi Aramco, Simmons asks:

'Given all the uncertainties that still surround Ghawar's complex reservoirs, how can Saudi Aramco boldly advertise… that they can *accurately predict* Ghawar's performance for the next 50 years?' [94]

Simmons is right to be sceptical about forecasts prepared by Saudi Aramco. But poor forecasts are not just the handiwork of that firm, or even of an opaque OPEC, but also of the IPCC's Working Groups II and III (see Chapter 3). They are also the handiwork of Western firms and governments, on oil and on every other subject.

Simmons wants to catalyse 'urgently needed energy data reform', so as to reduce uncertainty about the future of oil. [95] But about one thing he is certain:

'… the outlook for the future that Saudi officials broadcast for all to hear is simply too sanguine for the realities that are now emerging. As Saudi Arabian oilfields age and the world's need for oil steadily rises, the probability increases month by month, year by year, that we are approaching an oil-curtailing twilight in the desert kingdom that has provided the greatest single contribution to the world's oil supply at the least expensive cost.' [96]

In Chapter 3 we saw that climate zealots highlight scientific uncertainty as a means of confirming the certitude that disaster could strike at any time. With peakoilers, the same stretched logic is evident.

We'll spend no more time on conspiracy theories about oil and the Saudis here. We'll deal with oil's dangers to the environment in our later discussion of *transport* – the key application for oil, and, as we saw in Chapter 1, a key contributor to the growth of CO_2 emissions. For the present, we turn to *oil security and peak oil*, by way of a look at oil's economic status.

Is oil a commodity like any other, or is it the key factor in the world economy?

Of all the goods in the world, oil is the most traded. Like any material thing, it has its idiosyncrasies. But just as it's wrong to believe that oil is a commodity like any other, it's wrong to believe that people now live in a 'petroleum economy'. [97] Like labour, services, IT and weapons, oil plays a very important role in many modern economies. An *all-determining* role, however, escapes it.

Oil is too often picked upon as a commodity with a special force to it. For Greens, resources such as oil are a gift from nature. They lose sight of the actions that have gone into turning oil into a true resource for humanity. As a result, they imagine that it's oil itself that has created modern economies.

We've already mentioned that the West's recession of 1973-4 was too deep to blame simply on the OPEC oil embargo of that time (see Chapter 4). Serious recessionary trends had afflicted the West anyway for some years before that episode. After it, making a scapegoat of oil sheiks for the economic downturn was a useful device for Western elites; but as *Financial Times* journalist Toby Shelley acutely points out, there's more than a little evidence that the US in fact *supported* OPEC's campaign to raise oil prices in the early 1970s. [98]

The argument that an abundance of oil for an underdeveloped country is bad news – that it creates a 'paradox of plenty' and a 'resource curse' – rests on a wider exaggeration of oil's all-conquering powers. [99] The millionaire currency speculator George Soros has been particularly assiduous in promoting this point of view. In *Oil Wars* (2007), London School of Economics professor of global governance Mary Kaldor, a colleague of Soros, joins with Stanford University's Terry Lynn Karl in arguing that the reliance of states such as Iraq and Nigeria on oil money creates a culture of 'rent seeking', in which an authoritarian elite can bypass any need for a social contract with the people. [100] The contention is that the 'free' bounty offered by oil wealth undermines real economic activity and so creates dysfunctional societies.

In fact an oil-rich nation cannot be cursed with too many natural resources, for the simple reason that oil is not a natural product. To discover and extract oil from the ground requires sophisticated technology – which is why it was only available in small quantities before the 20th century. After extraction, oil requires more work to turn it into something useful. When Hurricane Katrina hit New Orleans and its environs in 2005, it was refinery capacity as much as pumping capacity that, after the disaster, limited the supply of useful fuel.

Nations such as Saudi Arabia have become reliant on oil. But this is a symptom of their prior underdevelopment and their position in the world economy. The problems experienced are not *caused* by oil. That's why the discovery of oil was not a catastrophe for economies such as the US or Britain.

Oil and war, 1898-1951

Access to oil has been important both as a *resource* for fighting and as a *motive* for fighting. Generally speaking, the gravity of war and the significance of oil to mobilisation have tended to make commentators overestimate its significance in peacetime, and even during 'peaceful' years of preparation for war.

Oil's significance as a military resource first became apparent during the First World War. It is well known that Winston Churchill moved the British navy to oil power in 1911 in a bid to stay ahead of Germany. Oil made possible new technology in aircraft, submarines, and in the tanks that broke the stalemate of the trenches.

How far the clashes on the fields of Europe were driven by the motive of trying to expand imperial control of oil is a rather different matter. When the Bolsheviks seized power in 1917, they made public secret treaties on war aims. Most famously, the Sykes-Picot Agreement mapped out a carve-up of the Ottoman Empire, dividing the territory of what is now Iran and Iraq between Britain, France and Russia.

Clearly part of the First World War was about the whole commercial future of the Middle East and, within that, the prospect of *profit from oil*, as well as further Western industrialisation and motorisation through oil. Even so, in those days coal was arguably more important than oil. After the war, France's occupation of the Ruhr confirmed coal's centrality to European economies and to foreign policy. Indeed it can be argued that only the European Coal and Steel Community, signed in 1951, finally resolved European tensions around coal.

In the lead-up to the Second World War, America's embargo on exports of oil to Japan played a decisive part in the latter's attack on the US Navy at Pearl Harbour, as well as in its imperial expansion into the oil fields of the Dutch East Indies. Here Japan's motive was less profit from oil, and more oil as an *essential technical material* for industry, motorisation, and the military.

Just as Churchill's commitment to oil is well known, so do writers on oil like to cite how, in the Second World War, Hitler would not relieve his tanks at Stalingrad, which were short on fuel, with troops summoned from oil-rich Baku. [101] Here, however, oil was much more resource than motive for fighting.

What comes out of these different episodes is that oil has been *a* motive, but certainly not *the* motive, for major conflict. Was it much of a motive in the Spanish-American war of 1898, the Boer War of 1899-1902, or the Russo-Japanese conflict of 1905? No. Did it play a bigger role in later years? Yes. But in every case the *specificity* of a war's origins and conduct, plus oil's role as motive and resource, has to be taken into account.

There is no law at work that suggests that war is immanent within oil reserves, refineries or pipelines.

Oil and war, 1946-1989

Since the Second World War, the Middle East has seen its share of conflict and intervention by outside powers. But in this period, oil has actually played a rather minor role. The ideological rhetoric of Cold War 'security' against the Soviet menace, together with national independence movements, were more important. Military methods and NATO were also critical, although as direct rule by colonial powers fell into disrepute, so military occupation became a poor way to ensure control over resources.

According to the distinguished American dissident, Noam Chomsky,

> 'If the Middle East didn't have the major energy reserves of the world then policymakers today wouldn't care much more about it than they do about Antarctica.' [102]

Perhaps Chomsky was merely being flamboyant, but he still doesn't convince. Nobody lives in Antarctica. By contrast, parts of the Middle East boast not just oil, or even construction and financial services, but also a fledgling renewables industry (see Chapter 6). When the West plans for war around the Middle East, it considers the Gulf Cooperation Council, which has begun to establish a regional market big enough for its six members to hope to establish the production of consumer goods.

During the Cold War, conflict centred on struggles for national independence waged against disintegrating empires. These struggles were played out against a background of animosity between the US and Soviet Union. At the same time, the US also asserted leadership over rivals in the capitalist world. Integrating Germany into the Atlantic Alliance, for instance, made NATO's first secretary general, Lord Ismay, famously remark that the organisation's purpose was 'to keep the Americans in, the Russians out, and the Germans down'.

Some of the most significant conflicts of the Cold War pitted the West against insurgency in Algeria, Angola, Cuba, Kenya, Korea, Malaya, Nicaragua, Vietnam and Zaire. It's this broader context, too, that is essential for understanding Cold War conflict in oil-rich countries.

In 1953, Iran's nationalisation of the British Anglo-Iranian Oil company led to the US and Britain overthrowing the country's prime minister, Mohammed Mossadeq. In

1956, Egypt's nationalisation of the Suez Canal also led to intervention by Britain and France. The US opposed the action. The significant difference in outcome between the two cases had little or nothing to do with the fact that Iran, unlike Egypt, is endowed with oil. Nor did it have to do with the fact that much of the world's trade, including that in oil, ran and still runs through Suez – not Iran. In both cases, too, the West felt that Middle Eastern expressions of nationalist independence were intolerable. Yet what proved decisive was that with Egypt's President Nasser, Eisenhower preferred to confirm the status of Britain and France not as victors, but as decidedly junior partners in the Atlantic Alliance.

Was oil a motive in Churchill's 5 March 1946 declaration about an Iron Curtain between East and West? Was it a motive in Harry Truman's presentation, to Congress on 12 March 1947, of the Truman doctrine on containing communism? Was oil the key factor in America seeing the Eastern Bloc brought to its knees in 1989? In every case, oil was *not* the key factor.

Today's 'ethical' wars have had little link with oil

Was oil, then, an American motive in prosecuting the first Gulf War of 1990-91? It was *quite* an important motive: together, Iraq and Kuwait accounted for a fifth of world oil supplies. But in retrospect, what was of greater importance was the need for the US, in triumphalist mood after the Cold War, nevertheless to assert its supremacy to the whole world. Oil or no oil, it had to do that, for it no longer had the powerful and global *legitimising* framework of the Cold War. What always escapes those who reduce war to oil is the importance of war to *politics*, and in particular its importance to the credibility of governments in the eyes of their electorates.

Even before 1992 and America's ostensibly humanitarian mission to Somalia, the supremacy sought in the first Gulf War wasn't really a material one. Philip Hammond, reader in media and communications at London South Bank University, notes that George W Bush's New World Order was one in which the US and its allies would obey not the amoral dictates of *realpolitik*, but the *moral* and *ethical* imperative to intervene against illegitimate regimes. [103] Nevertheless, those who dissented from the first Gulf War were largely oblivious to this. On the streets, they reduced the war to variants of the slogan 'Hell No! We won't go! We won't fight for Texaco!'.

It took 9/11 for there to be a real explosion of critical publishing about oil and war. After that, oil was more and more discussed, not just as vital to economic affairs, but also as *prime mover in international and military conflict*. The US invasion of Afghanistan in 2001, and of Iraq in 2003, only deepened this trend.

In many ways today's fingering of oil as a source of war is an ironic development. Today, war is more the continuation of fearful politics by other means than it is conscious, rational strategy based on material interest. At just this moment, however, a widely approved but cynical determinism suggests that most if, not all, military manoeuvres originate in oil.

Just between 1993 and 1999, Bill Clinton's administration fired more than 900 cruise missiles in anger, or an average of one every three days. Fewer than 500 were used in Iraq, 79 in Afghanistan and Sudan, and more than 300 in Kosovo. [104] While Clinton promised to focus 'like a laser' on domestic matters, he spent plenty of time pursuing ostensibly humanitarian interventions abroad, most notably in Yugoslavia, but also in Haiti, Rwanda and Somalia.

All these American adventures – even those in Iraq – were not propelled by oil. They started primarily from a restless, self-conscious, vapid and technocratic quest for national and ethical validation on the international stage. [105] Subsequent Western attempts at state-building, in Iraq and elsewhere, were of a piece with this. At Westminster University, professor of international relations David Chandler writes in his *Empire in Denial: the Politics of State-Building* (2006) that such attempts were about Western governments trying to *disassociate* themselves from power and *evade accountability* for its use. [106]

Altogether, the flexing of Western military muscle today is usually based on fear, and an accompanying search for a sense of purpose. It is not much based on a search for precious commodities.

The anxieties about Al Qaeda that arose after 9/11 explain much of George W Bush's desire to remove Saddam Hussein. No doubt, too, it's the inability of leaders such as George W Bush and Tony Blair to provide a coherent account of their motives for a second foray against Iraq that has led liberals to searching for a hidden motive – US desperation to secure oil from Iraq.

Life would be simple if Bush had gone to war in Iraq just because family money demanded it, or just because America needed oil. In reality, though, the motives for war in the 21st century differ from those of the 20th. Saddam Hussein could not have withheld oil from the world market for long. It was not oil that made Bush go to Iraq, but wider considerations – the first of which was fear.

Oil and tension in Central Asia

For its critics, oil's omnipotence doesn't stop in the Middle East. It also applies to Central Asia.

In May 2001, Hampshire College professor Michael Klare popularised the idea of wars based on a *struggle for resources*. [107] Demand for key materials, Klare argued, was up; resource scarcities had emerged; and disputes about resources had multiplied. Whether it be oil in the Gulf and the South China Sea, or water, minerals and timber elsewhere, resources were irrevocably associated with war.

Central Asia also loomed large in Klare's account. Back in 1997, he pointed out, the Clinton administration had engaged seriously with Central Asia. It had declared a need for the US to diversify its supplies of energy, and courted Azerbaijan in that cause. In southern Kazakhstan, it had held war games with Kazakhstan, Kyrgyzstan and Uzbekistan, ostensibly to show US support for these three countries remaining sovereign states.

Just a few months after Klare's book, 9/11 happened. Then

the US sent troops to Afghanistan, set up bases in Kyrgyzstan and Uzbekistan, and sent advisers to Georgia to train the military there. As if taking their lead from Klare, therefore, two authors now took these developments to indicate a search for and desire to guard the oil of Central Asia.

In 2002, refreshing a late 19th century category used by British diplomats, German journalist Lutz Kleveman published *The New 'Great Game': Blood and Oil in Central Asia*. Here, after extensive travels, Kleveman focused on Central Asia not as an arena in which Britain and Russia jockeyed for access to the riches of India, as they had done in the old 'Great Game' of the 19th century, but on Central Asia as a site for oil and gas. Kleveman also devoted a chapter to Georgia, insisting that the aim of president Eduard Shevardnadze from 1993 onward was

> '… nothing less than Georgia's reestablishment as the centre of a new Great Silk Road, linking Europe with Asia, as it did in the Middle Ages.' [108]

In the next year, Washington journalist Paul Sperry, in his aptly named *Crude Politics*, concurred. Like Kleveman, Sperry attributed the war against the Taliban to the Bush administration's desire to open up… a new Silk Road, in the shape of a gas pipeline through Afghanistan.

The concern with Central Asia did not stop there. In 2004, Michael Klare refined his thesis. *Blood and Oil: How America's Thirst for Petrol is Killing Us* reworked Kleveman's title to present hydrocarbons as sanguinary once more. In the same vein, London-based journalist Dilip Hiro followed up later with *Blood of the Earth: The Global Battle for Vanishing Oil Resources* (2007). [109]

A new game, a new road, and more blood: the same themes keep on reappearing. In his earlier book, Klare drew attention to

- Clinton's stated desire to make US foreign policy 'econocentric', and to include oil and gas within that approach

- The US military's playing up of its capacity to protect the flow of vital resources, a capacity readily grasped by the American public. [110]

For his part, Kleveman also drew attention to Vice President Dick Cheney's earlier *National Energy Policy* report, which identified the Caspian – along with Africa and parts of the Western hemisphere – as an opportunity to 'lessen the impact of a supply disruption on the US and world economies'. [111]

Yet just as it's foolish to read off military US dispositions from the oil underneath the boots of occupying troops, so is it foolish to take at face value those statements of Clinton and Cheney that upheld the significance of Central Asian oil. While Cold Warriors used to bang on about the communist menace in the past, few today would take that as a convincing account of why the West was so militarily active in the Third World during the Cold War.

No doubt the top politicians and military figures who have since stressed the importance of oil have meant what they said. That, however, doesn't imply that oil was really either the main impetus to war, or the enduring motive of military strategy in practice. What warriors say about oil or anything else is important; but what they end up *having to do* is even more important.

To his credit, Klare differentiated between America's posture in the Cold War and its posture since. But for him the change amounted to 'what might be termed the *economisation* of international security affairs'. In the Cold War, America had subordinated its national interest to far-flung global alliances and the containment of communism. By contrast, since the Cold War, self-interest, economic and technological dynamism and, above all, supplies of vital resources had predominated. [112]

This is wrong. Oil no doubt remains a factor in military deployment, both as resource and even as motive. But if anything, international security affairs have recently been dominated more by humanitarian rhetoric and, increasingly, by discussion of global warming.

With oil and war, history does not repeat itself

However much they protest otherwise, critics of the Bush years do tend to see today's wars as a repeat of the past. Yet as we've seen, if oil played a weighty role in conflicts before the Cold War, it was a role that was also circumscribed.

Writing in 2005 on 'Geopolitics Reborn', Klare argued that competition between the US, Russia and a newcomer – China – in the Gulf and around the Caspian Sea,

> '… comes under the rubric of *geopolitics* – that is, the struggle between rival powers for control over territory, natural resources, vital geographic features (harbours, rivers, oases), and other sources of economic and military advantage.' [113]

Klare observed that the US and Russia had established military bases in the Caspian, and that China had conducted joint military exercises with outlying Kyrgystan. On top of military power in the Gulf and the Caspian, US neoconservatives were, Klare observed, apt to 'acknowledge the energy dimensions' of major geopolitical contests. Recognising the importance of oil, the three great powers had bolstered political and military ties with key oil suppliers, so increasing mutual suspicion and setting off a new round of geopolitical competition. Klare went on:

> 'This competition touches practically every major oil-producing area in one way or another, but it is in the Persian Gulf/Caspian Sea region that it will almost certainly assume its most explosive form. With some 70 per cent of the world's known petroleum reserves and a vast potion of its natural gas reserves, the region is destined to become, in Brzezinski's words, the grand chessboard on which Washington, Moscow and Beijing will play out their struggle for primacy.' [114]

In these two passages, Klare revealed much.

For Halford John Mackinder (1861-1947), the English Fabian, geographer and inventor of the idea, *geopolitics* embraced not just natural features, but also national equipment and organisation, military power on land, and military power on sea. In particular, Mackinder felt that trans-continental railways were 'now transmuting the conditions of land-power.' [115] But for resource determinists, natural features tend to have the last word – so much so that Zbigniew Brzezinski, whom we earlier saw failing to make any mention of oil or gas, is hauled up, along with his grand chessboard, to vindicate the idea that oil and gas networks are really what war is all about now.

In all of these hip accounts of oil as a wellspring of armed combat, the specificities of each engagement are smudged over, and the general propositions made are ahistorical. Those propositions are well summarised, therefore, by the sentence with which British journalist Andy Stern opens *Who Won the Oil Wars?* (2005):

> 'Since the birth of the modern oil industry in the middle of the 19th century, the pursuit of oil has brought out three characteristics of mankind: greed, corruption and belligerence.' [116]

Yes, and love of money is the root of all evil too! In general, Stern links war to oil in a timeless manner. Unsurprisingly he finds that Standard Oil, Texaco and Texan oil billionaire John Paul Getty were all in league with the Nazis. [117] Today, he goes on,

> 'The Middle East, with two-thirds of the world's oil, has tensions high. So the US, Europe, Japan, and increasingly China and India, are seeking new sources. In a rerun of the race to colonise the developing world in the 19th century, these major powers are now competing for access to the oil of West Africa and Central Asia. Another flashpoint is the South China Sea, where six countries have laid claim to the oil-rich Spratly Islands...' [118]

In this all-too-monolithic reading of history, oil is always present and somehow, through human nature, causes violence. As Terry Lynn Karl insists in *A Crude Awakening*, oil is a magnet for war; it starts war; it prolongs and intensifies war. The conflict around Darfur in the Sudan is, she says, about oil. Likewise, oil was the reason for the First World War, and secure cheap oil has been *the* basis of US foreign policy since 1945.

Altogether, repeats – with differences – of imagined past military joustings over oil tend to mesmerise Western commentators. Inevitably, after new Great Games, new Silk Roads and Grand Chessboards, there are new flashpoints and chokepoints around oil.

Oil in Sudan and Nigeria

Terry Lynn Karl is not alone in attributing war in Sudan to oil. Andy Stern repeats her stance, saying that 'Chinese companies supply the Sudanese military with arms, tanks and helicopters to clear civilians from the oilfields, as well as fighting anti-government militias'. [119]

But John Ghazvinian, a journalist who has travelled around Africa, takes a more nuanced view. Despite the unwavering title of his book, *Untapped: The Scramble for Africa's Oil* (2007), Ghazvinian does offer some resistance to 'yellow peril' rhetoric about China in Sudan. China relies on Sudan for nearly 10 per cent of its imported oil, he notes; but unlike Western oil firms, Chinese ones are only just starting out in Africa, with oilfields that are mostly depleted. Of the debt relief, scholarships and infrastructure that China has offered Africa, he observes,

> 'China's ability to turn relatively small amounts of cash into tangible results, along with its strict regard for state sovereignty, are, for Western governments who still prefer to attach painful conditions to almost any interaction they have with African states, perhaps the most galling aspects of the rising strength of the Middle Kingdom in Africa.' [120]

Sadly, even these reasonable remarks rather miss the point.

The main conflict in Sudan lies in the west of the country, in Darfur. Sudan's oil lies in the south. What makes war in Sudan, however, isn't oil, or, as Stern imagines, the self-evidently outrageous fact that China 'has opposed a UN resolution for sanctions against the Sudan.' [121] What makes war there is the *absence of a coherent state*. It's the lengthy failure of Khartoum to gain legitimacy with the Sudanese that explains both the trouble around oil in Sudan, and the country's descent into war.

Oil didn't bring about the conflict in Sudan. Neither did it by itself incite separatist elements in Nigeria to force, in June 2008, the closure of Bonga, Shell's flagship deepwater oil production and storage facility 120km off Nigeria's coast. As Carl Mortishead, world business editor of the London *Times*, points out, Nigeria's navy is ill equipped and unable to patrol the seas around Bonga; and though NATO has yet to take action on deepwater platforms off West Africa, 'officials have proposed seconding warships to the area.' [122]

Wealthy Nigeria is not basket-case Sudan. But while oil provides separatists in both countries with a convenient target, it's the lack of popular support for the central authorities that turns the distribution of natural resources into a question of violence.

No peak oil yet

The idea that conflict is driven by oil is reinforced by the idea that oil is now becoming scarcer.

In April 2001, America's Council on Foreign Relations concluded that though the world would not run short of hydrocarbons 'in the foreseeable future,' there was a grave problem of underinvestment. [123] In 2001 that was still the consensus. For years, petroleum geologist Colin Campbell and petroleum engineer and consultant Jean Laherrère had predicted serious difficulties; but even as late as 1998, their conclusions were still fairly cautious:

> 'The world is not running out of oil – at least not yet. What our society does face, and soon, is the end of the abundant and cheap oil on which all industrial nations depend.' [124]

However, Kenneth Deffeyes cracked the consensus in September 2001 with his book *Hubbert's Peak: The Impending World Oil Shortage.* Since then, fear of peak oil has certainly been climbing. By 2005, as oil prices began to rise, the theory went mainstream.

In February 2005 the Hirsch report, commissioned by the US Department of Energy, warned that a peak could occur within 20 years, and that coping with it would be 'extremely complex, involve literally trillions of dollars and require many years of intense effort'. [125] Among oil companies, Chevron went furthest in acknowledging peak oil. Under the slogan 'Will you join us?', a Chevron advertising campaign warned that

> '… many of the world's oil and gas fields are maturing. And new energy discoveries are mainly occurring in places where resources are difficult to extract, physically, economically and even politically.' [126]

People must, said Chevron, 'start by asking the tough questions.'

In April 2007 the international Society of Petroleum Engineers held a conference in San Antonio, Texas. Promoting the conference, the SPE announced:

> 'We are now producing the second trillion barrels of oil, and in about 50 years we will have consumed all of that… how are we going to get the next trillion?' [127]

It was, and remains, a good question to ask. In fact, things could turn out more urgent than the SPE's 50-year timescale. At present rates of consumption, the world will consume a trillion barrels of oil in more like 30 years – sooner, if demand revives after the Crash of 2008.

According to theorists of peak oil, the next trillion barrels of oil are all there is in the world, and the crisis is *now*.

Such pessimism isn't appropriate. What *is* appropriate is to start investing in technologies that will help human society find and exploit the third trillion barrels of oil that the SPE was searching for.

Four techniques will get us there. In increasing order of significance, these techniques are:

1. Extraction of more oil from existing fields
2. Discovery of more oil
3. Unconventional oil, such as oil sands
4. Biofuels.

Partisans of peak oil are correct that none of these techniques, on its own, is a full solution. However, peak oilers also underestimate all four – especially biofuels.

Can new technologies assist, first, in the extraction of more oil from existing fields? For Simmons, the world is reliant on old supergiant fields such as Saudi Arabia's Ghawar, which has been producing since 1951. Moreover, technology has failed to bring about a real rejuvenation over the past 30 years:

> 'Horizontal drilling, multilateral well completion systems, intelligent wells, 3-D seismic, and computer-generated reserve simulation... have in some instances allowed 20 to 30 percent more of original oil-in-place to be ultimately recovered... None of these technical breakthroughs created an "oilfield fountain of youth".' [128]

Simmons is right that new technology cannot stave off depletion forever. It's also true that while new technology will squeeze more out of existing fields, diminishing returns set in: the more a field is depleted, the harder it is to squeeze the remaining oil out of it.

There's no reason to believe, however, that technological progress must be slower in the next 30 years than it was in the past 30. If rates of recovery from oilfields can be raised from 35 to 45 per cent, almost a third more oil will be produced.

Second, new technology will aid in finding altogether new oil. For many peak oilers, all the big discoveries of oil are firmly in the past and prospectors have looked everywhere on Earth worth looking – or at least have looked in so many places, it's possible to make an accurate estimate of how much oil remains to be discovered.

Yet there continue to be large finds. In 2007, those included Jidong, off the coast of China (up to 1.2 billion barrels), and Tupi, 160 miles off Rio de Janeiro (up to eight billion barrels). The Brazilian find is not just a lucky accident. Back in the 1980s, Brazil began to look for oil deep offshore, at a time when nobody else had thought to look. The first big find opened up brand new prospects for exploration. [129]

Most well known are the finds off the coast of Angola and in the Gulf of Mexico. Now, with Tupi, deep water drilling technology could pay off for Brazil, too.

There are also places to look closer onshore. A talking point in the 2008 presidential election was that, mostly for ecological reasons, and despite eight years of George W Bush supposedly in the pocket of the oil industry, the US has voluntarily placed much of Alaska, the western Gulf of Mexico, and the Pacific and Atlantic coasts off limits for exploration.

The latest technology has not been applied in countries such as Iran, Cuba or Sudan. On top of that, the limits to production and exploration in Iraq can hardly be said to be geological.

Third, recent years have seen an increase in the production of hydrocarbons from unconventional sources. As we noted in our discussion of coal, if there is a shortfall in oil, coal to liquids offers some redress. On top of that, there is the oil available from oil sands (see below). In 2003, the US Energy Information Administration added 180 billion barrels of sand-based oil to

Canada's reserves of conventional oil. [130]

Fourth, biofuels could substitute for oil very profoundly. That's why we give them a more extensive treatment later in this chapter.

There's oil in them there hills

Oil sands and their less important cousins, oil shales, are geological deposits of organic material that have not experienced enough heat and pressure to transform them into oil. That work of transformation has to be finished off by human industry.

Oil sands, sometimes known as tar sands, are made up of a mix of bitumen consisting of the thickest component found in conventional oil. Between them, Canada's Athabasca, East Alberta, and Venezuela's Orinoco contain greater deposits than all known reserves of conventional oil. Already, of the 87 million barrels of oil produced each day in 2007, about 1.6 million came from Canada's sands, which have seen increasing production since the 1960s. About 600,000 barrels also came from Venezuela's sands, in which investment is more recent. [131]

While there's plenty of them, oil sands are hard to get out of the ground, and need much more refining than conventional oil. Oil sands can sometimes be mined from the surface. The heavy tar-like bitumen then needs separating from the sand it's mixed with. Deeper deposits take even more work. Too viscous to flow directly out of the ground, oil sands need to be extracted by a process such as heating by steam injection. Then, once out of the ground, they need much more refining than conventional oil if they're to produce usable petroleum products.

All this makes oil sands a more expensive source of energy than conventional oil. Greens worry that, more than conventional oil, producing workable fuels from oil sands and shales requires vast amounts of water – and of energy in the form of natural gas. They also worry that oil from oil sands generates more CO_2 than conventional oil.

These are not insurmountable problems. Moreover, two points need adding to this. First, the existence of vast unconventional resources shows that peak oil will not be a sharp cut off. Estimates suggest there could be more unconventional oil than all the conventional oil in the world.

In the long term, then, oil will not vanish. It will just become harder to extract. [132]

Second, the more complex it becomes to pursue unconventional oil, the more attractive alternatives such as biofuels will become.

Hubbert's peak: an old theory makes its mark in new times

M King Hubbert, a geologist with Shell, was the first to develop a theory of peak oil. In 1956 Hubbert predicted the future course of US oil production. [133] Forecasts of the demise of oil have a long history of being wrong, but Hubbert turned out to be right: oil production from the continental United States did peak in the early 1970s, and then began to fall. It should be noted, however, that Hubbert failed to anticipate the discovery of oil in Alaska, and offshore in the Gulf of Mexico. Both of these finds have partially offset the decline in oil production for the US as a whole.

Hubbert's theory starts by trying to estimate the total amount of oil that will ever be recovered from the ground. It then makes the assumption that when the halfway point is reached, a decline in the rate of production will set in. Even if there is plenty of oil remaining in the ground, the theory states that it will be too difficult to find and extract for production to be expanded at all easily. The decline of the world's available oil resources is, finally, predicted to have dire consequences, especially if economic growth expands demand for them even higher.

Many economists are sceptical of peak oil, claiming that estimates of the total amount of oil in the ground are determined by how much money it has been worthwhile to invest in searching, rather than by comprehensive geological knowledge. Peak oilers reply that this shows that economists have never studied geology. In 2005 Deffeyes claimed that peak oil would occur later that

same year, suggesting that the following five years could, as a consequence, bring war, famine and death. [134]

Obviously, the cult of Hubbert today cannot be put down to his original forecast for the US eventually coming true. What has happened, rather, is that passing the year 2000, and the general discussion of tipping points (see Chapter 3), have contributed to the sensation that mankind is going over a hump of some sort – that it is *downhill all the way from now*. Until very recently, the soaring price of oil also contributed to the feeling that oil's moment of true scarcity had arrived.

Dr Richard Pike, chief executive of the Royal Society of Chemistry and a man with 25 years' experience in the oil industry, adds another factor that explains the popularity of theories of peak oil today. When oil companies estimate their future reserves, Pike says, they do so on a very conservative basis – namely, that the reserve has been discovered, is recoverable, and is 90 per cent certain to be exceeded. But to demand this level of certainty of every oilfield, as statutory bodies such as the US Securities and Exchange Commission do, is to exclude many oilfields that, on the balance of probabilities, could well yield useful results. To add up all the estimates of reserves in a simple arithmetical manner does no justice to the bell-shaped curve of probabilities that more accurately describes fields that are proven, probable and merely possible.

Thinking about what fields will yield in this more generous but actually more realistic manner, Pike believes that there could be two trillion barrels of conventional oil left on the planet, rather than the 1.2 trillion that is usually taken as given. [135]

Prophet of peak oil: Marion King Hubbert, a Texan, early on during a long career at Shell. Contemporary society's sense of loss has given this man a cult following long after his death

Summing up on oil

From the late 19th century onwards, the price-fixing practices of Rockefeller and his allies made American opinion see oil as a dishonest conspiracy, worthy of state regulation. But as we have seen, today's hatred of oil is very much linked to popular perceptions of the dangers that surround the Middle East. Before 9/11, there were certainly fears about what the US was getting into. But after 9/11, the personal connections between the Bush family and Saudi oil gave the Democratic Party an easy and popular means of attacking the Republican Party.

Despite its manifest weaknesses, many now take the link between oil and war as self-evident. Thus, writing about the Georgia-Russia clash of 2008, Rafael Kandiyoti, professor of chemical engineering at Imperial College and author of *Pipelines: Oil Flows and Crude Politics*, contended that 'from the beginning, oil and gas transmission has been at the centre of this conflict'. [136] Yet as Kandiyoti's book rightly observed, the spread of NATO membership into the Baltic States, Eastern Europe and Central Europe has also been a big factor in Russian strategic considerations in recent years. [137] No doubt Georgia is one place where oil and gas can move from East to West; but to Moscow, NATO and its missiles in Poland, as well as the policies of Georgia's leader, were probably just as important as oil, if not more so.

It's sad that the Middle East and Russia, sources of oil supply, are so much cast as bogeymen. Yet China and India are also now set to become villains – because of their demand for oil. The rise of state-backed national oil companies (NOCs) in Russia, China, India and elsewhere has very much unsettled the old international oil companies (IOCs).

The IOCs – oil 'majors' such as Exxon Mobil, BP, Total and Shell – rather resent NOCs such as Norway's StatOil, Algeria's Sonatrach, Mexico's Pemex, or Brazil's Petrobras. They are also nervous about the alliances firms like these have struck with oil services companies such as Schlumberger and Halliburton.

But in relation to Russia (Lukoil and Yukos), China (PetroChina, CNPC) and India, feelings run higher. The first section of *The New Competition for Global Resources* (2008), a report by the Boston Consulting Group (BCG) and Wharton Business School, University of Pennsylvania, contains important pointers for the international tensions that are likely to surround oil in future. [138]

Rick Peters, a senior partner at BCG and a former leader of its energy practice, points out that, given the advent of Chinese and Indian competition against Western companies, nations which own oil are now able to secure more favourable terms for their resource than those which obtained in the past. [139] From Wharton, management professor Mauro Guillen is sanguine about the market opportunities for Western engineering firms out to supply Chinese and Indian NOCs on their home turf in Asia. But Guillen's fellow management professor, Witold Henisz, is scathing about Chinese oil companies in Africa, saying that, compared with IOCs, they 'have worse labour practices, invest less in the community and are more tolerant of corruption'. [140] While Exxon Mobil, Shell and BP 'have made great efforts to improve their external stakeholder relations,' Henisz says, 'the Chinese and Indians,' he insists, 'have a lot of catching up to do.' He goes on:

> 'They need to think about an oilfield as not just something to take, but about their need to be partners and their need to take a more inclusive and holistic view about what they are doing in a place like Angola or Sudan, and how to manage that process. They are far behind Western companies in those realisations and in the development of those capabilities; that will be a big struggle for them in the medium term.' [141]

Oh really? For ourselves, we don't doubt that new Asian oil firms can be as rapacious as Western ones. But we also think that BCG's Peters is nearer the mark when he hints that oil owners in Africa, Latin America and the former Soviet Union welcome the fact that they face a wide choice of oilfield developers nowadays. By itself, that won't ensure that Russian and Asian oil firms win every contract. But the chances are that oil-owning nations won't roll over and play ball with Western IOCs the way they did for many decades. They will demand, and have been demanding, that oil developers contribute to local economies, engage in technology transfer, and have something to offer all the way from exploration through to retailing. [142]

So there is even more for the West to fear, if it wants to. Not every oil-owning state is run by a regime akin to that organised by Hugo Chavez in Venezuela. But we can expect that oil, like coal and gas, won't just be a useful source of energy for many years to come, but will also serve – again like coal and gas – as a stick with which the West will try to beat the East.

Demand for liquid fuels will continue to grow – driven, as we discuss next, by transport. In the short and medium run that demand will be met by oil. Oil will therefore continue to be big business.

To find and extract continuing amounts will require greater and greater investment. Oil from the Arctic or deep offshore won't come cheap. Beneath the ups and downs of market sentiment, that will tend to put upward pressure on the price of oil. Of course, high costs will not be the rule in locations where, as in Saudi Arabia, oil is still easy to extract. In such places, then, large profits will be made.

New technology will continue to be applied to finding oil and getting it out of the ground. A continuing increase in raw computer power will help in finding oil, as seismic data are used with greater and greater sophistication to reconstruct images of underground reservoirs. As for drilling for oil, new technology is already being applied so that extractors can sense the environment of the drill bit, better steer the drill, cut through rock more efficiently, clear wells, seal off unwanted leaks and generally stimulate the flow of oil.

For all these advances, 'conventional' oil production may expand only weakly over the next few decades. More oil may be pumped; but those nations with easy access, such as members of OPEC, show little interest in flooding the market.

In 2005, oil production stood at 84.3m barrels per day. Of that, 81.9m were conventional oil, 1.7m extra heavy oil and bitumen (for example, oil sands), 0.1m consisted of coal to liquids, and 0.5m were biofuels. [143]

By 2030 the world could be using an extra 30-40m barrels of oil a day. Much of that growth will come from gas-to-liquids, coal to liquids, oil sands and shales. But even more will come from biofuels.

Transport is why oil matters

Oil's principal use is in transport. From the end of the 19th century, oil spread to become a source of heat, light, electricity and motive power for transport. By the mid-20th century, it had become the key to modern economic development. But from then on, as electrification began its ascent more fully than before the Second World War, new sources of energy such as nuclear power and natural gas became more significant. The use of oil became concentrated in the transport sector: it was used to make fuels for motor vehicles. In 1950 the share of US oil used for transport was 54 per cent. It rose steadily to 60 per cent by 1980, and by 2001 reached 69 per cent. [144] The underlying reason is not just that carbon fuels – petrol, diesel and jet fuel – pack a lot of energy, but also that they pack it into a small space, with a small weight.

While much of the world's energy supply has gone electric, that hasn't happened with aeroplanes, ships or – above all – cars. The reason is simple. Where electricity can easily be delivered, it's by far the most convenient source of energy. The problem with electric transport is that vehicles that are meant to go anywhere need to carry their energy supply with them; but electricity is hard to store.

To see the difference that the capacity to deliver electricity makes, think about electric trains. An electric vehicle probably calls to mind something like a golf buggy, which carries its power on board in a battery. But a train such as the Eurostar, which runs from London to Paris, has electrical power delivered to it. It weighs 815 tonnes and travels at 300 km/h (186 mph). A container truck driven by a diesel engine carries its own fuel. It need not stick to a fixed electrified route. It typically weighs only 44 tonnes and travels at less than 100 km/h (60 mph).

The Toyota Prius shows that new transport technology will be about getting more from carbon fuels, not abandoning them altogether. The Prius uses a battery based on a material called nickel metal hydride (NiMH). But the Prius isn't an electric car. It's a hybrid, meaning that as well as battery-driven motors, it has a petrol-driven engine which it uses to charge up its battery. The Prius still takes on board all its energy in the form of petrol. The electrical system makes the fuel go further by allowing it to be used more efficiently. Regenerative breaking, for example, means that instead of throwing away energy when you put on the brakes, batteries get recharged. You can also get a lot more energy out of the same fuel just by running an engine at its optimum rate. By adding back or sucking in power, an electric motor can keep the engine running at its sweet spot. That dramatically increases efficiency, even though the battery doesn't add any new energy overall.

To avoid using petrol, a hybrid car would need large enough batteries to store energy for an entire journey. The snag is, however, that current battery technology just isn't up to the task. The NiMH material stores energy at a density of 250 kJ/kg (measured by weight), or 360 kJ/litre (measured by volume). That contrasts very vividly with petrol, which has an energy density of 44,400 kJ/kg or 34,800 kJ/litre. A battery to store as much energy as an equivalent fuel tank would have to be a hundred times as large and heavy.

The lesson is that oil, in its various refined forms, is an excellent store of energy. Whether we burn carbon-based fuels in internal combustion engines, jet engines, or fuel cells, they will allow us to carry more people and freight, further and faster.

In transport, as elsewhere, efficiency isn't enough

Individual cars will become more efficient, but in aggregate will demand more fuel. Amory Lovins complains that, in 2003, the average new American light vehicle had

> '... 24 per cent more weight, 93 per cent more horsepower, and [a] 29 per cent faster 0-60mph time than in 1981, but only 1 per cent more miles per gallon... America's light-vehicle fleet today is nearly the world's most fuel-efficient per ton-mile, but with more tons, it uses the most fuel per mile of any advanced country.' [145]

What Lovins sees as a problem is, in fact, Americans being able to travel around more. As we have already mentioned (see Chapter 1), increases in energy efficiency tend to be accompanied by increases in energy use. Better performance is a good thing. Most of the time, perhaps, people don't need the extra space and power of an SUV. But if they have it to hand when they do want it, without having to take special measures, that's a good thing, too.

The other great hope of the Greens is public transport – buses and rail. Public transport undoubtedly has an important role to play, especially in cities, where it can often be the most convenient means of getting about. But it's just impractical to think that more public transport will mean less consumption of energy. As statistics show, the convenience of the car is enormously popular.

In the US, the number of passenger miles travelled on the highways rose from 2.6 trillion in 1980 to 4.9 trillion in 2005. Meanwhile, the number of passenger miles travelled by train rose much more modestly – from 44 billion to 55 billion. This cannot be blamed entirely on a lack of service, even in the US. Between 1984 and 2006, the number of train stations served there rose from 1,822 to 2,975. [146]

Even in the UK, with all its European inclination toward the train, the picture is similar. Between 1980 and 2006, passenger kilometres travelled by cars, light vans and taxis rose from 388 to 686 billion, or 77 per cent. For rail, which has recently enjoyed a bit of a boom, the comparable figures were 35 billion and 55 billion, or an increase of 58 per cent. [147]

It's hard to avoid the conclusion that much of the enthusiasm for public transport today is motivated simply by hatred for a thing – the car. In Britain, Green campaigning journalist George Monbiot, for instance, puts Margaret Thatcher's road-building program on a par with the breaking of the unions and the dismantling of the welfare state. He muses that 'it is strange to see how the car has been overlooked as an agent of political change'. [148]

For Monbiot, the car has brought about an unwelcome shift to individualism. But this is to give an inanimate object the power of social forces. After all, making transport convenient through cars might more accurately be seen as overcoming distance and bringing people together. There are road hogs, and possibly even a little road rage; but lack of consideration for others is also sometimes evident on public transport. In fact, the level of cooperation on the roads can exceed that on a train or a bus, and is generally admirable.

While public transport deserves proper investment, most of the time it isn't a serious competitor to the car. The rise of the car should not be feared, but rather celebrated. Of course, transport planners can make mistakes with roads, traffic lights and all the rest. But the activities that surround driving tend to reflect, rather than cause, the moods of society.

The future of the car: internal combustion, fuel cells and the role of hydrogen

As options contrasted with the internal combustion engine, fuel cells are often spoken of in the same breath as hydrogen. And with both fuel cells and hydrogen, the assumption is that transport will become clean.

The truth is more complicated.

The key technology in a car takes the chemical energy in a fuel and converts it into the kinetic energy of forward motion. The Internal Combustion Engine (ICE) that made motor cars practical was developed at the end of the 19th century, most significantly in Germany by Nikolaus Otto, Gottlieb Daimler, Wilhelm Maybach and Rudolph Diesel.

In an ICE, fuel is mixed with oxygen and ignited. The energy released is captured in the motion of a piston, which is driven by expanding hot gas. Transmission and gearing take the mechanical energy of pistons and deliver it to wheels.

The most common fuels used with the ICE are standard gasoline and diesel. In both areas, ICE technology continues to improve. Homogenous compression charge ignition (HCCI) is a good example. It's based on precise control of conditions, so that fuel ignites uniformly and simultaneously at just the right moment. That can reduce nitrous oxide pollutants and increase efficiency by 10 per cent or more.

Many fuels can be used in an ICE, including ethanol and even hydrogen gas. After a century's concentration on petrol and diesel, some of these fuels don't get a second look today, which is a pity. Instead, fuel cells are now being discussed as never before.

We have already explained that electrical batteries suffer from poor energy density. But if an appropriate source of high energy density fuel can be converted into electricity, then turning wheels with electrical motors, rather than with controlled explosions, might have a future.

Like ICEs, fuel cells convert energy in a fuel into motion. Again, the fuel is combined with oxygen. However, with fuel cells there's no flame of combustion. Instead, the fuel is chemically broken apart as it passes through a special plastic membrane, and the resulting energy is captured electrically. The electricity can then run through wires to electrical motors powering a car's wheels.

Known membranes work best with hydrogen as a fuel. However, fuel cells have been constructed that work directly with methanol and ethanol: indeed, these direct methanol fuel cells are presently being developed as replacements for batteries in phones, cameras and portable computers. Another option is to convert a fuel such as methanol into hydrogen before feeding it into a fuel cell. Such indirect fuel cells have also been built, although they involve added complexity.

Vehicles based on fuel cells may come into serious use in a decade or so. At present, however, it looks likely that they will be outclassed by hybrids. But even with fuel cells, hydrogen is unlikely to be the fuel of choice. Compared to other gases, hydrogen is hard to handle, both for storage on board a vehicle and in pipelines and filling stations. Liquid hydrocarbons are much easier to deal with.

It is also sometimes forgotten that although hydrogen is carbon free, and so releases only water at the point of consumption, it has to come from somewhere. Hydrogen is a manufactured fuel that's only as clean as the energy that goes into producing it.

For fuel cells used in transport, manufacturing a fuel such as methanol would make more sense than manufacturing hydrogen. That's the problem with all those alluring visions of a *hydrogen economy*. [149] Using nuclear or solar power to make hydrogen will be important. But that hydrogen will mostly be used industrially in combination with carbon, as in the upgrading of oil sands, or in the production of fertiliser.

As a fuel or energy carrier for consumer use, hydrogen is a poor choice.

Greens fear the growth of cars in the East

The whole world needs more petrol. January 2008 saw the launch of India's cheapest ever car: the Tata Nano. Ratan Tata, responsible for seeing the car to market, explained that although the initial concept was for an improved scooter, he realised that

'… if we want to build a people's car, it should be a car and not something that people would say, Ah! That's just a scooter with four wheels or an auto rickshaw with four wheels or not really a car... people wanted a real car and not something that someone would say was not a car, this is half-a-car or three-fourths of a car.' [150]

So they do. 'My first reaction when someone says they need to buy a car is to say don't buy it... But people are buying cars, I cannot stop them', lamented Greenpeace India's Soumya Brata Rahut. [151]

For energy writer and consultant Andrew McKillop, there are natural limits not just to human population, but also to the number of cars on the planet. Denouncing what he is pleased to call 'The Chinese car bomb,' McKillop argues that the increase in cars in China and India means that these two nuclear-armed powers, along with another in the shape of the US,

'… are ever more likely to fight among themselves, or confront EU importers, including two nuclear-weapons states, for the last oil reserves on the planet.' [152]

Dramatic stuff – but having accused China of presiding over an explosion just waiting to happen, why not add more pejoratives? Somehow the growth of annual car sales in China, which slowed to fewer than six million in the wake of the Crash of 2008, is supposed to lead to a world war. No matter that sales of six million represent just half a per cent of China's population: for the enlightened Green, this is half a per cent too many.

It's much more likely, in fact, that China will be making cars in volume to export to Europe; already some GM Buicks are designed in China and exported to the US. India also plans to begin its first major car exporting effort to Europe. Finally, after spending $2m on a controlling stake in a Norwegian maker of polymer lithium ion batteries, Tata Motors hopes to bring its Indica EV electric car to the EU.

Asia will export cheap cars to Europe, and UK car workers' jobs could go. Meanwhile, China plans to build an extra one million kilometres of road before 2020. [153]

As transport becomes more Chinese, so China can expect to meet with more criticism.

Aviation: progress portrayed as child abuse

Worldwide, the number of air passengers more than doubled between 1985 and 2005: from 896 million to 2,022 million. The average length of those journeys rose too, from 1500km to 2000km. Freight – including all those food miles – grew even faster, from 40 billion to 510 billion tonne-kilometres. The International Civil Aviation Organisation expects growth over the next 20 years to continue at a similar rate, with the fastest happening in Asia. [154]

In the 1950s, the jet set consisted of the rich and famous. In the 1990s, easyJet and Ryan Air opened up air travel to millions. For some, the objection to cheap flights seems to be that they put the wrong sort of people aloft. George Monbiot found an alternative objection: flying across the Atlantic was as unacceptable as child abuse. [155]

Aviation will see advances in efficiency. For Rainer von Wrede, director of environmental affairs at Airbus, a new aircraft should have 15-25 per cent less fuel consumption than the plane it's going to replace. Better engines, improved aerodynamics and lighter strong materials are in the works. [156]

These goals are achievable. At the Farnborough Airshow in July 2008, General Electric launched a development of a new engine for jets carrying 150 or fewer passengers. Using advanced materials, new cooling technology, fuel mixing and aerodynamics, GE aims for a 16 per cent efficiency increase over the GE90-115B engines that power the Boeing 777. [157]

Meanwhile, Pratt & Whitney announced that its new engine, the Geared Turbofan PW1000G, had successfully taken flight, with an increased efficiency of 12 per cent. [158]

As with cars, increases in efficiency will be outweighed

by more flying. When told that new airport runways will make for more efficient flying, Greens are right to point out that more flying means more CO_2 overall. [159]

Though aviation adds to climate change, mankind should not take that as a pretext for abandoning flying, one of the noblest advances of the 20th century. What is needed here are some technological fixes. Nevertheless, Friends of the Earth has already made up its mind about aviation. It holds, very simply, that 'the fast growth in air flights can not be maintained without causing climatic disaster'. [160]

It's wrong to write technology off the agenda in this way. When Virgin made a test flight from London to Amsterdam, running on oil from coconut and the babassu palm tree, one Greenpeace activist complained that it was a 'massive piece of spin and greenwash'. [161] But a month later, Continental Airlines announced that it was working with GE and Boeing on a biofuel flight for 2009. [162] In May 2008 Airbus got together with JetBlue Airways and Honeywell to suggest that biofuels could provide 30 per cent of aviation fuel by 2030. [163] Inevitably, there is also keen interest from the US military. [164]

Fuel producers are also working on biofuels for aviation. Based in San Francisco, Solazyme makes kerosene from algae. Today, that's the only biofuel to meet the standards for regular Jet A fuel – but other high-flying biofuels are on the way. [165]

Biofuels: from Beauty to the Beast and back again

At one time, Greens favoured biofuels over fossil fuels. Fossil fuels are formed from the fossilised remains of dead plants and animals that have been exposed to heat and pressure in the Earth's crust over hundreds of millions of years. They produce billions of tonnes of CO_2 each year when burnt. Biofuels, on the other hand, are derived from raw plant matter and add no net CO_2 to the atmosphere when burnt. In fact, the photosynthetic plants used in the production of biofuels remove as much CO_2 from the atmosphere as biofuels release when they're burnt.

Solid or unprocessed biofuels include wood, sawdust,

grass cuttings, domestic refuse, charcoal and dried manure. When raw biomass is in a suitable form such as firewood, it can be burnt directly in a stove or furnace to provide heat or raise steam. This is known as 'traditional biomass'. However, the World Health Organisation has rightly decried the burning of traditional biomass, saying that, around the world,

> '… indoor air pollution from solid fuel use is responsible for 1.6 million deaths due to pneumonia, chronic respiratory disease and lung cancer… In high-mortality developing countries, indoor smoke is responsible for an estimated 3.7 per cent of the overall disease burden…' [166]

Biomass is also increasingly burnt in modern power stations, alongside coal. But this practice is unlikely to have a long-term future. Biomass isn't the most efficient way of capturing the sun's energy. As the first table in Chapter 6 shows, area for area, the energy in biomass is much less than that contained in sunlight. Instead of converting biofuels to electricity and losing energy in the process, it makes sense to use solar technologies to generate electricity.

The virtue of biomass is that, once processed into biofuels, it stores energy in the form of hydrocarbons – and it is only these light but punchy kind of fuels that can be used in road or air transport. Processed biofuels are very similar to existing liquid fuels such as petrol or diesel. All of these fuels were originally organic, and their basic carbon molecules are the same.

The main biofuel in use today is ethanol, which is known as a 'first generation' (1G) biofuel. Ethanol is produced by growing crops that are high in sugar. In the US, ethanol is produced from maize, whereas in Brazil, where production is more efficient, it's made from sugarcane. In both cases the ethanol is produced using yeast fermentation, similar to the way in which alcohol is produced in beer.

The other main biofuel in use today is biodiesel. Derived from vegetable oil, this is very similar to the diesel derived from rock oil. Europe gets most of its biodiesel from soybeans grown in sunny Indonesia.

A second generation of ethanol is planned. This *cellulosic* ethanol uses non-food crops or inedible waste products to produce the same end product, but first breaks down the cellulose or lignocellulose that makes up the body of plants into sugar for conversion to ethanol. That makes the process far more efficient than 1G ethanol.

As a fuel, ethanol is fine, but not ideal. It has a slightly lower energy density than petrol and it mixes well with water, which is a disadvantage for a fuel. It is also corrosive to pipelines.

However, ethanol does have a higher octane rating than petrol, meaning it burns more easily. Indeed, it has been regularly blended into much US fuel in the 1990s for its 'better burn', and for its ability to meet clean air standards. [167] What's more, car engines can easily be adapted to run any combination of ethanol and regular petrol.

One alternative to ethanol is butanol. Butanol produces more energy than ethanol and allegedly can be burnt 'straight' in existing gasoline engines – without modification to the engine or car. It is also less corrosive and less water-soluble than ethanol.

Many investigative groups are now working on biotechnology-assisted biofuels – 3G. In January 2008, Professor James Liao's group at the University of California, Los Angeles, showed that the fast-growing bacterium *E. coli* could be engineered to produce a variety of possible fuels, including butanol. [168]

The biofuel industry is still in its infancy. It continues to rely on plants that have been developed for food production. Still, research is beginning on non-food plants that are adapted to produce cellulose rapidly, and that will grow on marginal land with minimal inputs of water and fertiliser. Also, notwithstanding the improvements in genetics and agronomy that are surely to come, miscanthus, a tall growing grass, has been identified as a promising candidate for cellulosic ethanol production.

Beyond all this, biotechnology will contribute to both the

growing of biomass and its fermentation into fuel. This may go beyond improving the growth and utility of plants and, separately, improving yeast. Process improvements may well be combined to produce algae or bacteria that photosynthesise light in a way that directly produces fuel.

Craig Venter: a modern Prometheus

J Craig Venter knows how to shake things up. While many innovators have lost their sense of urgency and vision, Venter is not among them. He is clear where his urgency originated: drafted to Vietnam as an 18 year old in 1967, he worked as a medic in a US navy hospital. 'I had to learn in real time what triage actually meant', he has explained. 'I dealt with the death of thousands of men my age. It was a life-altering experience.' [169]

Venter entered medical research, but, frustrated with the slow pace of advance, he founded a private institute and then a private company, Celera Genomics, to lay out the sequence of the human genome.

Celera attracted furious accusations of privatising genetic knowledge. Perhaps some were justified, but it did light a fire under publicly funded scientists. When the complete human genome was unveiled in 2000, the breakthrough came at least three years ahead of schedule.

Venter has continued to make spectacular scientific advances – and lots of money. His grandest ambition, however, is to solve the world's energy problems. His plan is not just to modify an existing organism, but also to build a synthetic organism, a bacterium, from the ground up. That way, the capacity to produce fuel – or food, or medicines – can be designed in to organic materials.

Genius of genetics: the entrepreneur Craig Venter. His hope is to bring biology to the making of new fuels

'For the first time, God has competition', complained Green campaigners on hearing of the plan. [170] Venter replied:

'We are not afraid to take on things that are important just because they stimulate thinking... We are dealing in big ideas. We are trying to create a new value system for life. When dealing at this scale, you can't expect everybody to be happy.' [171]

Transport will get its fuel through high-tech farming

The plausibility of biofuels depends on raising the efficiency of agriculture. Because they make use of so much land, biofuels are attacked for causing deforestation and, above all, for pushing up the price of food. The only people in favour of biofuels are said to be farmers eager for new subsidies.

Critics have even claimed that when you add up all the energy that goes into growing a crop, you get out less than you put in. [172] There may be a bit of truth in this, especially with regard to 1G ethanol production from maize. But *done right*, with more sophisticated agricultural technology, there's no doubt that biofuels will provide plenty of energy in the future. [173] And as we show below, there's enough land around to accommodate biofuels.

Mechanising world agriculture will help motorise world society. Mechanisation both saves labour and makes most efficient use of other inputs. In Brazil, cane is still cut by hand with machetes. Even when fields are burnt before harvesting, to clear them of snakes and destroy razor-sharp leaves, the work isn't just difficult, but unpleasant and dangerous.

In 2007, a Brazilian government team freed 1,000 labourers from servitude at the hands of an ethanol producer. At the time, the company defended itself on the grounds that its workers were paid 'good wages by Brazilian standards.' [174] But given that its economy is relatively dynamic, Brazil's boom in biofuel production still promises to change sugar cultivation fundamentally. Brazilian sugar cane is gradually being mechanised. Between 1990 and 2000, the number of workers cutting cane fell from 1.2 million to just 700,000. [175]

Over the next few years, machines will substitute for hundreds of thousands of cane workers. In São Paulo state, for example, around 144,000 cutters will be replaced by machines that cost about $600,000 each and harvest the crop more quickly and cost effectively. Each will do the work of about 100 cane cutters, who earn $500 per month. [176]

Many suggest that mechanisation is being driven by the worries that ethical consumers in the West have about poor labour conditions. [177] But it seems more likely that mechanisation is being driven by concern for the bottom line. That's why investment is happening all the way through the process of production, from the development of new varieties of cane and new planting conditions, through to refineries and the commercial exploitation of waste products. [178]

Agricultural technology is important to the developed world. Global positioning systems and computer control have given rise to the concept of precision agriculture, which treats crops and soils within an accuracy of one centimetre. [179] With the right technology, however, tropical countries could still become more productive than the developed world. In the sunlight they receive, they have an edge in natural conditions that cannot be replicated by artificial means. This is the path that Brazil is travelling.

In the least developed countries, by contrast, improving basic transport is still the main means by which to increase agricultural production. Without roads, refrigeration and other methods of food preservation, food goes to waste before it can reach market.

Greens don't just attack the use of such techniques in the Third World. They also campaign against genetically modified seeds, pesticides, factory farming, food packaging, the preservation of food by irradiation, and increasing land use. On every front, they prefer to go organic.

In fact, though, *industrialised agriculture* will both abolish hunger and provide abundant fuel for transport.

Land enough to grow both food and fuel

It's myopic to accuse biofuels of squeezing agriculture off the landscape. There is plenty of land around for both kinds of crops to prosper.

The world already has more high quality food than ever before. Per head per day, world consumption of food increased from 2300 kilocalories (kcals) in 1961-3, to 2800 in 2001-3. In the developing world, it rose from 1950 kcals to 2650 kcals. Importantly, it was not just the calorific content of food that changed, but also the *types* of food consumed. In the developing world, diet moved away from cereals, pulses and tubers, toward meat, dairy and fruit:

The changing composition of world food intake [180]

	per cent calories, 1961-1963	per cent calories, 2001-2003
Cereals	60	52
Pulses	6	2
Roots and tubers	9	6
Sugar	4	6
Vegetables	3	5
Meat	3	7
Milk	2	3
Other	13	19

Although the world has seen a small absolute fall in the consumption of cereal per head, that's a sign of the advent of more balanced diets, not a food crisis. [181]

Despite all the progress in food, far too many countries have been left behind. But the problem in those countries is conflict, not shortage of resources. The places that have seen the largest increases in the number of malnourished are in the war-torn regions of Africa and the Middle East. Between 1990/2 and 2001/3, the world's largest national increase in hunger, by far, occurred in the Democratic Republic of Congo, where malnutrition tripled from 12 million people to 36 million.[182]

There have been increases in agricultural output. Yet once again these have been achieved more by increasing yields than by using additional land. America's yield of soybeans, for instance, rose by 16 per cent from 1979 to 1995, and then by another 15 per cent in the shorter period between 1995 and 2007. Its yield of corn for grain rose by 58 per cent between 1974 and 1990, and then by another 29 per cent between 1990 and 2007. [183] At present, the US is so productive that it's a net exporter of food.

There's no *technical* reason why productivity in the developed world should not continue to increase, or why the developing world should not match it.

In developing Asia progress is already happening; but, thus far, sub-Saharan Africa is being left behind. Between 1960 and 2005, the yield of cereals in the developed world rose from 0.84 to 2.16 tonnes per hectare. In East Asia, yields rose from 0.48 to 1.8 tonnes per hectare, closing most of the gap between the developing and developed worlds. In sub-Saharan Africa, yields did less well, rising from just under 0.4 tonnes to just over 0.4 tonnes per hectare. [184]

The most important technological input into food production is *fertiliser*. Nitrogen is most critical, followed by phosphorus and potassium. High technology can ensure not just the right nutrients, but also the right proportions.

Improvement of crop varieties had especially good results in Asia's green revolution. But in many other developing countries, figures confirm that crops have not yet been created especially to suit local soil and climatic conditions:

Application of farming technologies in the Third World [185]

	per cent irrigated land, 2002	per cent of area planted with improved crops, 2000	kg of nutrients per hectare, 2002
Sub-Saharan Africa	4	24	13
Latin America	11	59	81
South Asia	39	77	98
East Asia	29	85	190

Pesticides, herbicides, growth hormones and antibiotics are all carbon derivatives of oil. They all still have a greater contribution to make to high-productivity agriculture.

Even without the continuing advance of productivity in agriculture, there would be no shortage of land in the world – land to grow both food and biofuels. At the end of June 2008, the EU abolished the compulsory set aside of 10 per cent of farmland. [186] Environmentalists warned that the decision 'could be one of the worst disasters for wildlife for 40 years'. [187] Similarly, in the US, Greens have lobbied for strict enforcement of the Conservation Reserve Program (CRP), which keeps land out of production. [188] Indeed, in July 2008, the National Wildlife Federation won a lawsuit that blocked the Department of Agriculture from opening 24 million acres of CRP land to additional haying and grazing. [189]

In reality, releasing more land should open up a more dynamic future for the countryside. A large part of that future will be the production of biofuels.

Summing up on transport, biofuels and agriculture

Transport will be harder than electricity to make carbon-neutral. For that reason it will continue to attract Green hostility.

Energy efficiency, and especially that offered by hybrids,

will slow the transport-related growth of emissions, but will not reverse it. As prices fall, hybrid technology will become standard, first of all in the developed world and, in time, globally. But it will take several decades before hybrids make up a majority of new sales, let alone of the whole vehicle fleet.

Plug-in hybrid cars have much to recommend them. Though they still rely on liquid hydrocarbons for long journeys, and for quick refuelling, plug-in hybrids can be recharged from the grid. They can therefore benefit from electricity that's clean and cheap. But while modification from existing hybrids should be relatively straightforward, plug-ins are only now beginning to enter production. Their first real test looks likely to be as a fleet of 200 taxis in Shenzhen, China. [190]

Rising emissions are not a reason to cut back on transport. Eventually, the use of biofuels and electricity will allow transport to have little net impact on emissions of CO_2.

Driving, shipping and air travel will demand liquid hydrocarbon fuels. At present, demand is met by fossil fuels, and it will take decades for the new biofuel industry to gear up. But as the world heads toward 2050, biofuels will supply an increasing proportion of liquid fuel.

Use of biofuels has been rising. In the US, the Renewable Fuel Standard programme required that, by 2007, the proportion of fuel ethanol deployed in cars be at least four per cent. [191] Before the scare, in 2008, about biofuels driving up the cost of food, the EU adopted a target of 10 per cent biofuel by 2020. [192] India has adopted a target – possibly too ambitious – of using 20 per cent biofuel by 2017. [193]

The final pace at which biofuels will be adopted will depend on the rate at which the technology develops, but at present penetrations of about 10 per cent by 2020 look reasonable. By the middle of the century, biofuels could grow to cover a majority of transport fuel.

On this scale, the growth of biofuels will transform the Earth's landscape. They will affect the already productive Western world, but, in addition, vast tracts of abandoned land in Eastern

Rating different sources of biofuels

Source	Intermediate product	Processing method	Period of take-off
Cereals (maize / corn, wheat, etc)	Sugar	Breakdown of starch	2005-2010
Sugar beet	Sugar	Sugar refining	2005-2015
Sugar cane	Sugar	Sugar refining	2005-2025
Cellulosic cereals (maize / corn, wheat, etc)	Sugar or lignocellulose	Sugar from breakdown of cellulose in whole plant	2010-2030
Grasses (switchgrass, miscanthus)	Sugar or lignocellulose	Sugar from breakdown of cellulose in whole plant	2015-2035
Woody plants (willow, poplar)	Sugar or lignocellulose	Sugar from breakdown of cellulose in whole plant	2015-2035
Genetically modified plants	Sugar or lignocellulose	Could be genetically programmed to break itself down into useful products	2020-2040
Soy, rapeseed, palm oil	Vegetable oil or lignocellulose	Pressing seeds for oil	1995-2005
Jatropha	Vegetable oil or lignocellulose	Pressing seeds for oil	2010-2020
Algae	Vegetable oil or lignocellulose	Pressing seeds for oil	2025-2045
Agriculture and industrial woody waste	Sugar or lignocellulose	Sugar from breakdown of cellulose in whole plant	2010-2030
Domestic and food organic waste	Vegetable oil or lignocellulose	Chemical and thermal processing	Not practical
Genetically modified algae or bacteria	Sugar or vegetable oil or lignocellulose or direct production of fuel products	Combined chemical and biological processing. Could be genetically programmed to break itself down into useful products	2030-2050

Advantages	Disadvantages	Generation	Rating now / 10	Rating for the future / 10
Builds on existing technology Stepping stone to future technologies	Inefficient. Plant optimised for food, not fuel	1G	2	1
Can be grown in wide range of climates	Less efficient than cane	1G	3	2
Very efficient source of sugar	Requires tropical sun	1G	6	5
Stepping stone from first to second generation	Plant optimised for food, not fuel	2G	3	5
Can be optimised for fuel rather than food	Technology under development	2G	2	7
Efficient production through coppicing	Long term plantations inflexible	2G	3	7
Greater flexibility and optimisation of grasses or trees	Technology under development	3G	1	8
Builds on existing technology. Stepping stone to future technologies	Plant optimised for food, not fuel	1G	2	1
Can be optimised for fuel rather than food	Underdeveloped technology. Long term potential not clear	1G	4	4
Easy to handle	Very different from existing agriculture. Energy from algae will require serious innovation	2G	2	5
Usefully recycles existing waste	Limited supply	2G	2	3
	Limited supply and hard to centralise	1G	1	1
Potentially most flexible and efficient	Technology still under development	3G	1	8

Rating different biofuels as end-products

Biofuel	Starting material	Production
Ethanol	Sugar	Biological fermentation
Butanol	Sugar	Biological fermentation with use of genetic engineering
Synthetic petroleum	Sugar	Biological fermentation with extensive genetic engineering
Biodiesel	Vegetable oil	Chemical processing
Methanol	Lignocellulose	Chemical processing
Dimethylether (DME)	Lignocellulose	Chemical processing
Fischer-Tropsch biodiesel	Lignocellulose	Chemical processing

Advantages	Disadvantages	Rating now/10	Rating for the future / 10
Builds on existing technology in beverage industry	Corrosive, mixes with water, low energy density	4	3
Compatible with existing distribution infrastructure. Mid energy density	New production technology needed	4	3
High energy density	Complex production technologies, still to be developed	2	8
Use in existing engines. Potentially adaptable for use in aircraft	Limited supply of starting material	4	3
Can be used in fuel cells. Application in consumer gadgets	Toxic. Low energy density. Needs development of fuel cell vehicles	2	5
May work well in existing engines	Low energy density. Development needed and potential unclear	2	5
Use in existing engines Potentially adaptable for use in aircraft	Development needed and potential unclear	1	3

Europe look set to be drawn back into production. [194] The most dramatic shift could come in the parts of Africa that, like Brazil, enjoy tropical sun. If those regions can begin to industrialise, that could open up a whole new era in biofuel agriculture.

Altogether, as the productivity of food production increases alongside a booming biofuel sector, it's possible that, by the middle of the century, rather more land will be used for fuel than for food. One method that would cut the amount of land needed to make biofuels is to create other biofuels through the offshore farming of aquatic bacteria and algae. [195]

Either way, biofuels will help the world become more mobile. The two tables above summarise our view of the prospects facing different sources of biofuels, as well our view of the prospects facing different kinds of biofuels as end-products.

Toward a New Carbon Infrastructure

Of more than 100 elements in the periodic table, only carbon has an entire branch of chemistry – organic chemistry – devoted to it. It's no accident that the chemistry of life is based on carbon. In creating living organisms, evolution has shown the remarkable uses to which carbon can be put.

Tomorrow, human beings will go further in exploiting carbon, both in making new carbon fuels and in gradually taking control of the natural carbon cycle.

The petrol in a car mostly consists of molecules based on a backbone of between five and 12 carbon atoms strung together. Diesel and jet fuel are based on chains of between 10 and 15 atoms. All these compounds are found in crude oil, which formed the basis of most transport fuels in the 20th century. But in the 21st century, transport fuels will come from a much wider variety of sources, including the

- breakdown of the large bitumen molecules in oil sands
- breakdown of even larger coal molecules (CTL)
- conversion of oil shale to oil
- building up into oil of small molecules from natural gas

('Gas To Liquids')
- conversion of sugar, cellulose, lipids or other biomass to liquid fuels (biofuels).

It's a similar story with natural gas. Methane is a single carbon molecule with four hydrogen atoms attached. In the 20th century, it was obtained by refining raw natural gas. In the 21st century, however, methane will also come from

- breaking down large coal molecules (coal gasification)
- converting sugar, cellulose, lipids or other biomass (biogasification).

How, though, will liquid fuels and gas participate in a New Carbon Infrastructure?

Today, by far the biggest part of oil and gas is consumed for energy. But oil and gas also provide the raw materials for the production of plastics, as well as the production of fine chemicals such as pharmaceuticals. For this reason, the two fuels together are rightly known as the petrochemical industry.

The rise of biofuels, however, will turn *petrochemicals* into what might possibly be termed *petroagrichemicals*. That's why the chemical giant DuPont has teamed up with BP working on biotechnologies – and why the giant agricultural producer Cargill has moved into plastics. [196]

The greater integration of the energy sector with the chemical and agricultural industries will open up opportunities for new fuels beyond petrol, diesel, methane and coal. Ethanol is the first of these new fuels, and biobutanol will be the next. The development of fuel cells may see compounds such as methanol put to work, too. As a result, future refineries will have to become more and more flexible in their outputs.

The commonality between oil, gas, chemicals and biofuels is that each is built on carbon – judiciously combined with other elements. Yet with climate change comes the need to manage a fifth form of carbon: CO_2.

At first sight, prospects here do not look very promising. It's true that CO_2 is, in the first instance, a waste product. But when two of the planet's most eminent scientists recently penned a reminder to the world not to forget long-term fundamental research in energy, they identified the *chemistry of CO_2* as a key area for work. 'For decades,' they wrote, 'there has been little research, whether fundamental or exploratory, in this area; it was considered a solved problem.' [197]

One of the first results of renewed work on CO_2 will be improved technology for its capture. That will allow it to be sequestered away from the atmosphere. Beyond the CCS technologies that this chapter has already discussed in relation to coal, there are other possibilities. One is to process CO_2 into rocks of calcium carbonate – chalk, or marble – that can easily be stored, or even used as construction materials.

These technologies will help human beings manage the planet's carbon cycle. However, CO_2 will not just be stored, but will also be put to work. It will be used to fertilise the growth of new plants for food and fuel.

The full intricacies of the biology that can 'fix' or extract carbon from the air have yet to be unravelled. The process will rely on

- molecules that can harvest light and focus energy into a reaction centre that splits up water to produce hydrogen
- a complex system of membranes and enzymes that can use the energy produced to manufacture hydrocarbons from CO_2.

By understanding that biology better, it will be possible to make plants, bacteria and algae produce more and better biofuels.

Improved methods of producing hydrogen, together with insights into the chemistry of joining carbon atoms from CO_2, may however allow the same process to be done better, and without any biology at all. The energy to power such artificial photosynthesis might come from the sun, in the same way as plants engage in photosynthesis. But other forms of energy, such as nuclear or geothermal, would work just as well.

In any case, such a system would give human beings simultaneous control over both the carbon cycle, and the production of carbon-based fuels. Far from heading toward a low carbon economy, therefore, the world will see an emerging web of industrial uses for carbon, building up into a New Carbon Infrastructure.

Can algae save the world?

That was the question asked by an exhibition at the Science Museum, London, in Spring 2008. The exhibition drew on the work of Carol Turley, a microbial ecologist at the Plymouth Marine Laboratory, and an earlier paper by Turley gave a scholarly preface to it. [198]

When plants take up carbon from the atmosphere, the carbon is usually returned to the air when the plant dies and decays. An exception occurs when plant material is buried, which is what happened with the ancient forests that form today's coalfields.

The natural process of carbon burial, however, is slow. To remove the carbon that burning fossil fuels would generate in a century would take millions of years. Yet it may still be possible to increase the rate at which biomass is buried.

That's where algae come in. We have already mentioned the promising possibility that marine algae might be farmed and harvested for biofuels. The plan here is different: algae on the surface of the ocean first absorb CO_2, then sink to the bottom of the ocean, taking their carbon with them.

The Southern Ocean around Antarctica has been identified as an area in which only a single nutrient, iron, appears to be missing. Simply adding a small quantity of iron may vastly stimulate algal growth. Experiments have been carried out on a small scale, showing that there is a real effect. [199]

There are, however, big question marks over whether such

a scheme could be made to work in practice. Could the system scale up? Will dead algae sink, or will they be consumed by other marine life so that the carbon stays in circulation? Will there be knock-on effects, reducing carbon uptake elsewhere in the ocean? At present all that can be said is that experiments are cheap relative to the potential long-term gains.

The fertilisation of the world's oceans is particularly galling for Greens. After all, it's being pursued by commercial enterprises – most notably Climos, of San Francisco. [200]

Any scientific results produced by a commercial organisation should be scrutinised with a sceptical eye. Typically enough, though, Greenpeace and others seem keener on stopping experiments in ocean fertilisation than on keeping an open mind about such experiments. [201]

Summing up this chapter:
Obama, cheap oil, and New Carbon

Around the negotiations on climate change in Copenhagen in December 2009, coal will certainly dominate the West's arguments with China and India. The *interpretation* of national and general security in Western Europe will undoubtedly revolve more around Russian gas. Last and most importantly, the presidency of Barack Obama has long been billed as one that will take climate change seriously. For that reason, and – even more – on grounds of energy security, Obama will pay special attention to fossil fuels and, in the first place, Middle Eastern oil.

Obama's will be the energy presidency as much as the one that has to mend the US economy. Obama has already promised that energy efficiency and conservation around fossil fuels will be a major feature of his programme for US economic revival. For him, energy efficiency is 'the cheapest, cleanest, fastest energy source'. [202] His appeal on energy will be a personal one, in the sense that it will ultimately be more about Americans' everyday conduct than it will be about technological innovation. *Obama will make great rhetoric on innovation, but his focus will be on*

America and Americans cutting back on energy use.

The starting-point for Obama's strategy in energy isn't energy innovation, but a federal system for capping and trading CO_2 emissions in the US. [203] More broadly, his efforts will be spent on

1. 'Inclusive' international diplomacy aimed at regaining international prestige for the US on climate matters
2. Driving a hard bargain on carbon-based fuels made in, used by, or generally sought by, Brazil, Russia, India and China and their NOCs
3. Humanitarian interventions as a perceived means of shoring up US energy security in the Middle East, Central Asia and elsewhere
4. A domestic programme that will devote 'substantial resources' to repairing US roads and bridges, but will also call for 'significantly more attention to investments that will make it easier for us to walk, bicycle and access other transport alternatives'. [204]

Obama's election platform on energy, *New Energy for America*, covered the emissions levels he would like to see. The platform referred to *carbon*, rather than CO_2, as the thing needing to be reduced. It was only a slip, and one that is made quite commonly, too; but, in an election platform for the US presidency, one might expect more precision.

Reflecting today's zeitgeist, Obama has some fundamental disagreements with carbon-based fuels, for all their merits.

He has admitted that the prospect of 'no coal' is an illusion, and has said that CCS is right in principle. But he has added, rather notoriously, that with a cap and trade system on US GHGs, the rigours of the market mean that 'if somebody wants to build a coal-powered plant they can; it's just that it will bankrupt them, because they're going to be charged a huge sum for all that greenhouse gas that's being emitted'. [205]

Joe Biden, Vice President, has gone further. On the stump

in Ohio before the 2008 election, he charged that China's coal burning had brought pollution and even some deaths to The US. He said that 'we' – Americans – should figure out how to clean 'their' coal up. For good measure, he added: 'No coal plants here in America'. [206]

Many commentators have remarked upon Obama's protectionist tendencies. These certainly apply to gas and oil: in those sectors, he wants to promote the supply of domestic energy, rather than the foreign sort. Obama takes a particularly belligerent line on oil. He has said:

> 'We cannot sustain a future powered by a fuel that is rapidly disappearing. Not when we purchase $700 million worth of oil every single day from some the world's most unstable and hostile nations – Middle Eastern regimes that will control nearly all of the world's oil by 2030. Not when the rapid growth of countries like China and India means that we're consuming more of this dwindling resource faster than we ever imagined. We know that we can't sustain this kind of future.' [207]

Few know whether, when, by what means and by how much Obama will really be able to reduce US 'dependency' on oil from the Middle East. But we do know that he is so fearful about energy security that he wonders out loud about America being 'held hostage to the whims of tyrants and dictators who control the world's oil wells'. And we do know that, in the simplistic manner of peak oilers, Obama believes the world's oil to be 'rapidly disappearing'. [208]

It's over oil that Obama has most comprehensively adopted the environmentalist credo.

In his election platform, Obama said that a small portion of the receipts generated by auctioning allowances to emit would be used to 'support the development of clean energy, invest in energy efficiency improvements, and help develop the next generation of biofuels and clean energy vehicles'. [209] But given that this list of causes included efficiency measures, renewables and carbon-based fuels, exactly how small was Obama's portion of cash?

The answer was $15bn a year; not much, compared to George Bush's $700bn bailout for US banks. Some of the money would go, Obama's platform said, on helping put one million plug-in hybrid vehicles on America's roads by 2015. But converting the White House's fleet of cars to plug-in – to 'show government leadership' – would take priority over the mass market. [210]

To assist the retooling of US car plants so that they can build new, fuel-efficient cars, Obama's platform promised them just $4 billion of tax credits. No doubt, since his election and the crisis in Detroit, Obama will now be more expansive. Still, his platform already made clear that, as with oil, the dynamic of his policy was broadly toward a more autarchic America. The old $4 billion, after all, was meant to ensure that the new fuel-efficient cars could be 'built in the US by American workers rather than overseas'. [211]

Obama may or may not understand the difference between CO_2 and carbon. But whatever he thinks, *carbon will remain central to the world's energy system – and especially its transport – for a long time to come.*

Astronomical flows of energy

Greens underestimate the potential of wind, solar, water and geothermal power. Even more than nuclear or fossil fuels, flows of energy around and inside the planet need working on at scale. Do that, and enormous quantities of energy will come on stream

Worldwide, more wind turbines

were installed in 2007 than existed at all in the year 2000. The 94 GW of wind power in operation in 2007 provided about 1.3 per cent of the world's electricity – a perceptible and growing part of global energy production. [1]

By contrast with wind power, however, the shares of global energy production held by the other three main kinds of renewable energy – solar, water and geothermal – are truly insignificant.

Set against the *results so far achieved*, the current interest in renewables might be considered misplaced. Measured by *future potential*, however, things stand differently. As the insightful Vaclav Smil, distinguished professor at Manitoba University, points out, even a fraction of the two per cent of solar energy that drives the world's winds would be enough to power human civilisation. [2] Hurricanes unleash more energy than the bombing of Hiroshima. Each year, the sun floods the Earth with more than 50,000 times the energy that was generated by the world's electricity power stations in 2000.

The gigantic potential of wind and solar doesn't mean that putting them to work is easy. But it does show that humanity's problem isn't one of limited natural resources. The energy is out there. *Renewable energy is limited only by humanity's political will and engineering talent to capture it.*

It's time 'renewables' grew up. In this chapter, we'll sometimes use a better word for them: *astronomicals*.

Environmentalists often group renewables together by what they are *not*: not nuclear and not fossil fuels. But this is a limp logic with which to greet a clutch of terrific possibilities.

The word 'renewable' is also limp. Nobody is trying to renew the sun, like mankind will renew supplies of carbon-based fuels through the New Carbon Infrastructure.

Wind turbines near Palm
Springs, California

So what benefits, exactly, do renewables offer?

Wind, solar, water and geothermal sources of energy have one thing in common: they're about *capturing continuous flows of energy over the Earth as an astronomical entity, not mining it as a stock of geological fuels*. Flows of energy from the sun drive the Earth's winds, which also supply waves with power. The sun is the force behind the water flowing to hydroelectric dams, while the moon's actions on the Earth's seas give the tides power that can be captured. As the heat of the Earth's deep radioactive rocks and molten mantle escapes into the cold of outer space, it can be drawn upon to provide geothermal power.

The physics of wind, sunlight and water also *produce no waste* – giving them a slight advantage over nuclear power, where waste is tricky but can be handled, and a big plus over fossil fuels, which necessarily produce CO_2. And geothermal power? Chemically, the hot water it produces does contain dissolved minerals, so it needs a bit of waste management.

The most important benefit of all four techniques, however, is that *each can generate enormous supplies of energy*. The energy available from the Earth as a member of the solar system is truly astronomical.

Collecting that energy, however, can be difficult. Apart from its stupendous size, what characterises the energy contained in the wind, sunlight and water is its *diffuseness* – the fact that it is spread so thinly compared with the needs of mankind. It's useful, first, to compare the amount of energy generated by different techniques over a square metre: the energy available from coal fields, oil fields and power stations is far more concentrated than that available from renewable sources. Then it's useful to compare the different energies generated per square metre with the energy demands, per square metre, made by different kinds of buildings and activities within those buildings. The basic figures are striking:

Energy and surface area: selected sources, Watts generated per square metre [3]

	W/m^2
Coal fields, oil fields, power stations	1000-10,000
Solar in, say, the Sahara desert	250 or more
Solar, global average	168
Tides, upper parts of rivers	10-50
Wind	5-20
Lower parts of rivers	just above 1
Biomass, geothermal	below 1

Energy and surface area: selected final uses, Watts consumed per square metre [4]

	W/m^2
High rise buildings	up to 3000
Steel mills, refineries	300-900
Supermarkets, office buildings	200-400
Homes, modest manufacturing/service facilities	20-100

The sun can muster quite a lot of energy per square metre, but even in tropical zones it lacks the concentrated clout of fossil fuels. With what Smil calls 'indirectly harnessed solar flows', the energy available for a given surface area is weaker still.

That's an important result. Because humanity needs *a lot* more cheap energy, what's available from wind, sun and water must be *collected over a wide area* and *linked to major grids* if it is to be put together into a decent sized supply – let alone transmitted somewhere else.

It's true that geothermal energy contrasts with the energy in the wind, sun and water. Geothermal sources of heat, like deposits of uranium or of the fossil fuels, are rather concentrated: they can be drawn on in locations as tight as 100m^2. In all four cases, however, the amount of energy that can potentially be put in mankind's hands invites a unique slogan: *scale is beautiful*.

We saw in Chapters 1 and 2 that changing personal habits in the home, even across every household in society, provide few savings in energy and in CO_2 emissions. We saw in Chapter 5 that the energy density of carbon-based fuels will continue to make them a very big part of the future of transport. The same harsh facts apply to attempts to engage in the *microgeneration* of energy in the home through wind or solar power. Leaving aside all those people who live in blocks of flats with small roof areas per head of population, *capturing astronomical flows of energy at a personal level is a fool's errand.*

Technologically, too, wind turbines have already gone way beyond the individualist gesture to neighbours and the planet, as this table shows:

Progress in grid-connected wind turbines, 1992-2008, kW

Date	Specification
1992	50-200 peak capacity; 300-750 was best economically [5]
Late 1990s	1000 was the peak on the market; most models rated at 500-750 [6]
2003	General Electric: 3600 offshore. Diameter 110m [7]
Early 2008	Germany's Enercon: rated at 6000, may reach 7000 or more. Diameter 126m [8]

In IT and elsewhere, miniaturisation can be a wonderful thing. But with renewables, things are very different. The miniature approach, for example, of British Conservative Party leader David Cameron and his wind turbine would be laughable if it wasn't so sad (see panel).

Renewables are astronomical flows of energy. They make grand amounts of power available, but very often over expanses of the Earth that, though not of astronomical size, are grand grand in scale as well

To help humanity, renewables need investment and they need space.

A wind turbine for every home? Forget it

With wind power, bigger is better. In 2007 Conservative party leader David Cameron installed a D400 Stealthgen wind turbine as part of a £150,000 green makeover of his house in Kensington, West London. The £3,000 turbine was estimated to save him £72 a year on electricity. [9]

Used afloat, the application for which it was designed, the D400 has been highly rated by *Practical Boat Owner* magazine.[10] But with boats at sea, wind turbines have two competitive

Setting a dishonest example: Tory chief David Cameron goes Green, April 2006. The chauffeur-driven car was just behind

advantages. First, there are no electricity grids on water; second, there's often a good supply of wind.

Over London, by contrast, wind speeds only reach about 20 kilometres per hour. [11] However, the D400 is rated for optimal output of 400 Watts in a wind of just less than 60 kph. So because wind power rises with the cube of wind speed, the D400 will produce a measly 40 Watts in London – barely enough to power a light bulb. [12]

In 2006, a Green photo opportunity of Cameron on a bicycle unravelled when it emerged that a chauffeur-driven car had followed him so as to carry his papers. A Conservative spokesman explained that 'if he could carry all of the boxes of documents for work on his bike, then he would'. [13]

Cameron's home windmill turned out to have just as little to do with the practicalities of reducing CO_2 as his bike trip. The machine was removed after less than a week because planning permission had specified that it should have been attached to his chimney rather than to the adjacent wall – where it was actually installed. [14]

Within cities, where the majority of the world's population now lives, wind energy on a domestic scale can only be described as a folly.

The geometry of renewables

Coal, oil and gas tend to be concentrated in particular fields, and the energy in them is distributed widely. Apart from geothermal, renewables, as we've seen, are different. Before they can distribute energy, they first have to take it in over large expanses of land or water.

In some cases, it's true, natural geography *can* be used to concentrate renewable energy. Wind farms in special locations can whip up a fair bit of electricity. However, solar water heating, concentrated solar power and photovoltaic panels can only take advantage of *latitude*.

With hydroelectric dams, location is more important than

it is for wind. With dams, energy originates from the evaporation of water by sunlight. That evaporation lifts water high into the sky, where it falls as rain, joins rivers flowing down toward the oceans, and so gives up energy. A dam works by intercepting the flow. It makes use of the way sunlight lifts water above sea level. By channelling a flow through turbines, a dam sets to work the energy given up by water as it falls back to Earth. And by strategically placing dams across suitable valleys, water can be collected from an entire river basin. The water that falls as rain right across the basin will, in turn, have been evaporated over an even wider area. In this way a dam concentrates at a single point solar energy that originally fell on open water or forests.

Tidal and wave power also take advantage of natural geography. Inlets and estuaries can give the tides and the waves a special force.

Hot springs, geysers, and volcanoes are traditional, concentrated sites for geothermal energy. In enhanced or engineered geothermal, by contrast, a special hole is drilled and rocks are cracked using hydraulic pressure. Cracking allows water to flow through several cubic kilometres of rock, picking up heat.

Then, a second well is drilled, allowing water pumped underground to return heated to the surface; so in this more artificial form, geothermal energy is also somewhat concentrated.

Prospects in brief

While wind will shortly become a significant part of the world's energy supply, solar technology is not as mature as wind turbines. Only in a decade's time will it start to provide a significant part of energy supply.

Among renewables, hydroelectric power has developed the furthest. In the form of dams, then, it's the Green source of energy most disliked by Greens. Hydroelectric still has considerable scope for expansion; but Greens turn the usual practical difficulties around big hydro into the escapist chimeras of micro or even 'pico' hydro.

Because the moon moves the sea from high to low tide and back again, that motion can be used to turn turbines. As for waves at sea, which are mostly driven by wind, a variety of mechanical devices has been constructed to use their motion as power to turn an electrical generator. Even more than tidal power, though, wave power is still at an experimental stage.

What environmentalists see in renewables

Environmentalists don't really see wind, solar, water and geothermal as massive sources of energy. Their 'renewables', rather, are meant to renew the world *morally* – by leading it away from industrialism and modernity.

In 1987 the World Commission on Environment and Development, led by former Norwegian Prime Minister Gro Harlem Brundtland, internationalised debate on sustainable development. The Commission wrote that most renewable energy systems

> '… operate best at small to medium scales, ideally suited for rural and suburban applications. They are also generally labour-intensive, which should be an added benefit where there is surplus labour. They are less susceptible than fossil fuels to wild price fluctuations and foreign exchange costs. Most countries have some renewable resources, and their use can help nations move towards self-reliance.' [15]

In fact these characteristics are not inherent in astronomical sources of energy, but represent Green value judgements about what is ethically correct. Facilities for harnessing astronomically sized amounts of energy don't necessarily work best when decentralised, and they're no more inherently labour-intensive than other sources of energy. Just like cars, windmills can be built manually, or by robot. In use, too, renewables facilities should demand as little attention as possible.

Environmentalism prefers renewables to be labour-intensive because it would rather make people toil than see automation use up more natural resources. Because renewables are an emerging technology, they provide environmentalism with a blank canvas upon which to project its fantasies. But for us, progress is measured by the saving of human labour, by convenience; and as it happens, astronomical flows of energy can be *less* labour intensive than other energy technologies – provided they are organised at scale.

The World Commission was right about the price of renewable energy fluctuating less than that of fossil fuels. This is because most of the investment renewables require is made upfront. But this outlay is itself subject to price fluctuations. In 2008 Shell passed off its withdrawal from the 1 GW London Array wind farm by blaming rising materials prices. [16] Just between 2005 and 2008, those prices may have risen 48 per cent for offshore turbines, and 74 per cent for onshore ones. [17]

The World Commission's final plaudit for renewables – that they move nations toward self-reliance – has emerged as one of the most significant.

While grids are seen as dangerous, renewables are backed as good for energy security

Many environmentalists want households to do enough microgeneration of renewable energy to *go off grid*. They believe that such a tactic represents a challenge to the status quo. In fact, it's not just them who entertain foolish hopes of energy independence; the government, the military and businesses all do the same.

Chapter 5 discussed how governments now mistrust long supply chains for gas and oil, believing them a threat to geopolitical equilibrium. The same localist apprehensions about energy security do much to explain both official nervousness about grids, and official enthusiasm for renewables.

In Chapter 1 we met Amory Lovins. Back in 1976, Lovins put down a marker against grids – they were, he said, 'a likely target for dissidents'. This is what he wrote:

'The scale and complexity of centralised grids not only make them politically inaccessible to the poor and weak, but also increase the likelihood and size of malfunctions, mistakes and deliberate disruptions. A small fault or a few discontented people become able to turn off a country. Even a single rifleman can probably black out a typical city instantaneously.' [18]

In 1982 Lovins followed up with a book, *Brittle Power: Energy Strategy for National Security*, in which he pioneered the idea of the *vulnerability* of the US energy system – grids very much included. [19]

After 9/11, Lovins republished the book, and many others came to see grids both as targets for terrorists, and as a means of spreading disaster. Meanwhile, the RAND Corporation warned about *Networks and Netwars*. [20] Then, reflecting on the North American blackout of 2003, one physicist explained:

'The power grid could be a metaphor for our modern scientific world…[it] has become so complex that no one fully understands it.… Complexity leaves us vulnerable to natural disasters and simple human blunders, as well as low-tech terrorist attacks.' [21]

Not content with feeling prone to victimhood through the foreign manipulation of gas and oil, establishment experts join with environmentalists in fearing that wide-area grids within a nation such as the US merely increase risks.

In 2005, when thinking about terrorism, the European Commission proposed the need for EU-wide Critical Infrastructure (CI) protection, identifying energy and transport as key areas. The Commission noted that damage to infrastructure in one country could have consequences elsewhere, claiming that such an untoward turn of events

'… is becoming increasingly likely as new technologies (eg

the Internet) and market liberalisation (eg in electricity and gas supply) mean that much infrastructure is part of a larger network. In such a situation protection measures are only as strong as their weakest link.' [22]

In the US, too, military experts like to magnify the 'knock-on' problems around infrastructure, using this argument as a device to talk up renewables. At the Naval Postgraduate School in California, Ted G Lewis enthuses over wind and solar power at the level of every home and office, and indicts grids. He argues by analogy:

> 'Just as the personal computer has reduced dependence on large, centralised mainframe computers, distributed generation may be able to decrease and perhaps even eliminate dependence on the unreliable and vulnerable middle of the grid.' [23]

The convergence, around energy security, between environmentalism and the US defence establishment was apparent even before Obama's election consecrated the marriage. Robert James Woolsey Jr is one example. Director of Jimmy Carter's Central Intelligence Agency between 1993 and 1995, Woolsey became an enthusiast for any source of energy other than oil, favouring ethanol as alternative as early as 1999. Today he has photovoltaic panels on the roof of his home and drives a Prius with a bumper sticker saying 'Bin Laden Hates This Car'. [24]

In fact, localist dreams aren't yet likely to trump the need to maintain national and international supplies of energy. As Woolsey concedes elsewhere, his policy is not national energy independence for the US, 'because,' as he puts it, 'there's nothing wrong with our importing natural gas from Canada'. [25] In the same way, nobody at all rational argues that it's time to abandon grids.

Take Texan billionaire T Boone Pickens. An oilman all his life, and now in his 80s, Pickens nevertheless launched a nationwide advertising blitz on 8 July 2008, attacking America's addiction to overseas oil as a danger to national security. Pickens wants 22 per cent of America's electricity to be generated by wind, and the gas it displaces put to work in cars. He's completely against America continuing to entangle itself in international grids of oil pipelines and tankers.

It's all very patriotic. But Pickens is himself building the world's largest wind farm in the Texas Panhandle; so though he's against international grids, he's all in favour of the US electricity grid. Thus, to ensure profits on his venture, he wants it expanded and modernised. [26] And he's already had a result. Within two weeks of his advertising campaign, the Texas Public Utilities Commission decided to put $5bn into transmission linking wind farms in West Texas and the Texas Panhandle to the electric grid, and then to Austin, Dallas and Houston. [27]

No man is an island. The desire for insularity in personal energy supply is as silly as that for autarchy in the national sort. Talking up the local is a very backward argument for tapping flows of energy that are of astronomical proportions. It will not encourage the scale of investment that renewables demand.

Renewables as freedom: a bogus idea

Hermann Scheer is a Social Democrat member of Germany's Bundestag. He argues that renewables can provide people with 'autonomy,' and freedom from subordination to outside forces:

> 'The goal must be to make energy available in a way that is self-determined... energy must be free and independent of external constraints, free of opportunities for blackmail and outside intervention, used according to decision making criteria of one's own. In the long run, all these dimensions of energy autonomy are possible only if renewable energy is used.
>
> 'The counter-plan to energy autonomy would be to integrate renewable energy into the existing energy supply

system in order to contain it and keep it under control.

'Renewable energy facilitates an independent way of life that corresponds bet to human needs for individual and social self-determination and, thereby, to the "programme" of liberal democratic societies.' [28]

Here choice of technique in energy supply is made into a moral matter.

Scheer argues that what he calls a 'specialised energy business' – commercial energy *delivered from somewhere else* – is required least of all for heating and cooling buildings, and need only be mounted on a regional scale so as to provide electricity on top of that made through 'self-production'. [29] But in fact the reality of energy production and use is very different from what he says.

Taking advantage of the world's division of labour to deliver energy from somewhere beyond the home, village or town opens up *more* freedom for self-development, not less. Spending days collecting firewood might eliminate 'opportunities for blackmail' by energy firms, OPEC and Russia; but it would also eliminate the opportunity to spend time doing something more worthwhile.

Anyway, self-determination and dependence on others can be mutually reinforcing, not mutually exclusive. People are often at their strongest when they are linked to others.

Wind: a long gestation, followed by impetuous growth

Wind power has been used for centuries: first for sailing, then for grinding corn in windmills. To help form the Netherlands, wind power was set to pumping water. In the 20th century, wind has been used to power electrical turbines.

In 1923, the left-wing British biologist John Haldane could already see the potential of wind power. For Haldane, it was axiomatic that the exhaustion of the world's coal and oil fields was 'a matter of centuries only'. But Haldane took issue with those who assumed that this exhaustion would lead to the collapse of civilisation:

> 'Four hundred years hence the power question in England may be solved somewhat as follows: The country will be covered with rows of metallic windmills working electric motors which in their turn supply current at a very high voltage to great electric mains. At suitable distances, there will be great power stations where during windy weather the surplus power will be used for the electrolytic decomposition of water into oxygen and hydrogen. These gasses will be liquefied, and stored in vast vacuum jacketed reservoirs, probably sunk in the ground. ... In times of calm, the gasses will be recombined in explosion motors working dynamos which produce electrical energy once more, or more probably in oxidation cells.' [30]

Haldane was an all-too-rare visionary. He saw that, particularly on an island like Britain, wind turbines could play an enormous energy-generating role.

In the first half of the 20th century, windmills were still commonly used in rural areas; but as grid connections spread, they fell out of use. After the Second World War, cheap oil and then hopes in nuclear power pushed wind off the agenda. With the oil embargo of 1973, there was renewed interest; but soon wind was forgotten in the West's 'dash for gas'.

For 20 years development took place, but application was weak. Different materials for blades were tried. Steel proved too heavy, aluminium suffered from metal fatigue, and eventually carbon fibre turned out best. Different designs were also tested, such as turbines with vertical axes and, instead of the familiar three-bladed horizontal sort, turbines with one, two, four and many blades. But throughout the 1970s and 1980s, reasonable contributions by wind to power supply were confined to places such as Denmark and California.

The UK's first commercial wind farm opened in 1991 at Delabole in Cornwall, generating just 4 MW. [31] Over the past decade, however, worldwide wind power has expanded 10 times. Even 10 years ago, wind was little more than a curiosity in the global energy mix. But today it has become a dynamic, high-tech industry. That's no mean achievement – and it backs up the optimism we have about astronomicals.

In February 2007, the Braes O'Doune wind farm opened in Scotland, taking Britain's capacity above 2 GW. Putting in the first Gigawatt of wind energy capacity took the UK 14 years: putting in the second, just 20 months. [32]

If wind is to make a significant difference in future, the Gigawatts will have to come faster still. Right now in the UK, they're being added quite slowly – at the rate of about 1 GW every 27 months. [33] Worldwide, however, the accelerated pace of change with wind today is clearer.

World wind energy capacity, GW [34]

Year	Capacity	Rise in capacity
1997	7.475	
1998	9.663	2.187
1999	13.695	4.344
2000	18.039	6.280
2001	24.320	6.844
2002	31.164	6.844
2003	39.290	8.126
2004	47.693	8.403
2005	59.033	11.340
2006	74.153	15.120
2007	93.849	19.696

Next to Denmark, Germany has been the historic leader in wind power since the 1970s. In 2007, it had more than 22 GW, and the US had 15.5 GW. However, the US, Spain and China respectively added 5.2, 3.5 and 3.3 GW in 2007, while in Germany new wind installations, at 1.6 GW, fell. [35] Indeed, in

September 2008 it was reported that the US, taking advantage of the fact that it has much higher wind speeds than Germany, had for the first time surpassed Germany in the amount of wind energy it generated. Total installed capacity would also rise by 7.5 GW to more than 24 GW by the end of that year. [36]

Spinning more strongly – with more snags fixed

Turbines are now bigger and punchier; but the future will see even larger ones. In April 2008 the UK Crown Estate, which owns almost all the UK territorial seabed out to 22km, announced that it would buy the world's largest offshore turbine, a 7.5 MW, 163m high prototype from Clipper Windpower, California. [37] The turbine is being developed in Blyth, Northumberland, where Clipper will centre its European operations – supported by a grant of £5m from the One Northeast regional development agency. [38]

Installing turbines offshore certainly demands sophisticated engineering. But at present, the limit to bigger turbines comes from the need to transport blades to the site of installation – something that's much more easily done at sea than it is over land. In deliberating where to build new facilities for testing blades, the US Department of Energy's National Renewable Energy Laboratory noted the importance of access to waterways in its decision to build at Boston harbour in 2009 – as well as at Ingleside, Texas, which enjoys access to shipping routes through the Gulf of Mexico. [39]

The world's installed base of wind turbines is big enough for lessons to be learned from it. Detective work has tracked down the origin of mysterious lapses in performance of turbines after heavy winds: the fouling of blades by insects. [40] Improved blade design and materials have since been put to work on the problem.

Accumulation of ice also disrupts aerodynamics, as do collisions with particles carried in the wind. Here too, solutions such as heating blades and new blade coatings are being pursued – but much remains to be done. [41]

Beyond the aerodynamics of blades, there remains room

for progress in gearboxes and generators, and in controlling the yaw and pitch of the blades so as to keep them facing the wind and running at the right speed. Still, the power electronics that can cope with fluctuating demand and cleanly integrate wind into the grid has also been developed. With the first wind farms, the easiest solution was to build a switch that would cut off the grid connection at any sign of trouble. As wind has built up on a larger scale, the need to keep wind online has emerged. Fulfilling that need means building electronics that can smooth out peaks and troughs in energy – whether these originate in the turbines or in the grid. [42]

Scale, grids and factories are vital to wind turbines

With wind engineers gaining more and more experience, operations and maintenance costs are falling, and technologies are moving forward. The real question is whether, after the Crash of 2008, wind can continue to grow at a rapid pace. Consider the statistics on wind power capacity provided by the World Wind Energy Association:

Forecasts of world wind energy capacity, GW [43]

Year	Capacity	Capacity forecast for next year	Capacity forecast for 2010
2005	54	70	120
2006	74	90	160
2007	94	109	170

So far, paper forecasts for future expansion have risen year on year. But it's a moot point whether wind turbines can keep up with these forecasts in the real world. What principles should energised, clear-eyed enthusiasts for wind seek to have followed?

The total sum T Boone Pickens will invest in wind power will be $10bn. Over 81,000 hectares, his 2,700 turbines will have a capacity of 4 GW. 'Don't get the idea that I've turned green,' Pickens told the London *Guardian*: 'My business is making money, and I think this is going to make a lot of money'. [44]

In 2007 Texas overtook California as the leading US producer of wind energy. Planting windmills in its open spaces has created a flow of royalties to ranch owners, and has pushed up the price of land. [45] But California is not to be outdone: in March 2008, Southern Edison California announced the construction and upgrading of more than 400km of high voltage transmission lines. This will connect up Los Angeles and the California power transmission grid to 4.5 GW of wind power generated in the Tehachapi mountain area. [46]

Whether or not he eventually turns a profit, Pickens confirms in practice the significance of scale. He is right to make his initiative on wind a truly massive one – equivalent to a square 28km x 28km, with 100 turbines in a line for each kilometre travelled along on a side of that square. That kind of investment is necessary if the major amounts of electricity that society needs are to be made available without too much expense.

Second, both Pickens and Southern Edison are right to highlight the importance of grids. Without grids, wind-based electricity has no chance of getting to the skyscrapers and industries that will need it. Indeed, Britain alone may need to invest about £10bn in pylons and cables by 2020. [47] One of the major reasons: the government says it wants 14 GW of onshore wind and 14 GW of offshore by that time. [48]

By the M74 motorway near Abington, south Lanarkshire, Scotland, Scottish and Southern Electricity will be adding to the transmission tasks of Britain's Grid. It will be building Europe's largest onshore wind facility – a £600m, 152-turbine farm generating 456 MW. Meanwhile, offshore wind farms will need entirely new grid connections.

There is, however, a third thing that wind needs, if it's to keep up its impetuous rate of growth. Wind needs advanced factories capable of manufacturing large numbers of turbines, complete with different transformers for different national grids.

Turbine manufacture is a high tech, aerospace-style affair.

Ditlev Engel, CEO of Denmark's Vestas, the world's biggest player, says of his company and of each turbine it makes: 'we are no stronger than the last delivered component out of the 8,000 components'. [49]

Some of the operational failures of large turbines have been put down not to design, but to human error in the lay-up and checking of composite blades. More automation should fix that. [50] The production of wind turbines, like the production of wind energy, is a matter of scale and technological advance. And just as wind energy depends on grids, so wind turbine production depends on long, coordinated supply chains.

Suzlon: India's wind energy giant

Wind turbines have become a matter of heavy engineering. As a result, Indian and Chinese manufacturers have become important to the future of wind power.

Suzlon Energy is an Indian company which not only manufactures wind turbines, but also buys the land for them, installs them and operates them. In 2007, it had a 10.5 per cent share of the world market for wind power.

The world's top wind turbine manufacturers, 2007 [51]

Company	World market share, per cent	Entry to wind business	Country of origin
Vesta	22.8	1978	Denmark
GE Wind	16.6	2002	US
Gamesa	15.4	1993	Spain
Enercon	14.0	1984	Germany
Suzlon	10.5	1995	India

Back in 1995, Tulsi Tanti, a textiles manufacturer based in Gujarat, western India, founded the firm in 1995 because of his dissatisfaction with his country's power shortages. Tanti entered the wind business just to supply to his own operations; but by 2007, aided by Indian tax breaks and other incentives for renewables, wind power had secured him a $10 billion fortune and the number 10 spot on the *Forbes* list of richest Indians. [52]

Suzlon gained its first contracts within India, but soon began to make a global impact. In 2001 it formed its first overseas subsidiary in the Netherlands. In 2002 the firm obtained its first order from the US, and in 2003 opened an office in China. In 2006 Suzlon took over Hansen Transmissions, Belgium – the world's second largest manufacturer of turbine gearboxes. The following year the firm outbid France's Areva to buy Germany's REpower. [53] By 2008 Suzlon was operating in 20 countries. [54]

Manufacturing in the East and acquiring established firms in the West are two strategies that have allowed Suzlon to apply new technologies and consolidate market share. The company has also brought the manufacture of blades and nose cones to Minnesota in the US. [55]

Greens against wind

The rapid growth of wind power has led to controversy. Those who instinctively mistrust the Green agenda feel uncomfortable with the spread of giant wind turbines across the countryside. The turbines appear a physical manifestation of Green politics, towering overhead. It's a testament to the power of the environmentalist agenda that even its opponents feel that their arguments against the wind turbine must be couched in terms of it wrecking the landscape and killing birds.

The beauty, or otherwise, of wind turbines, and of the new pylons that will connect them up, is a subjective question. Greens find them graceful and elegant; others object that they're ugly. This is a ridiculously narrow way to evaluate a new energy technology.

As for birds, if wind really is the solution to human energy needs, should they really have a veto? The more powerful turbines are, the more slowly their blades turn, and so the more modest is their threat to birds. Wind only acquired its reputation for killing birds because of the poor record of weak, badly positioned Californian machines back in the 1980s and 1990s. The latest evidence, collected around two wind farms in the fens of East Anglia, is that farmland birds are not affected by wind installations – even if birds in upland and coastal locations are. [56]

In fact, objections to wind go deeper than aesthetics and ornithology. In April 2008 Scotland's government rejected plans for what would have been Europe's largest wind farm on the Isle of Lewis, the largest island in the Outer Hebrides. The formal reason given was that the land was protected under the European Community Birds Directive, and that birds could suffer. [57] However, Jonny Hughes, head of policy at the Scottish Wildlife Trust, had something to add:

> 'We welcome renewable energy, but the moor will be lost forever. We have looked after it, grazed it, dug it, walked it and known it for centuries. We are tied to the land. We are inseparable from the moor.' [58]

In feudal times, what Hughes calls being 'tied to the land' wasn't a choice. Today, crofters in the Outer Hebrides are not compelled to submit to a Lord; they *volunteer* to submit to nature. Since they actually live on the Isle of Lewis, perhaps it's right that they should decide what's done there. But they're hardly peasants being thrown off the land by the Terrible Tartan Turbines. After all, they've turned crofting into a lifestyle, and persist in speaking Gaelic.

Their rejection of wind is a microcosm of wider attitudes that exist all over Britain.

Ultimately the reason that Greens are hostile to big wind is that it holds the promise of progress, right now. When wind power is made available as a reality, environmentalist enthusiasm tends to fade, and all sorts of reasons are offered for why wind turbines – and indeed astronomical sources of energy generally – are a bad idea. That's the reason Hughes fell back on the excuse about people being tied to the land.

Hughes added that a 'staggering' 800,000 hectares of Europe's land was converted to artificial surfaces between 1990 and 2000. [59] But Europe is a big place: it covers more than a billion hectares. Even if Hughes' figure were taken at face value, growth of 'artificial surfaces' would take up, over the next century, less than 1.3 per cent of Europe's land.

Of course there is even more space available offshore. It's more expensive to build and maintain turbines out at sea than on land, and also more expensive to transmit electricity from sea to grid. Yet the sea doesn't just offer more space, but also stronger and steadier winds than those on land.

It's likely that wind turbines will eventually move so far offshore that, instead of being mounted on the sea bed, they will float, anchored to the seabed by tethers in a similar fashion to oil and gas platforms. Today's offshore turbines typically stand in water 20 metres deep or less. The development of floating systems will open up the continental shelf for turbines, out to distances of hundreds of kilometres. [60]

The idea of floating turbines was first suggested in the 1970s, but is only now receiving renewed attention. Paul D Sclavounos, Professor of Mechanical Engineering and Naval Architecture at MIT, has worked with colleagues to design hurricane-proof floating steel and concrete blocks that can be anchored as deep as 200 metres of water. [61]

The fact that turbines are big is bound to bring practical problems. Britain's Ministry of Defence has objected that offshore wind farms interfere with its radar. But this is really nothing more than a dispute over who pays for the necessary upgrading of the MoD's system. [62]

What large infrastructure project does *not* have a multitude of such issues to deal with?

Next generation wind: high altitude kites

Turbines have developed incrementally, becoming larger and moving out to sea to take advantage of stronger, more consistent winds. But the next generation of wind energy offers a more radical innovation. The really strong, consistent winds are available at high altitude, kilometres above sea level, where no turbine tower can hope to reach. Here speeds are three or more times faster than on land, and power outputs are 30 or more times higher.

Attached by tether to the ground, kites will one day harness that energy.

The force generated by a kite pulling on a tether can be used to turn an electric generator. A variety of systems are under development: Makani Power in California, Laddermill at Delft Technical University in the Netherlands, and KiteGen, in Milan, are already active.

The most promising configuration in the short term appears to be similar to a yo-yo. A 100 MW generator could have 50 kites attached to the top 5km of a 6.5km cable. The power is then generated as the kites heave the last 500m of the cable up. The clever part of the design is controlling the kites into a pattern that continually loses and then gains altitude, generating power with each gain. [63]

Solar: in the right places, a great long-term future

Solar energy doesn't yet match even the modest scale of wind. Yet the activity around it, the basic research already undertaken, and the sheer amount of energy available in sunlight suggest that *solar could one day outdo wind.*

Today, investment in solar isn't driven by commercial imperatives. Government subsidies and obligations on suppliers have created a market, especially in Germany and Japan (see panel). The challenge for solar is to break out of subsidies – in the first place, in those parts of the Earth that are near the equator and have clear skies.

As it becomes commercial, solar's applications in Arizona and Spain are likely to overtake those in Germany and Japan.

In familiar style, solar already has environment-minded detractors.

In May 2008, the US Bureau of Land Management (BLM) put on hold 130 applications for leases covering 4000 of the 500,000 square kilometres it manages in the Western US – much of them for patches of desert ideal for solar power. The BLM proposed a two-year assessment of solar's impact on vegetation and wildlife such as the desert tortoise and Mojave squirrel. It also worried about how the relatively tiny amounts of land round power stations could be reclaimed after those stations reached the end of their working lives. [64]

Just a few weeks later, however, the BLM bowed to pressure and lifted its moratorium... while still demanding a site-by-site assessment. As Rhone Resch, president of the Solar Energy Industries Association, bitterly noted, the BLM had 'yet to lease a single acre of land to the solar industry'. [65]

It's likely that, as solar scales up to make a real contribution, it will come under further attack.

Renewables and the proper role of the state

If it's ever to be cheap, the scooping up of astronomical flows of energy requires both state-supported basic research, and risk-taking private development. Yet modern capitalism seems to prefer to support renewables through the roundabout, ineffectual methods of state subsidy and regulation.

* In 1995, Japan provided a 50 per cent subsidy for domestic PV installations on roofs. The subsidy gradually fell to zero in 2005. The intention was that, by then, Japan's PV industry would be self-sufficient; and solar installations did drop from $11,500 per peak kW to $6,000. [66] However, installation numbers fell in 2006, and there's now talk of reintroducing the subsidy

- Under what are known as feed-in tariffs, electricity utilities in Germany have, since 1999, been compelled to pay a fixed price for renewables supplies from anyone who produces them. A similar regime applies in France and Spain

- The UK's Renewables Obligation (RO), made nearly five years after New Labour's election manifesto commitment to 'a new and strong drive to develop renewable sources of energy,' demands that utilities obtain 9.1 per cent of their electricity from renewables today, and 15.4 per cent by 2015/16. Along with exemption from a government Climate Change Levy, the RO should provide renewables suppliers with 'up to' £1bn a year by 2010. [67] That's less than one thousandth of the UK's GDP

- In the US, the federal Production Tax Credit, instituted under the 1992 Energy Policy Act, pays an income tax credit of two cents per kilowatt-hour. After the PTC expired in the past, US renewables production faltered in 2000, 2002, and 2004 – which is why the putative expiry of the PTC at the end of 2008 saw renewables suppliers mount a desperate defence of it. [68]

There are three outstanding features of state intervention in renewables:

1. How much the cash involved means to suppliers
2. How little the cash involved represents relative to the national economy
3. How much the Precautionary Principle has made governments avoid their proper responsibility to plan ahead and finance fundamental science and technology.

The thoughts of Angel Gurría, Secretary-General of the OECD, are interesting here. Gurría rightly calls for governments to increase their energy R&D, 'because pricing carbon will likely not be enough to bring about the R&D spending we need'. Yet to boost renewables, he has a whole list of other proposals: investment incentives, tax measures, preferential tariffs, quantitative targets or obligations, and tradable certificates for meeting such targets. And Gurría insists that governments should use these policy tools 'carefully', and avoid picking technology 'winners'. [69]

Perhaps the government should indeed not pick technology winners. But in retrospect, Concorde looks like it was a better horse to pick than those financial institutions rescued by the US Federal Reserve Bank in 2008. The Fed picked proven financial losers, not likely high-tech winners.

Today Western policy on energy supply is completely dominated by the idea that governments should never do very much in technology. Even Obama's enlarged, post-election programme of Green investments is more about creating jobs and avoiding imports of oil than it is about a careful selection of renewables on their own merits.

So long as the mantra of the state not picking winners is allowed to go unchallenged, it ought to be clear that that's only an irresponsible get-out from putting *any* kind of government money anywhere near renewables research.

The minimum size of a solar unit is considerably smaller than that of a wind turbine. Solar technologies are much more modular than other renewables, so the concept of local generation can *sometimes* make sense. The evidence from Beltsville, Madison in the US, however, is that solar installations are set to get bigger.

From that town, North America's largest solar energy services provider, SunEdison, has built a fast-growing business in the supply, installation, operation, finance and maintenance of solar power. The scale on which the firm works is far larger than a domestic dwelling. SunEdison will build, own and operate a 1.2 MW electricity generating plant near Wilmington, North Carolina. It will rent four hectares of land from, and sell the energy to, Progress Energy, a utility. [70] The company is also trying to integrate no few than 3300 utility industry standards into its work. [71]

Beyond utilities, SunEdison's typical installations supply several hundred kilowatts to 'big box' stores or car parks. However, General Motors will put a 1.2 MW Sun installation – 8,700 panels over nearly three hectares – on the roof of its transmission assembly plant in White Marsh, Baltimore, Maryland. [72]

GM's roof-mounted solar panels are just one example of a wider trend. Anxious to show off their Green credentials, many organisations have tried to rally staff around a programme of *energy conservation in buildings*. The GM example suggests that facilities management around large buildings, at least, might do better to install renewables-based generators of 1 MW or higher (see panel).

Renewables in the workplace

In this book we have concentrated more on energy issues around households, consumers and personal transport than on energy issues around employers and employees. That's for a reason: facilities managers, employers and the state have tended to approach workplace energy in a way that turns it into a matter of employee consumption and responsibility. In other words, the agenda at work is made into something very akin to the agenda in the home.

Interest in the workplace *generating* energy is much lower. In commercial and public buildings, the whole discussion on energy is about conservation, efficiency, regulation, tracking and metering energy use and GHGs, and changing employee behaviour in line with all of this. Insofar as workplaces are thought about as sites for energy generation, this is done in terms of the need for a stand-by supply should the grid for any reason fail.

In the UK in 2008, as part of the implementation of the EU's Energy Performance of Buildings Directive of 2003, both dwellings and workplaces for the first time were required to have Energy Performance Certificates (EPCs) accompany their construction, sale or rent. All public buildings larger than 1000 square metres were similarly required to have Display Energy Certificates, covering their operational use of energy, on public view. As the government puts it, 'Acting on an EPC is important to cut energy consumption, save money on bills and help to safeguard the environment'. [73]

In the future, space planners and facilities managers will use IT-based building management systems to try to coordinate the human utilisation of buildings in time and space with the energy taken by workstations, meeting rooms and lighting. The tendency will be to trace how individuals, teams or departments use energy – and it's already mooted that our old friend, carbon footprints (see Chapter 2), may be traded accordingly.

In this petty, penny-pinching conception, the highest goal is for workstation users to be able to check their personal energy profiles on control panels that are nicely designed into their furniture.

It is all very well – but a more constructive alternative is already available. Although there is every case for large-scale renewables projects to be done in special locations by specialised companies, workplaces of a certain size and scale may also have a role to play in working with and supplementing the national grid with renewable energy that they generate on site.

In an open-handed spirit, *Energise!* is prepared to entertain any workplace-based generation of energy above a very modest 1 MW. Actually 10 MW is more sensible, and 100 MW would represent a real contribution; but we will settle for 1 MW.

Here the news could have been good. In Britain in October 2007, the telecommunications firm BT offered electricity generating companies the chance to build 250 MW of wind farms on its many industrial sites, and sell the electricity back to BT. The aim was, by 2016, to generate electricity worth up to 25 per cent of BT's existing UK requirements.

At the time, Ernst and Young, which advised BT, was convinced that UK energy companies would be very interested. For investment in renewables, an E&Y spokesman enthused, 'There is a wall of capital out there'. But since the announcement, there has been no news from BT. [74]

Whatever the final fate of this promising project, the sad thing about it was brought out by Hanif Lalani, BT's group finance director. BT decided to shoot as high as 250 MW not because of the intrinsic merits of harnessing astronomical flows of energy, but because it was coming under pressure about – you guessed – its carbon footprint. 'Our customers are very clear', said Hanif Lalani. 'They want us to develop products and services using clean energy'. [75]

So far, renewable energy for workplaces has shown two things. First, the preference for conservation and for cutting costs is larger than that for innovation in renewables supply. Second, around GHGs there are already in place strong elements of duress that force organisations such as BT to make strenuous adaptations to those 'customers' – in the first place, the state – who behave with, and demand of others, the due amount of Precautionary Principle. [75]

Within certain limits, it can make sense for large organisations to learn how to generate their own energy. But renewables in the workplace can never substitute for grid-delivered electricity, nor should they try to. Their contribution cannot be about assuring business continuity: even before the Crash of 2008, the mere survival of business dominated the management mind far too much. Neither should they be ineffectual baubles, trying to prove corporate social responsibility, strengthen the company brand and all that.

No, the contribution of renewables in the workplace is worthy in its own right: to expand energy supply.

Passive solar: architecture as passivity

With 'passive' solar energy, humans use the sun's rays directly, without converting them into another form of energy. Passive is the least promising solar technology: it's about energy efficiency and conservation more than energy supply.

The case of pure passive solar occurs when buildings are made with windows that face south. That makes them warm up in winter. Usually, however, such buildings are also designed so that the natural airflow around them – solar power at one remove – provides good ventilation and cooling. These ideas have been developed furthest in Darmstadt, Germany, and have led to the PassivHaus standard.

As architecture, passive solar energy has acquired the status of orthodoxy. Uruguayan architect Rafael Vinoly's proposed redevelopment of the Battersea Power Station site in south London, which includes a giant chimney to obviate the need for air conditioning, is based on passive principles.

Those principles are fine in themselves. In the hands of Green architects, however, they all too often impose unnecessary and clumsy design constraints. The shape and orientation of a building, the locations of its windows, the thickness of its walls, the construction materials in its roof – these should not be dictated by the need to save energy.

The more extreme variants of passive design are anyway far from passive in what they demand of a building's occupants. The UK government's endlessly complex, endlessly updated Code for Sustainable Homes, first published in 2006, wants new buildings airtight, so that heat cannot escape from them. [76] But what if people feel like opening a window? And what about the extra expense? [77]

The amounts of energy involved with passive solar are small, and hard to bring under control. With passive solar, you get no thermostat.

Heated water is fine, but Concentrated Solar Power is far superior

Above passive solar stands solar water heating. While high temperatures are needed to generate electricity, warming water and buildings is relatively easy. Even in temperate zones, solar water heating, a simple, low-tech solution, has been a success. It will continue to be deployed, especially for applications such as swimming pools.

Greens tend to get over-excited about the spread of solar water heating – particularly to China. It's true that in Rizhao, a city of three million in northern China, a third of the population uses solar water heating. [78] But Rizhao is poor; with greater wealth, the popularity of solar water heating will likely decline there.

Solar water heating will remain a niche affair. But concentrated solar power, or CSP, is different. Because it generates electricity, CSP has the long-term potential to be a universal application, even if only about 500 MW is installed worldwide today. [79]

The main kind of CSP is Solar Thermal Energy Generation (STEG), in which heat is used to drive turbines. Steam-powered turbines are key to electricity generation, so the obvious way to turn sunlight into electricity is to use it to heat water into steam. The snag is that even in the tropics, the midday sun doesn't generate much steam.

Everyone, however, has seen a convex lens concentrate sunlight to set paper alight at Fahrenheit 451°. The large-scale concentration of solar energy is a little different from this. It's most practical and cheapest if done with giant concave mirrors. Unlike lenses, mirrors need not grow thicker as they get larger. Engineering a support underneath a large mirror is easier than holding up a giant lens around its edges.

With mirrors as with large reflecting telescopes, mountings can track the sun as it moves across the sky, all the while focusing energy where heat is needed. There are three main ways of configuring the mirrors in CSP:

1. **Solar troughs** look like pipes cut open lengthways. A long straight tube runs along the focus of parabolic mirrors; down it runs heat-absorbing oil. By the time the oil hits the far end of the tube after exposure to the sun's rays, it can reach temperatures of 400° C
2. **Solar dishes** have the same geometry as satellite dishes. They follow the sun in two dimensions across the sky. A generator placed at the focus of a solar dish can make use of temperatures of up to 750° C
3. **Power towers** are large arrays of mirrors focusing energy on to a single central point set on a high tower. Europe's first opened in Spain, near Seville, in 2007. There, more than 600 mirrors focus on a tower 115m high, generating 11 MW. [80]

As with wind energy, CSP enjoyed a burst of investment in California in the early 1980s, followed by a lull. Nevertheless, between 1984 and 1990 the Israeli-American company Luz built a network of nine parabolic trough power stations north of Los Angeles, with a total generating capacity of 354 MW. [81]

Today there is renewed interest in CSP. Torresol Energy, a 60:40 joint venture between Spanish engineering group Sener and Masdar, Abu Dhabi's state-backed renewables fund, will build two 50 MW parabolic solar trough plants in Seville and Cadiz, Spain, as well as a 17 MW power tower near Seville, at a cost of €800m. The aim is to standardise technologies so as to globalise CSP. [82] Meanwhile Stirling Energy Systems, Phoenix, Arizona, has filed an application to build a 750 MW, 30,000-dish CSP plant across 2600 hectares of desert 160km east of San Diego, California. [83]

Tower of power: Solar Two, an experimental Concentrated Solar Power plant in the Mojave desert, California. Backed by Boeing, Bechtel and others, Solar Two generated a modest 10 MW of electricity during its life from 1996 to 1999. Heat storage was performed through the use of molten salt. New, bigger designs are now planned

The main limitation of CSP is that unlike plain photovoltaics (PV), it can only bring *direct* sunlight into focus. While PV will work even on an overcast day, clouds are bad news for CSP.

Even so, that leaves a large fraction of the world where CSP has high potential. There are opportunities in Mexico; parts of Chile, Argentina, Bolivia and Brazil; Southern Africa as far north as Democratic Republic of Congo and Tanzania; a swathe across all of North Africa, the Middle East, Pakistan, parts of India and Western China; and in Australia. As for Southern Europe, some hope to see 30 GW of STEG installed there by 2020. [84]

Desertec: CSP in the deserts of North Africa and the Middle East

Perhaps the most ambitious plan to develop CSP is Desertec, a scheme drawn up for the Sahara and the deserts of the Middle East by the Trans-Mediterranean Renewable Energy Corporation (TREC). In 2003 the Club of Rome, the Hamburg Climate Protection Foundation and the National Energy Research Center of Jordan founded TREC. In 2007 Prince Hassan bin Talal of Jordan outlined the Desertec concept to the European Parliament. The idea is to install CSP in North Africa and the Middle East to generate power for these regions, and, by 2050, to transmit no fewer than 100GW of electricity by high voltage direct current to southern states within the EU. Waste heat from CSP could be used for desalination, and wind turbines in Morocco and around the Red Sea would also contribute to the total output.

First proposed by French President Nicholas Sarkozy, the Union for the Mediterranean, a group of EU member states and countries with Mediterranean coastlines, could provide the framework for realising Desertec. At a technical level, the project looks promising: if solar energy is ever really to fulfill its promise, it will need to be implemented on the sort of scale envisioned by Desertec.

Interestingly, TREC has already tried to protect itself from accusations that Desertec will end up as a means for Europe to exploit Africa. Its argument, however, is weak: because solar energy is unlimited, its owners can't be exploited. [85]

Things are not so simple. European interests could well dominate not just financing, but also installation and operation. The terms of trade for the northward export of electricity will also be critical. But despite these potential difficulties, and despite Desertec's rather disparate backers, *Energise!* wishes it well.

Photovoltaic panels: the solar power with the greatest potential

With PV, human beings have arrived at the solar technology with the best long-term prospects. PV panels are devices that convert light directly into electricity. [86]

Demand for PV has been growing for 30 years. Between 1977 and 2007, it rose at a compound rate of 34 per cent per year. [87] That adds up to a big difference between visions of a solar future in the 1970s and real investment today. The industry today is thousands of times larger, and on the verge of contributing significant amounts to world energy.

It's easy to grow fast from a base of nothing. Growth in the 1980s amounted mostly to watches and calculators. But in 2004, through another feed-in tariff, German legislators guaranteed a minimum price for PV-generated electricity. [88] Then PV began an even bigger boom, growing by 41 per cent in 2006 and by 55 in 2007. [89] Historically, first-generation (1G) solar cells arose from two sectors: IT and satellites.

In 1940, at Bell Laboratories, Russell Ohl discovered the first PV silicon cell purely by chance while trying to build a detector for radio waves. [90] Bell Labs quickly saw the significance of Ohl's work, and it was through building on it that Walter Brattain, John Bardeen and William Shockley invented the semiconductor transistor in 1947. From the beginning, then, solar cells have developed in parallel with electronics.

The first real application of solar power came, typically enough for post-war American history, in 1958. The sun powered

Vanguard 1, the fourth artificial satellite to be launched in the Sputnik era. In satellites, cost was little problem, and only small quantities of cells were needed.

How do PV cells work? To make chips, silicon has to be 'doped' with small amounts of impurities. *P-type* impurities, such as aluminium or boron, carry extra positive charge. Impurities such as phosphorus or arsenic, which carry a negative charge, are called *n-type*. The switches in computer chips that can turn flows of electrons on and off are built from clever combinations of *p* and *n* type regions.

PV cells are also built from doped silicon. A solar cell, in fact, is essentially nothing more than a *p-n junction*. As a result, the PV industry has largely grown out of the semiconductor industry. The world's leading manufacturer of PV is Sharp, a company that developed its expertise in the process of making memory chips.

Second-generation PV

With 2G solar cells, things are different. Still based on p-n junction technology, they're built around solar power rather than computer power. 2G cells still use doped semiconductors; but instead of crystalline silicon as a base material, they depend on options more suited to solar power.

A first set of 2G devices comes under the general name of *thin-film* photovoltaics. A key advantage of thin films is that they use less silicon than 1G devices. Until 2004, PV could survive using discarded silicon from the computer industry. Since then, the boom in PV has created a major shortage of new silicon, which is refined by stripping sand of oxygen. [91]

One approach to thin films uses amorphous silicon. In the crystalline silicon that's used in microchips, all the atoms are arranged in a very regular pattern. In amorphous silicon, by contrast, atoms are more jumbled up, but the material can more easily be laid down in a thin film on glass.

The world's largest semiconductor company, Applied Materials, has leapt into the market, adapting machines that lay down the pixels on to liquid crystal display screens so that they deposit amorphous silicon instead. CEO Michael R Splinter has conceded that in 2007, the solar panel part of the company's business was 'pretty much nothing' In 2008, however, he thought it would be worth between seven to 10 per cent of Applied Materials' worldwide business. 'We think the market for solar panels is very big, around $2.5 to $3 billion in business for us by 2010', he said. [92]

In 2008 other leading semiconductor firms seemed to share Splinter's upbeat view. Intel spun out a silicon cell company, SpectraWatt, and made a $38m investment in a German thin film firm, Sulfurcell. [93] National Semiconductor announced new technology that, it said, maximises solar cell output even when part of the cell is shaded or dirty. [94] In 2008 it was also anticipated that a new wave of investment in silicon refineries would soon come on line in China and South Korea. [95] Meanwhile, on the roof of its plant in Zaragoza, Spain, GM will install nearly 10 times as many solar panels based on amorphous thin film technology as it will with conventional solar in Baltimore. NASDAQ-quoted Energy Conversion Devices, Michigan, US, which specialises in amorphous thin films, hopes that its subsidiary United Solar Ovonic will generate up to 12 MW from the Zaragoza site, which at nearly 19 hectares is claimed to be the largest rooftop solar venture in the world. [96]

PV without silicon

Amorphous silicon is only the beginning for thin film technology. Although it's cheaper than crystalline silicon, it's relatively inefficient at converting light into energy. In other thin film regimes, silicon is abandoned altogether. By using different base materials on the two sides of the p-n junction, the divide between the two can be made sharper ('hetero-junctions' rather than 'homo-junctions'). That allows a greater proportion of light falling on the cell to be converted into electricity.

First Solar, which launched on NASDAQ in 2006, has developed cells based on Cadmium Telluride (CdTe). In 2008 it had manufacturing capacity to produce, annually, cells capable of generating 495 MW. By the end of 2009 it plans to raise this figure to 1 GW. [97]

Global Solar is the leader in Copper Indium Gallium Selenide (CIGS) systems. For some years, its manufacturing capacity was modest; but in 2008, its US facilities quadrupled in size, and the firm opened an additional factory in Germany.

One of the most impressive 2G technologies is Nanosolar's system for manufacture of CIGS cells. Founded in Silicon Valley in 2002, the firm's factories have adapted the roll-to-roll printing technology of the newspaper industry. In 2008 Nanosolar raised $300m from investors, including EDF and AES Corporation, another major power firm. [98] Also in 2008, IBM announced a deal with Japanese semiconductor equipment manufacturer Tokyo Ohka Kogyo to develop and license thin-film solar cells able to convert 15 per cent of incident sunlight into electricity, compared with current CIGS efficiencies of 8-12 per cent. [99]

Another 2G technology is multi-junction cells. These overcome the problem that each material used in a solar cell only efficiently extracts energy from a fraction of the spectrum. Light that is toward the red end of the spectrum is poor at generating current, while some of the energy in colours more toward the blue end is also wasted. Multi-junction technologies get around these limitations by stacking different cells on top of each other so as to make efficient use of the whole spectrum.

The main drawback with multi-junction cells is their expense. That problem can be mitigated, however. In Concentrating Photovoltaics (CPV), tracking mirrors focus a very large area of sunlight not to make heat to drive a turbine, as in STEG, but rather on to a very small area of multi-junction cells.

In northwest Victoria, Australia, Solar Systems is constructing the world's largest photovoltaic power station using CPV. It will concentrate light by a factor of 500 on to multi-junction cells. Rated at 154 MW, it will cost A$420m. A$295m will be invested by Solar Systems, and A$125m will consist of state subsidy. [100]

As of August 2008, the US National Renewable Energy Laboratory (NREL) holds the record for the multi-junction cell that extracts most energy from sunlight. The thickness is just a fifteenth that of a conventional design. The cell derives its capacity to absorb light from mismatches in the alignment of atoms between layers. In the laboratory, under a light intensity equivalent to 326 suns, the device converted 40.8 per cent of the light energy into electricity. [101] The technology is currently being commercialised by Emcore, based in Albuquerque. [102]

CPVs need cooling. In May 2008 IBM demonstrated technology adapted from the cooling of computer chips for use with concentrated solar. In a test, IBM used liquid gallium-indium metal to cool chips exposed to a light intensity equivalent to 2300 suns. [103]

A third generation of solar cells is presently coming out of the laboratory. 3G PV is built on a number of technologies: dye-sensitised solar cells, quantum dots and polymer systems.

In 2008 only 1 GW of thin film solar capacity was produced. But in the summer before the autumn financial crisis, at least, Travis Bradford, a leading industry commentator, raised his forecast of production in 2012 to an amazing 9 GW for thin film photovoltaics alone. [104]

If the world has any luck between now and 2020, solar energy will be able to add scores of GW each year, making it into a substantial contributor to global energy.

Hydroelectric power: time Africa had its own

Among all the technologies that take advantage of astronomical flows of energy, hydroelectric power is the only one that already makes a substantial contribution to world energy supply. It accounts for 90 per cent of all power generation by renewable sources. [105]

Hydroelectricity has been in use as long as the electrical grid has existed. In 1896 Thomas Edison's General Electric collaborated with George Westinghouse using Nicola Tesla's Alternating Current (AC) technology to feed power from the

Niagara Falls into the growing North American grid.

Perhaps the most spectacular pre-1939 construction was the Hoover Dam. Built in the desert on the border of Arizona and Nevada, it was completed in 1935. Roosevelt's New Deal undoubtedly accelerated construction. The Hoover Dam generates 2GW of power, exporting mostly to Los Angeles and other Californian cities.

Hydroelectric dams can generate spectacular amounts of energy. On the border of Brazil and Paraguay, the Itaipu dam is the largest power station of any sort in the world. Negotiations between the two countries in the 1960s led to construction beginning in January 1975. By 1978 the course of the Paraná, the world's seventh largest river, was diverted. The Itaipu's reservoir began to fill in 1982, with power production beginning in 1984. By 1991 all 18 electricity generators were online, with a total capacity of 14 GW.

In 2006-7, two more generating units were installed, allowing operation at full capacity even during maintenance. Altogether, the Itaipu provides more than 90 per cent of Paraguay's electricity and 20 per cent of Brazil's.

Environmentalism tends to prefer small-scale hydroelectric to dams on the scale of Hoover or Itaipu. 'Small hydro', as it's called, can produce anything up to a few megawatts. It's often promoted as a means to rural electrification.

We think the opposite. In hydroelectricity, the more ambitious goal of connection to a centralised supply should be pursued.

Big dams generate a lot of electricity efficiently. They demand big grids. More importantly, the urbanisation of developing countries makes both big dams and big grids essential.

In the big picture, cases where small hydro is truly appropriate will remain niche applications.

Where there is a real case for new hydroelectric power is Africa, whose rivers and waterfalls contain a vast and mostly unexploited energy. It has been suggested that a new dam on the Congo River – Grand Inga, alongside the existing, neglected dams

Inga 1 and 2 – could provide 39 GW. The African Development Bank has allocated $15.7m to study the idea. [106] The South Africa Development Community (SADC), the New Partnership for African Development (NEPAD) and the World Energy Council (WEC) have listed Grand Inga as a priority project. [107] But since it could cost $80bn, its future remains uncertain.

In June 2008 the WEC assembled possible financers in London. It concluded that while the Congo should get a 4 GW dam, Inga 3, quite soon, for the 40 GW Grand Inga, 'all actions need to be put in place to start the development phase by at least 2015, in order to get the first kWh around 2025–2030.' [108] That seems to us to be a thoroughly complacent schedule.

There is potential for large hydro beyond Africa. In Tajikistan, one of the world's poorest countries, the proposed Rogun dam would be the world's highest, at 335 metres.

According to the IEA, hydropower could double in output by 2050, reaching no fewer than 1700 GW of capacity. And according to the WEC, if every region exploited its hydroelectric potential to the same extent as Europe, then Africa would see a ten-fold increase, Asia a three-fold increase, South America a doubling, and the United States a 10 per cent increase. [109]

Greens and the social impact of dams

The industrialisation of China and India has been accompanied by the building of large dams. But projects such as the Three Gorges complex in China and the Narmada dam in India have proven unpopular with Greens. Dams displace local people, who see few of their benefits, say Greens. They destroy wildlife, or make animals die young. To add to their sins, dams create GHGs: when reservoir basins are flooded, rotting vegetation gives off methane.

With Grand Inga in Africa, the WEC has tried to deal with Green objections, and repeats again and again that it is about 'promoting the sustainable supply and use of energy for the greatest benefit of all'. The WEC concedes that the advent of Inga

The Hoover Dam, 48 km southeast of
Las Vegas, generates electricity equivalent
to two conventional electricity plants. By
modern standards of hydroelectric power,
however, its output is quite modest

1 and 2 has not seen local people properly compensated for being moved out. Nor have they been properly resettled, employed by dam companies, or even provided with electricity. That's why the WEC promises, in the case of future dams at Inga, to make a comprehensive social and environmental development and management plan 'in a participatory manner with the affected communities and all the stakeholders'. [110]

These words are none too convincing. Yet nor do Green arguments persuade, either. From Berkeley, California, International Rivers, a $2m environmentalist organisation, adds to the WEC's worries about Grand Inga. International Rivers says that Three Gorges, put forward as a model for Grand Inga, has been merely one of several large dams marked by huge cost overruns and corruption. At Inga itself private interests from overseas would benefit from local land giveaways. Transmission costs have been underrated, and maintenance has been non-existent. [111]

For ourselves, we've no doubt that most of the grievances and imbalances reported are completely genuine. After all, Chinese officials themselves have had second thoughts about Three Gorges. But problems with foreign investors, with the displacement and compensation of local people, with them getting a share of the electricity produced, and with maintenance – these are problems of politics and wealth. They are not at all intrinsic to the energy performance of dams. These problems are a reason to fight for more equitable politics, not a reason to oppose dams.

In its mission statement, International Rivers says: 'We oppose destructive dams and the development model they advance.' [112] Yet though it may seem churlish to say so, dams do not advance any development model. That tends to be something that people do. Of course, planners of big projects can wreak havoc, and not just with dams. But like coal (Chapter 5), a dam is what we make it. Loss of wildlife and the generation of GHGs are small penalties compared with the benefits of a dam – if, that is, it is well managed. As for the costs of long-distance transmission, that's a point in favour of building big dams in Africa wherever they makes sense.

In its pursuit of African oil, China frequently offers to build big dams as a sweetener. For instance, it plans to construct and pay for a 2 GW, $1.5 billion dam in the Mambila Plateau, Nigeria, so that, in return, it can import Nigeria's oil. [113]

As China strikes more new oil deals in Africa, expect to see a new level of Western hostility to dams being built there.

From the tides and the waves: good for some places, but no solution for the world

There's plenty of energy in the motion of water flowing around the oceans. It's estimated that the motion of the tides alone dissipates about 2500 GW worldwide. [114] But as with other renewables, the difficulty is converting this enormous quantity into energy suitable for human consumption.

Energy from the tides can be captured in two ways:

1. In *tidal barrages*, or *tidal ranges*, dams or artificial lagoons form reservoirs that get filled to a very high head at full tide. Energy is generated by turbines as the water vertically builds up ('flood flow' generation) and drains down in release back to the sea (ebb flow generation), in a manner similar to hydroelectric power

2. In *tidal stream generation*, or *tidal currents*, a watery version of wind power, fast, horizontal flows of water in narrow straits drive turbines directly. Although water currents move more slowly than wind, water is much heavier, which in terms of the energy that could be generated, more than compensates for the slowness.

In contrast with wind flows, the movement of the tides is highly predictable for years ahead, although availability does follow a natural cycle. Tidal power also demands suitable locations – not least, long lengths of coastline.

The world's only existing tidal barrage is on the River Rance, in northern France, generating 250 MW of electricity. Larger projects are certainly possible. A barrage on the Severn, one of the world's best sites, could generate up to 8 GW or more of electricity.

Tidal stream generation is even more experimental than tidal barrages, but first estimates suggest that the UK could derive 2-7 GW from it. In 2008 the world's first seabed-mounted tidal turbines were installed by OpenHydro, Dublin, in the Orkney Islands. A 1.2 MW facility was also installed by Marine Current Turbines, Bristol, in Strangford Narrows, Northern Ireland. [115]

According to a 2007 review of UK literature on tidal power, much of the UK tidal stream resource is concentrated in the Pentland Firth and the Channel Islands, and is mostly 40m deep or more. Once tidal streams around Rathlin Island, Northern Ireland and Mull of Galloway are added to Pentland Firth and the Channel Islands, however, perhaps five per cent of UK electricity demand could be met. Tidal streams in the South West, and especially in the Bristol Channel, would cater for another 10-16 per cent. In the Eastern Irish Sea, both tidal stream and tidal barrages might be able to handle another six per cent. [116]

The geography of the UK makes it one of the best prospects for tidal power. For that reason, the potential of tidal elsewhere has not been so comprehensively assessed. It also means that for the rest of the world, tidal can only meet much lower proportions of electricity demand.

Like tidal power, wave power needs long lengths of coastline; unlike it, wave power is driven by winds, and lies at a very low level of technological development. The first modern proposal for use of wave energy came from Stephen Salter, an engineer at Edinburgh University in 1974, in the wake of the oil crisis. [117] Salter's 'nodding duck' design featured a string of large canisters, each weighing hundreds of tonnes, which pumped oil through a hydraulic system as they bobbed in the waves. The pressure could be used to turn an electric generator.

Funding was killed, some hinted, after the UK nuclear industry lobbied against his proposal behind closed doors. [118] Perhaps this is true; but low gas prices in the 1980s and 1990s may also have made development of the Salter duck unattractive.

Wave power may eventually make a useful contribution in places that have lots of coastline – for example, Scotland, or Hawaii. However, it seems doubtful that it will make a much of a contribution on a world scale.

At present the most advanced design from a commercial point of view is the Pelamis, which is more snake than duck. Four 30m lengths flex at the joints as waves move past them. As in Salter's duck, the flexing is used to generate pressure and then electricity. Three Pelamis snakes are in operation off the coast of Portugal, and generate a total of 2.25 MW. [119]

More deployments of Pelamis are planned worldwide, and the device could be deployed alongside offshore wind, making use of common grid connections. The Anaconda, a snake-like rubber tube that bulges with water when a passing wave squeezes it, may also have potential. [120] But in the round, there's just not enough usable energy in the waves for them to make big difference to world energy supply.

Geothermal: bubbling under

Geothermal is one of the least developed energy sources, but has great potential. Finally, today it's beginning to get the attention it deserves.

The Earth's internal heat comes mainly from the natural radioactivity of its rocks. Although the radioactivity only exists at a low level, it's hard for heat to escape from underground, so high temperatures can build up. Without this heating, the Earth would long ago have frozen solid.

Averaged across its surface, the flow of heat from the Earth's is a miniscule 0.06 Watts per square metre. [121] That would be too thinly spread to make geothermal a practical source of energy. However, in specific locations, the flow of heat is much higher than average. Because of this wrinkle, geothermal energy has been used on a small scale for millennia, in the form of hot springs.

Electricity was first produced from geothermal energy at Larderello, Italy, in 1904, where heat from the rocks had been widely used several decades in the boric acid industry. The success was repeated in Japan in 1919. Then, in 1921, John D Grant drilled a well at The Geysers holiday resort, California. He

An eruption at Castle Geyser, Yellowstone National Park, US. The force of geothermal energy can be enormous, even without the enhancement of natural sources through clever engineering

produced 35 kW of electricity to power the place. [122]

As with wind, geothermal energy was little more than a curiosity until after the Second World War. Even then, it went through several decades of slow development rather than dynamic growth.

It was 1960 before the first commercial geothermal electricity entered operation, built by Pacific Gas and Electricity at The Geysers, California, and generating 11 MW. The 1980s and 1990s saw the sporadic construction of geothermal plants around the world – from Hawaii to Iceland. Between 1990 and 2005, installed generating capacity rose from 5.8 to 9 GWe. In 2005, the leading countries in geothermal energy were the US, at 2.5 GWe, the Philippines, at 1.9 GWe, and Mexico, at 0.9 GWe. For its small number of inhabitants, Iceland put in a strong performance at 0.3 GW. [123]

A complex of power stations at The Geysers is now the largest geothermal producer of electricity in the world, generating 0.9 GWe. [124] But geothermal provides even more power – 15 GWt worldwide – in the form of heat. Here China, Iceland, Turkey and the US are the leaders.

In Turkey, fossil fuels are rare but tectonic shifts are frequent. Geothermal heating and (especially for tourists) hot springs have appeal: already geothermal sources for such direct use amount to more than 1 GWt, and latest estimates put Turkey's geothermal capacity as high as 3.7 GWt. [125]

Beyond natural springs and geysers

To go beyond natural springs and geysers, wells must be bored so that pressurised water can be pumped down in to hot rocks. Heated water returns through a second nearby well. The heat can be used directly for district heating or industrial processes. Alternatively, it can make steam to turn an electric generator.

A 2006 study by a Massachusetts Institute of Technology team led by chemical engineering professor Jefferson W Tester has done much to put geothermal in the spotlight. [126] The report drew attention to what are known as *hot dry rocks*.

401 ASTRONOMICAL FLOWS OF ENERGY

Even where flows of it are low, heat can accumulate in rocks over millennia. In this case, heat can be extracted without waiting for the flow of geothermal energy to reheat those rocks from below. Power here comes from a stock, not a flow of heat.

To get at that heat, the MIT team enthused about enhanced or engineered geothermal systems (EGS), in which hydraulic pressure is used to create cracks in hot rocks, so that water to be heated can flow around in great volumes. The MIT study estimated that the energy that could be practically mined in this way in the US amounted to about 2000 times the country's annual consumption of primary energy in 2005. With technology improvements, the economically extractable amount of useful energy could multiply by a factor of 10 or more, thus making EGS possible for centuries. It's the prospects of enhancement that supported the MIT report's optimistic conclusions. By allowing access to larger and deeper volumes of rock, enhancement will free geothermal from the need to rely on those few natural formations in which heat comes near the surface.

Though geothermal is insignificant on a world scale, useful lessons have been learned from it. Hot fluids from geothermal wells, for instance, contain a cocktail of dissolved minerals: handling them well requires sophisticated engineering. But decades of experience have contributed to a gradual build-up of novel technologies. In 2000, Toshifumi Sugama, a chemist at the US Department of Energy's Brookhaven National Laboratory, won an award for a new kind of cement for well walls and a new plastic coating that protects steel heat exchangers from corrosion. [127] In California, technology now exists to extract dissolved minerals so as to stop pipes scaling up, and to gather useful items such as zinc and high purity silica. [128]

Geothermal can also benefit from the oil sector's techniques of geological surveying, drilling wells and pumping fluids – not to mention the sector's legal and logistical capacity to handle underground resources.

On the other side, as a relatively unexplored field, geothermal will benefit from new ideas that neither come from oil drilling nor can be applied to it. Potter Drilling aims to break up rock with spallation, which for many years has used drills and a supersonic flame jet in an air-filled hole to break up and remove crystalline rock – and to mine iron ore. In 2009, Potter aims to field-test a prototype of spallation based not on hot air, but on hot water. That should make for faster drilling, more stable drill bits and boreholes, and the ability to reach depths of 9km – 'ultimately allowing geothermal energy to be developed,' Potter says hopefully, 'anywhere on the planet'. [129]

Potter is backed by $4m from Google. The company claims that its rock-piercing capabilities could also come in handy in the disposal of nuclear waste and the sequestration of CO_2.

With the African Rift Geothermal Facility (ARGeo), the World Bank, Germany's Kreditanstalt für Wiederaufbau (KfW), the government of Italy, UNEP and others hope to bring geothermal power to the East African Rift Valley, which runs from the Red Sea to Mozambique. Kenya, Uganda, Tanzania, Djibouti, Ethiopia and Eritrea are interested. In Australia, too, there are hopes that geothermal will be able to provide 2.2 GW by 2020. [130]

How to overcome intermittency (1): energy storage

The most important sources energy – wind and solar – suffer from the drawback that they're not on all the time. They're intermittent. If astronomicals are to become an effective source of energy, this drawback will have to be solved. Several approaches are available.

At present intermittency can be dealt with by building 'spinning reserve'. This means keeping fossil fuel stations – typically gas – running and ready to take up the slack when inputs of renewable energy are weak. Critics of renewables point out that if you have to build enough fossil-powered plants to cover possible failures, then there isn't much point in renewables in the first place.

There are two fundamental strategies for a better solution. Either the energy from astronomical sources can be stored, or it can be fed to large grid systems.

The idea behind energy storage is simple. When energy is plentiful, it can be used to build up supplies. Then, when winds slow or the sun stops shining, the stored energy can be used.

As Chapter 5 showed with transport vehicles, storing electricity isn't easy. Luckily, storage is easier at a power station than it is in a car.

Whether generated from fossil fuels or from renewables, the main way that electricity is stored today is by using hydroelectric dams. When excess power is generated, usually at night, dams are put into reverse. That's done either with pumps, or by turning the generation turbines the other way. In this manner water is pumped back up into the reservoir. Then, when power is needed in the day, there's plenty of reserve for generation.

Beyond pumped storage, a variety of new technologies for electricity storage are under development, including flywheels, batteries, compressed air and superconducting coils.

The most promising of these is compressed air storage. In this system, excess electricity is used to pump air into a store such as an underground cavern or disused gas field. When electricity is needed, the air is released and used to turn a turbine, sometimes in combination with a gas burning system.

While the concept has been around for several decades, the rise of wind energy has now won it serious attention, as with the formation of Energy Storage and Power of New Jersey in August 2008. [131]

For concentrated solar thermal, there is another option. Although electricity is hard to store, heat can be retained for longer periods. One technique uses giant tanks of salt that can be heated into a molten state during the day and retain sufficient heat to produce steam for electricity generation at night. [132] The solar tower planned for Seville also does that. [133] Another technique, under research for more than 20 years at the Australian National University and now being deployed by Wizard Power, uses a closed-loop thermochemical system for ammonia dissociation and synthesis. [134]

Beyond pumped storage and thermal storage, a variety

of new technologies are under development, including flywheels, batteries, compressed air and superconducting coils.

How to overcome intermittency (2): bigger grids, more devices

The second strategy for solving intermittency is to *build bigger grids, with more devices to capture astronomical flows of energy*. The wind may not always blow for a single power station; but when it doesn't, it will be blowing elsewhere. Over a wide area, then, grids even out supply. The wider the area connected up, the more predictable average output becomes, as fluctuations in different places tend to cancel each other out.

Bigger grids also make demand for energy more even and easy to forecast. Demand from a single house is relatively unpredictable: it all depends when you decide to plug in the hair drier or kettle. Demand from a million houses, by contrast, is much easier to plan for, even though not every home, office or factory switches power on or off at the same moment.

Astronomicals call forth a need for bigger electricity grids for another reason. Like fossil fuels, they're often not strong where they're needed. But unlike fossil fuels, they cannot be transported by pipeline, railway truck or ship. Instead, wind or solar power must be converted into electricity – electricity that grids move to locations where it's needed.

We have shown that Greens prefer small, local systems. But rather than leading down the path towards energy security and personal self-reliance, investing in renewables properly will mean more interconnection between bigger grids – grids that cross wider areas and indeed more national borders.

As turbines become larger, offshore locations are likely to become more significant for wind. For solar energy, cloud-free deserts are good news if they are as close to the equator as possible. A future European electricity supply might partially rely on wind energy from the North Sea and Atlantic, and, as we have seen, on solar power imported from deserts in North Africa and the Middle East. The whole apparatus would best be continental in scope, and would likely also include massive facilities for pumped storage in mountainous Norway and Switzerland.

In the US, solar energy will be generated in the barren deserts of Arizona and New Mexico. As well as offshore, wind will be generated in a central corridor running from Texas up through Nebraska and the Dakotas. But that will need a new grid to link these sparsely populated areas to the large conurbations in which most people live.

New grids will benefit from new technologies. High-voltage direct current (DC) lines, for example, offer two advantages. First, less energy is lost in transmission. Second, DC is less susceptible to blackouts.

Greens and transmission losses

In their dogmatic desire to identify renewables with decentralisation, Greens like to point out that decentralising electricity supply has the advantage that less money needs be spent building transmission lines – and that transmission losses will also be reduced, since electricity will no longer have to travel long distances.

Up to a point, that's true. If the cost of solar power can be reduced to near that of grid-generated electricity, then decentralised solar systems ought indeed to save a few percent of electricity by having to transmit only to local, rather than national users. In turn, that ought to give solar a competitive edge.

In 2007, however, Sun Edison founder and solar enthusiast Jigar Shah went too far in this argument. Shah attacked the building of a $1.8bn transmission line from West Virginia to Maryland as a 'subsidy' to the US coal industry. [135]

That was unfair. Realistically, decentralised generation will never mean widespread independence from the grid. Local sources of wind and solar energy will still need grid connections for when the right weather isn't forthcoming, and for when localities need power on top of what renewables can provide. To the extent that decentralised generation has a role to play, the grid will be upgraded to make such generation more workable.

If anything, accommodating renewables will mean costly new investments in updating grids. That's fine. But when advocates of renewables calculate their output prices, they should take into account that these prices will be inflated by necessary investments in a centralised grid.

Renewables as a great technological fix – if built at scale

The growth of renewables has been profoundly influenced by the rise of Green politics. If it weren't for the Greens, as we saw in Chapter 4, it's probable that nuclear energy would by now have become the world's main source of electricity. However with things as they are, renewables – especially wind – have become significant industries. They've made substantial technological progress, fostered their own generation of engineers, developed professional supply chains and standards, and integrated themselves into the future of grid planning.

But now that a real renewables industry exists, many environmentalists are uncomfortable. As renewables are readied to become successful technological fixes, they undermine the environmentalist demand that people change their lifestyles and their whole relationship with nature.

As for us, we're adamant that the exploitation of astronomical flows of energy is something that energised citizens should support. Exploitation will depend on more investment than that which has come so far from government and industry. It will also mean abandoning a Green vision: that of the small-scale, independent production of energy as a kind of personal lifestyle choice. Finally, it will mean adopting a high-tech, capital-intensive approach to the production of renewables equipment.

Both physical and economic considerations point toward the need for scale. The sheer quantity of energy demanded by modern society requires it. In addition, scale in renewables

- tends to reduce the fraction of energy lost in conversion, to friction, heat and the like
- has a disproportionately positive effect. Larger systems can handle faster flows of wind and water – and energy is delivered in proportion to the cube of that speed.

On top of this, the intermittency of renewables argues for connections to large-scale grids and against attempts at local independence.

From the physical point of view, efficiency in renewable energy almost always points toward a need for scale. But the logic of scale from an economic point of view is even stronger and more basic. Both in and beyond energy, building at scale takes advantage of the most important wealth-creating *process: human cooperation*.

One of the first to appreciate this point was the Scottish economist Adam Smith. Smith was studying a very different society from today's, one where mechanisation was only just beginning to emerge. Some of his specific examples are therefore dated. But the growth of the extent of the market, together with the humanistic climate of the Enlightenment, did allow Smith to understand the importance of cooperation. In his famous discussion of the division of labour in Chapter 2 of *The Wealth of Nations*, he wrote:

> 'By nature a philosopher is not in genius and disposition half so different from a street porter, as a mastiff is from a greyhound, or a greyhound from a spaniel, or this last from a shepherd's dog. Those different tribes of animals, however, though all of the same species, are of scarce any use to one another. The strength of the mastiff is not in the least supported either by the swiftness of the greyhound, or by the sagacity of the spaniel, or by the docility of the shepherd's dog. The effects of those different geniuses and talents, for want of the power or disposition to barter and exchange, cannot be brought

into a common stock, and do not in the least contribute to the better accommodation and conveniency of the species.' [136]

Some might object that Smith sees human cooperation only in 'barter and exchange.' That's true, but doesn't negate his basic point. In highlighting cooperation, a uniquely human capacity, Smith undoes contemporary prejudice, which sees people as not much better – if not a whole lot worse – than jumped-up dogs.

For Smith, ships and barges expanded the market and thereby the division of labour. In energy, then, today's ships and barges are grids, and in particular electricity grids.

Grids will be essential, but so will the social forms of co-operation. Smith thought that the development of *markets* was a natural product of technology such as shipping and roads that brought people together. Today, a subtler picture is drawn: markets and other institutions of co-operation are thought to be more variable, and to develop historically. [137] Nevertheless, in all cases it's cooperation, and through that *centralisation*, that's allowed productivity to develop.

Without a division of labour and cooperation, people would be scrabbling around all day trying to find the energy they needed to pursue their lives. At the same time the two factors create an opportunity and motivation for professional energy engineers to concentrate all their efforts on the improved production of more energy. Not only do they have the time and resources necessary for innovation centralised in their hands, but also, with production on a large scale, quite small improvements will be worthwhile, and large improvements are unlikely to outstrip national and international demand in the long term.

Centralisation, like technology generally, is not inherently anti-democratic. In 1976, Lovins wrote:

'In an electrical world, your lifeline comes not from an understandable neighbourhood technology run by people you know who are at your own social level, but rather from an alien, remote, and perhaps humiliatingly uncontrollable technology run by a faraway, bureaucratised, technical elite who have probably never heard of you. Decisions about who shall have how much energy at what price also be-come centralised – a politically dangerous trend because it divides those who use energy from those who supply and regulate it.' [138]

In this petit-bourgeois vision of neighbourhood energy as lifeline from 'people you know,' Lovins mistook the defects of capitalist democracies as defects of centralised energy supply. Of course many people find themselves thwarted by faraway bureaucrats. But that's a problem that grows out of politics, not out of energy technologies.

Some of centralisation's more elitist opponents fear it will enforce what they call the tyranny of the majority. That's their problem. Still, there can be no doubt that

- the planning of major renewables installations often proceeds without proper discussion, yet is also slow and bureaucratic [139]
- attempts by mandarins in Whitehall to 'nudge' people into microgenerating tiny amounts of energy through renewables make little sense, and are offensive.

The answer to these difficulties, however, is to energise democracy, not to oppose centralisation. There's no magic formula or technical solution to the problem of balancing the needs of centralised energy generation against the conservation of wilderness, or the objections of minorities. The only chance of success is to have an informed citizenry energetically engaged in democratic debate.

If that seems like wishful thinking, that's because an alliance of technocrats and Greens has presented energy as a set of incontrovertible choices imposed by science and nature.

But energy isn't like that.

Summing up this chapter

When thought through carefully and implemented on an ambitious scale, astronomical sources of energy will show just how clever 21st century human beings can be.

Naturally, politicians and environmentalists make a lot of noise about the need to make major investments – especially in wind. In the summer of 2008, Thomas Friedman (whom we chastised in Chapter 2) caught presidential candidate John McCain running TV commercials that featured turbines, whilst McCain failed to show up at no fewer than eight Senate votes on whether or not to continue with tax credits for investment in technologies such as wind. [140]

So long as the world's need for much more energy is downplayed, and the need for energy conservation, energy security and off-grid contrivances is played up, the real potential offered by astronomicals will not be realised. A particular joke in the 21st century is for fans of astronomical flows of energy to plead a nationalist case for it. Why? Because the renewables sector is *already* a globalised industry:

- Two Chinese wind turbine manufacturers, Goldwind and Sinovel, add a further 4.2 and 3.4 per cent world market share to Suzlon's 10.5 [141]
- In 2007, no fewer than four Chinese solar firms listed on the New York Stock Exchange [142]
- Abu Dhabi's interests extend beyond Spain. Masdar will operate a major thin film photovoltaics factory in Erfurt, Germany. Masdar's owner, Mubadala Development Company, has formed a partnership with General Electric around renewables research and finance in the UAE. [143]

Renewables cannot provide energy autarchy: they already amount to an international business. A worldwide and reasonably integrated division of labour exists with them, too.

Until the economy crashed in 2008, wind power looked

like a stable sector. It had few venture capital or private equity players, plenty of asset financing, and strong mergers and acquisitions, with deals worth $7.2bn between September 2006 and September 2007. [144] One year later, though, and America's First Wind could no longer rely on more than $200m of Lehman Brothers' financing. [145]

Lehman had also taken 10 and seven per cent stakes in Clipper Windpower and in Ormat Technologies, a specialist in geothermal energy. Certainly, renewables have had to rely on sources of finance that now look questionable. Yet until as late as the summer of 2008, hundreds of millions of dollars were still pouring into thin film solar energy: not just to Nanosolar, which we mentioned above, but also to Colorado's AVA Solar and California's Miasolé. [146]

It is as fruitless to speculate about the future financial shape of renewables as it is to guess how much electricity generated from them will eventually cost. What is certain, however, is that the new, post-Crash enthusiasm for investment in renewables has a strongly nationalist aspect to it. After his repudiation, during his election campaign, of oil tyrants and dictators, Obama went on:

> 'Or will we control our own energy and our own destiny? Will America watch as the clean energy jobs and industries of the future flourish in countries like Spain, Japan, or Germany? Or will we create them here, in the greatest country on Earth, with the most talented, productive workers in the world?' [147]

All the Greens' high hopes in decentralisation can only lead to a flag-waving finale.

What is also certain is that renewables have to work with each other, and with other sources of energy. At Clipper Windpower, CEO James Dehlsen points out that most wind energy is generated at night – and that though household demand is low at that time, charging the batteries of tomorrow's more electric-orientated cars would be an excellent application. Meanwhile, one consultant critic of the Pickens Plan insists that, without large-scale storage of wind power, only gas can meet America's demand for air conditioning on hot, still afternoons. [148]

Taking the long view on renewables

Wind energy will see explosive growth over the next 20 years. By 2030 wind could be generating one fifth of the world's electricity. The trend will be towards larger turbines, further offshore. After 2030 growth will likely continue but share of generation will stay constant. Further offshore, turbines will be mounted on floating platforms rather than on the seabed. Kites may also be able to offer vastly improved performance.

Solar energy will undergo rapid development over the next 10 years, with development and deployment of many real-world systems and billions of dollars of investment. This will lay the basis for massive expansion in the period 2020-2040. Generation of electricity by photovoltaics and concentrating solar power has even greater potential than wind. By 2040 it could be generating one quarter of the world's electricity.

Hydroelectricity will maintain its share of electricity generation even as world demand grows. By 2050 it may still be providing a tenth of much larger world demand. Africa has the potential to construct the world's largest power station using hydroelectric technology.

Wave and **tidal power** will continue to undergo development. Without unforeseen breakthroughs, they will be significant on a regional scale, but niche technologies on a global scale.

Geothermal may have great potential in the long term. Over the next decade much basic development and exploration remains to be done. If this results in wide commercialisation over 2020-2040, geothermal could be one of the largest sources of electricity. While the development of goethermal is furthest off and hence least predictable, by 2050 it could provide one quarter of the world's electricity.

Grids will continue to become more international. In the short-term, high voltage DC transmission lines and computer control will be critical. Superconducting cables are an example of an innovation that is entirely plausible in principle but rather unpredictable in detail.

Energy **storage systems** including batteries, thermal storage and compressed air will be developed and deployed in parallel with wind and solar energy.

In the table below, we present the third and last series of our ratings for selected energy technologies.

The ratings in the table should not give the reader a false sense of precision. Might wind provide 35 per cent of electricity by 2050, or solar 15 per cent? Quite possibly.

In the table, we have entered question marks when a particular technology is not yet advanced enough to permit a sensible evaluation.

Rating renewables technologies

Technology	Period of take-off	Top share of global electricity, 2050	Advantages	Disadvantages
Wind	2010-2030	20 per cent	Relatively cheap. Already developing and can be rolled out quickly	Need for energy storage to beat intermittency
Concentrated Solar	2020-2040	5 per cent	Relatively cheap and easy to roll out	Thermal storage better developed, but still needs progress. More reliant on locations with clear skies
Photovoltaic Solar	2020-2040	20 per cent	Abundant. Can be decentralised and expanded in modular units where appropriate	Need for energy storage to overcome intermittency
Hydroelectric	Already mature	10 per cent	Dual use for water management. Can be used for energy storage	Limited by the need for suitable sites.
Tidal	Could begin expansion now	probably less than 5 per cent	Useful where location allows	Limited locations
Wave	Unclear if it will take off	probably less than 5 per cent	Potentially useful where location allows	Technology not yet demonstrated
Geothermal	2030-2050	20 per cent	Power available on demand	Technology not yet developed

Potential	Resources required	Key disciplines for research	Rating now / 10	Rating for the future / 10
Continuous growth is size from kW to several MW. Will likely growth in size to >5 MW per turbine. Will move further offshore. Installation in water deeper than 20m likely	Steel, copper and concrete	Mechanical engineering, aeronautical engineering	4	7
Designs only incrementally improved since the 1970s. May converge multi-junction high temperature with photovoltaic solar technology	Steel for structures and mirrors	Mechanical and civil engineering	3	6
1G: Past and present. Crystalline silicon. 2G: now until 2020 or beyond. Amorphous and polycrystalline silicon, thin film. 3G: dye sensitised cells, organic materials, quantum dots, multijunction cells	Silicon processing, supply chain from mining to refining for tellurium, indium and gallium, handling toxic cadmium	Condensed matter physics, chemical engineering	3	9
Continued possibility for larger and higher dams	Concrete and construction materials	Civil engineering	8	8
Not yet clear. Projects such as the Severn barrage need to be constructed to obtain a clearer picture	Concrete and construction materials	Marine and civil engineering	2	4
Economic generation technology not yet demonstrated	Steel and other construction materials	Marine engineering	1	Unclear
Key to future takeoff will be Enhanced Geothermal Systems that fracture hot dry rocks allowing fluid to flow through	Water for operation	Geology	1	7

Energising humanity, humanising the planet

Human activity has unintentionally had an enormous effect on climate. Now it's time to uphold what intentional action could do

7

Technology as alien and malign: between the wars, Karel Capek's dystopian play *Rossum's Universal Robots* had a worldwide influence. Television production of *RUR* for the British Broadcasting Corporation, 1938

Mary Shelley's novel *Frankenstein, or the modern Prometheus* (1818) suggested that human power over nature could lead to tragedy. Then, a century after Shelley's lone doctor foolishly experimented with human body parts and electricity, Karel Capek's play *Rossum's Universal Robots* – performed in London in 1923 – mixed biology with large-scale industrial processes, and had the resulting worker androids turn against their manufacturer.

Movies soon developed the same theme. Man became subordinate to machine in Fritz Lang's *Metropolis* (1927), Stanley Kubrick's *2001* (1968), and James Cameron's *The Terminator* (1984). Still later, even software engineers in the world of IT had become nervous. As the 21st century began, Sun Microsystems founder Bill Joy argued that genetics, nanotechnology and robotics could conspire together to rid the planet of mankind. [1]

Today, environmentalism fears that continued industrialisation could warm the planet enough to annihilate it. Yet the world should resist such dystopias.

Unwittingly, a growing but chaotic civilisation on Earth has changed the planet's climate. Now, with a more conscious approach, *people can gain still more civilisation by adopting a less chaotic energy regime.*

An economic and technological programme for universal energy supply would have a broad political effect, too. It would *energise humanity*, by organising enough cheap energy for people to lead richer and freer lives.

We must put energy in perspective, however. Just as questions of energy cannot be reduced to climate change, neither can the fate of the world be reduced to energy. Innovations in energy supply need to accompany innovations in other sectors, if the root causes of backwardness around the world – and of genuine environmental degradation – are to be tackled.

Alongside other investments, though, energy can help *humanise the planet* and make it a delight to live on. Human beings are multi-talented.

By integrating innovations with other innovations,

human beings can do more than just survive. They can make the environment a place where they can better realise their potential.

This chapter shows that:

* Energy firms and the state have abdicated responsibility for technological advance, preferring displacement activities in *finance* and elsewhere.
* This abdication isn't peculiar to the energy sector, but general to capitalism today. Environmentalist theory, however, represents it as a virtue.
* Adapting to climate change isn't as desirable as developing a *30-50 year gale of new-generation technologies*, in energy and elsewhere, aimed at *transforming* the planet in a human direction.

To meet the world's needs means to recover a sense of human capabilities in innovation, both in and beyond the energy sector.

It's vital that people get serious about basic scientific research conducted entirely for its own sake. In March 2008, Britain's Science and Technology Facilities Council announced that it could no longer finance Cheshire-based Jodrell Bank, one of the world's leading centres of radio astronomy. As *The Times* noted, the proposed saving of £2.5m a year was equivalent to the grants and subsidies paid out to the Prince of Wales in 2007. [2]

In 2008, the heir to the British throne gets more funding than research into fundamental aspects of the universe. There's no predictable or measurable 'pay-off' for basic physics, so it's left to languish. In 2004, the Engineering and Physical Sciences Research Council (EPSRC) showed just how much nuclear fusion is valued in Britain. In its largest ever grant allocation, the EPSRC lavished £48m on the UK Atomic Energy Authority (UKAEA) at its Culham site in Oxfordshire. Over four years, this amounted to just £12m a year. [3]

People need to know how contemptuous Western elites can be toward scientific research. David King, whom we have already met, used the opening of the CERN Large Hadron Collider experiment in Geneva in 2008 to suggest that it was 'all very well' to search for fundamental particles, but that 'we need to pull people towards perhaps the bigger challenges where the outcome for our civilisation is really crucial' – in other words, toward climate change. [4]

Baby you must drive my car: Prince Charles, deep Green heir to the monarchy in the UK and no fewer than 15 other sovereign states, including Canada, Jamaica, Australia, New Zealand and Papua New Guinea. In 2008 Charles let it be known that, as King Charles III, he would depart from the practice of previous British monarchs and go 'speaking out' on matters of importance. Can he hope to address the US Congress so as to raise its awareness of global warming?

This is very wrong. *To get a great deal more energy, the world needs to do a great deal more basic science research – including particle physics research. In the long term, research driven by curiosity lays the basis for radical innovations, in energy just as much as in any other sector.*

It's tragic that King can see no higher aspiration for science than to assure survival.

Research, investment and the Enron paradigm

In 1969, Robert Rathbun Wilson, the US physicist who built Fermilab, the world's highest-energy particle accelerator laboratory, addressed the Congressional Joint Committee on Atomic Energy. Rhode Island Senator John Pastore asked Wilson to spell out what research into high-energy particle physics would do to improve the defence of the United States.

Wilson gave a reply that went down in scientific history. Fermilab, he said, had 'nothing to do directly with defending our country, except to make it worth defending'. [5]

Civilisation needs to recreate the pride in human curiosity that Wilson evoked. Research into the secrets of the nucleus, like earthbound or space-bound technology that improves mankind's grasp of the cosmos, is worth defending in its own right – even if it brings no benefits to energy supply.

That said, we've no doubt that such research will, over the course of the 21st century, help the world put together some impressive advances in the generation and transmission of energy.

In Chapter 1 we highlighted the West's slothful record on innovation, especially in energy. As the IPCC's Working Group III puts it, atechnology and R&D response to the challenge of climate 'has not occurred'. [6]

In the US, weak investment by general business is the context for dismal investments in energy R&D, whether public or private. Between 1959 and 2007, the ratio of gross private investment to America's GNP hovered around 15 per cent. By way of comparison, in continental Europe in the 1950s, the ratio of gross fixed investment to GDP mostly exceeded 25 per cent. In China in 2004, the ratio of total fixed capital formation to GDP was an astonishing 41 per cent. [7] Worse, from 1999 to 2005, US outlays on non-residential equipment and software dropped from 55 to 45 per cent of gross private fixed investment, while those on housing rose from little more than 25 per cent to nearly 40 per cent. [8]

Enduring example: Enron's preference for finance over actually supplying energy remains an inspiration to energy firms the world over. Here, workers carry out boxes from the company's Houston HQ after it declared bankruptcy – less than three months after 9/11

US investment has tilted toward housing and, since 2005, toward commercial property. The main innovations have been in finance, not engineering or energy. In the US but also beyond, 'new product development' often has little to do with technological innovation, and everything to do with new forms of finance. And nowhere has the broad financialisation of the firm and of industry been clearer than in energy.

The practices of Enron – once the world's largest firm trading energy and America's seventh biggest corporation – have proved more exemplary than exceptional. Just as Enron started out mainly in gas pipelines and ended up mainly in financial instruments known as derivatives, so the 'business models' favoured by energy firms today tend to focus on streams of revenue, not streams of fuel

Derivatives are ways of betting on the price movements of a thing rather than buying the thing itself. [9] During the post-war boom, one could bet on the future price only of basic commodities such as wheat. But in 1972, the collapse of fixed rates of exchange for currencies led the Chicago Mercantile Exchange to offer international businesses the chance to speculate and especially insure themselves through bets on currency futures. Then, in 1974 and 1978 respectively, gold and energy futures began to be traded.

Today's market for derivatives, however, only emerged in the 1980s, when the discipline of *risk management* expanded out from corporate finance departments into every aspect of business practice. [10] It was in the 1980s that the desire to *hedge* against the future price of everything – currencies, interest rates, raw materials, the stock market – established burgeoning derivatives markets not just on exchanges, but also among private parties.

In the late 1990s, Enron began to offer derivatives related to energy prices and future weather conditions over the Web. In this way it was able to capitalise on investors' fears. Meanwhile, the relative weakness of US business investment, again born of risk aversion, meant that there was plenty of cash around for

Enron to borrow – at low rates of interest.

Enron went bankrupt on 2 December 2001. But risk aversion and plentiful cash persisted. As a result, other energy derivatives traders filled Enron's place, while traditional energy concerns such orientated their operations toward finance in a big way. Indeed, as Vijay Vaitheeswaran, energy correspondent for the *Economist* presciently put it in 2003, 'the heart and soul of Enron's strategy (minus the fraud of course) is alive and well in energy circles'. [11]

When Enron's former chief executive and chairman, the late Kenneth Lay, was convicted of fraud and conspiracy in May 2006, Representative Michael Oxley, co-author of the bill that became known as the Sarbanes-Oxley Act of 2002, was forthright. The end of Lay's trial, Oxley said, would mark 'the end of a dark era'. But in fact the darkness continues today, because even genuine suppliers of energy now have the same impulses as Enron.

Enron's final CEO, Jeffrey Skilling, pioneered the doctrine that its old physical assets were not worth having. All that was important was revenue streams. And this remains broadly the doctrine of energy companies. Just as governments have wanted to break up monopolies in energy and insisted that energy utilities should not own power stations, so utilities have been happy to sell their assets – or through Special Purpose Vehicles, get them off the balance sheet. The result is that many firms in today's energy sector pay little attention to energy *supply* or *innovation*.

Energy assets are bought and sold as financial instruments, not as instruments for raising output, increasing productivity, or lowering carbon emissions. As electric power industry veteran Jason Makansi makes clear in his book *Lights Out* (2007), about 40 per cent of power generation in the US is not financed, built or operated by utilities. The upshot is:

> 'Financial engineering takes precedence over physical engineering. With respect to the electricity system, the impact is this: investors are making more money buying

and selling assets than they are investing in those assets for the long term. When you're playing for the big money in this game, short-term return on increased transactions always trumps long-term investment in upgrading – or even maintaining – infrastructure.' [12]

As Makansi also notes of the US, it isn't just power plants that change hands: so do gas and electric utilities. Since the 1980s, the energy sector has helped deepen capitalism's general preference for corporate acquisitions over innovation. [13] Even in Europe, where national energy champions remain strong, acquisitions have multiplied. [14] Meanwhile investment banks such as Goldman Sachs exchange billion-dollar titles to future revenue streams from energy; and, through energy traders, corporate buyers of energy play energy markets so as to exact the maximum cost savings.

Is all this financialisation of energy just a product of its *deregulation*? No.

In 1986, Margaret Thatcher privatised British Gas. There followed a decade of UK energy privatisations. After that, the EU gradually let independent power companies compete with state-owned energy firms for large clients, then small ones, and – most recently – for householders.

Yet privatisation, liberalisation and increased competition were never equivalent to deregulation. Many Europeans think of America, and especially Texan oil, as a free-market brawl. But US capitalism, its banks and its energy sector have long been regulated. Indeed, after the Crash of 2008, the regulation of banks went into overdrive. On top of the Keynesian 'mixed economy' of the Cold War, in which taxes on private employers helped fund state-run industries and services, the private production of energy now occurs in a context of growing state regulation, legislation, intrusions and enforcement.

In that kind of context, genuine innovations in energy supply come second to other, dubious kinds of innovations, named *energy services*.

A dry run for energy services: turning complaints about energy into... a state service

In *The Culture of Complaint* (1993), the Australian critic Robert Hughes brilliantly attacked a divisive trend: approaching politics as a plaintiff or a supplicant victim. [15]

Consumer complaints about energy utilities were not among Hughes' chosen targets. Nevertheless, but he implicitly anticipated how, in 1998, complaining about utilities became *institutionalised*. In New South Wales, Australia, six electricity suppliers and one transmission company set up an Ombudsman to handle complaints. [16] In Ontario, Canada, the Ontario Energy Board became officially responsible for investigating all complaints about energy. [17]

In 1999, Britain's Office of Gas and Electricity Markets (OFGEM) was formed. Today, it declares the protection of consumers its 'first priority'. In 2000, too, the British government also established Energywatch, a watchdog whose mission was 'to get the best deal we can for energy consumers'. [18]

Posturing politicians like to present the consumer as vulnerable to rapacious energy utilities, always out to rip people off on prices and skimp on customer service. But the purpose of this right-on populism is only to help the state more easily insinuate its policies into people's heads and homes.

Once complaining about energy is given state sanction, there's no stopping it. Thus in 2005, Britain's House of Commons Public Accounts Committee, chaired by rightwing Conservative MP Edward Leigh, complained about Energywatch. Only two per cent of Britons, Leigh protested, had heard of Energywatch. It hadn't supported complaints vociferously enough. [19]

Today energy in the UK builds on this cloying approach to social cohesion. Energy is more about the state 'protecting' helpless householders than it is about leaps in the productivity of energy production. Every aspect of the household's consumption of energy is held to deserve inspection and improvement.

Energy services distract from energy supply – and help the state control lifestyles

With energy services, the supply of power and heat to a home or organisation is nothing compared to the multi-faceted *service relationships* that electricity and gas utilities now want with their customers. Forget kilowatts and kilojoules; think billing, call centres, web sites and IT systems for customer relationship management. Utilities now believe that their task is to increase customer loyalty, stop customers defecting to other suppliers ('churn'), flaunt Green credentials, and build trust – not just with consumers, but also with state regulators.

Here, energy utilities are very similar to retail banks and telecommunications providers. In Europe,

> '... a product that was once a basic commodity – electricity – is getting a marketing makeover and is being sold with dynamic pricing, special offers and tailor-made deals. Some companies are offering the electricity equivalent of weekend minutes, special offers for Saturday and Sunday when overall demand is lower.' [20]

This is what energy innovation has now been largely reduced to – discounts for use at weekends.

Of course, consumer lobby groups and governments insist that energy utilities adopt particular price regimes, and call on them to display thoroughly ethical conduct, transparency and all the rest. But for householders, problems – though by no means insuperable – often lie elsewhere. Often, genuine differences between energy providers are hard to discern, bills impenetrable, call centres lethargic, sales calls legion, junk mail enormous, and meter inspections a hassle. In the UK, even those websites that claim to clarify the confusion can end up adding to it. [21]

In energy services, the ghost of Enron lives on. It's possible to be in the virtual, financial energy business more than in real energy supply. Worse, energy services embrace something more sinister than frilly marketing and pricing: a state intent on reforming the behaviour of the citizen at home.

In 2003, the British government crowed that, rather than simply selling electricity and gas,

> 'Energy services focus on the outcome the customer wants – such as warm rooms and hot water – and offer the most cost-efficient way of achieving it. Under an energy services contract a supplier might, for example, install insulation or a more efficient boiler in a customer's home, and recoup the investment through the quarterly bill over, say, 3 to 5 years. The householder uses less energy as a result, and the savings on the energy bill are used to repay the cost of the measures. So, worthwhile home improvements are installed with no upfront cost to the householder, who benefits from a warmer, more comfortable home and lower energy bills for years to come once the initial investment has been repaid. Some have called this approach selling "negawatts" instead of "megawatts".'

In sum, 'Energy services could help to overcome consumers' reluctance to invest in energy efficiency improvements'. [22]

In fact, it was Amory Lovins who, at the end of the Cold War in 1989, announced the idea of *negawatts* – counting up reductions in energy use as if these amount to increases in energy supply. [23] But what Lovins' negawatts mean today is that, instead of innovations in energy supply, there's a labour-intensive government drive to:

* patch up Britain's ageing housing stock to make it more energy efficient
* treat the myopic, wasteful consumer with *authoritarian CO_2 therapy.*

The ostensible purpose is to save money and household CO_2 emissions – although as Chapter 2 mentioned, less than 15 per cent of Britain's CO_2 comes directly from houses. The real purpose is to overcome the British consumer's rather proper 'reluctance' to suffer disruptive home improvements for little financial or environmental benefit.

Why UK householders won't go Green

In the summer of 2008, the UK householder's bill for energy was set to top £1300 – high not just for the poor, but for middle-class homeowners too. [24] In the autumn of the same year, Gordon Brown announced a £1bn initiative to go about 'helping people make long-term savings through cutting the cost of energy through insulating, draught-proofing, getting better heating in their homes and taking all the conservation measures that are necessary'. [25] Indeed Brown aimed at the insulation of all Britain's homes, where practical, by 2020. [26]

The problem is that going Green in home energy means more than just adding insulation. [27]

Insulating walls and roofs does indeed cut fuel bills. [28] But done without controlling moisture vapour and without improving ventilation, the result is condensation and mould. It's impossible to properly renovate homes without disruption and expense, as 15 years of German research into a housing concept free of central heating – the Passivhaus – shows. [29]

Nor will insulation alone meet the 'zero carbon' requirement of Level 6 of the government's *Code for Sustainable Homes* – an extraordinary regulation demanded of all new UK homes in 2016, and of all new commercial buildings in 2019. [30] To maximise energy efficiency means redesigning complete building envelopes to incorporate heat recovery through mechanical ventilation systems. [31] In practice, that would require making houses with integral air-conditioning, or 'active houses'.

Insulation is one thing on which British state policy is unconvincing. The microgeneration of energy in the home –

through wind turbines or solar panels, for instance – is another.

To achieve significant CO_2 savings through microgeneration, officialdom has conceded, householders must export the electricity they generate domestically at prices equal to those offered on the National Grid. For that highly unlikely prospect, each household will need three separate meters to collect data on electricity imports, exports, and generation. [32] More recently, one government energy quango has discovered that between 10 to 80 per cent of new homes may not be able to meet the current definition of 'zero carbon'. [33]

People are right to suspect that insulation and microgeneration may turn out to be a fool's errand. No wonder the government finds them 'reluctant' to change their behaviour, even when it insists that efficiency measures are 'demonstrably cost effective'. [34]

The British state wants to change the population's behaviour. It wants energy utilities to become expert not in energy innovation, but in loft insulation. [35] Utilities in the UK, therefore, now focus on new kinds of billing, 'smart' meters to measure home energy usage, and on assisting households in the microgeneration of energy. Strategies like this, a White Paper argues, help suppliers develop 'alternative business models'. [36]

What, though, is meant by this phrase? After all, the respect *Fortune* magazine paid to such models allowed it to name just one firm as America's Most Innovative Company in each of the six years from 1996 to 2001.

That firm was Enron. And what *Fortune* held to be so innovative about Enron was its alternative business model: the practice of trying to make money not through energy supply, but through IT-based derivatives markets, in energy and other sectors.

How could *Fortune* make such an error of judgment?

Today's theoretical crisis around innovation

According to *Fortune*, Gary Hamel is the world's leading expert on business strategy. He's certainly charismatic – and he's often acute. Back in 2000, he observed:

> 'Spin-offs, de-mergers, share buybacks, tracking stocks, efficiency programs – all these things release wealth, but they don't create new wealth. Neither do mega-mergers. These strategies don't create new wealth because they don't create new markets, new customers, or new revenue streams.' [37]

Hamel was right. The flatlands of finance have preoccupied Western multinationals – much more, we might add, than the stormy seas of the laboratory. But alongside this unfortunate trend, *theories* of innovation have also grown more conservative. Even Hamel has a little bit of form here, through his notoriously uncritical praise for the 'grey-haired revolutionaries' at a certain American company: you guessed it, Enron. [38]

To understand how someone as clever as Hamel could make the same judgment as *Fortune*, we must go back to the 20th century brain most associated with casting technological innovation in a progressive economic role: the Austrian economist Joseph Schumpeter (1883-1950). Beginning with him, we can hope to understand today's theoretical crisis around innovation – the cultural context for the contemporary paralysis of energy innovation in practice.

Guru's guru: Gary Hamel. Probably the most brilliant management thinker now that Peter Drucker is dead, Hamel nevertheless got in a muddle about business models

Schumpeter's 'gale of creative destruction' vs business models

Publishing his classic book *Capitalism, Socialism and Democracy* in the US in 1942, Schumpeter showed himself a stronger advocate of technology than Veblen. [39] It wasn't hard to be impressed by the technological advances achieved by America in wartime: for example, Schumpeter intriguingly noted that corporate 'safeguarding activities' such as insuring or hedging were not only about long-period contracts in advance, but also about *patents* and 'temporary secrecy of processes'. [40] In Schumpeter's day, one hedged with technological innovations, not just financial ones.

When, however, he famously described capitalism as being in a *'perennial gale of creative destruction'*, Schumpeter wasn't just reflecting on technological advance during the war. Competition with and *the destruction of commercial rivals* now occurred, Schumpeter argued, not by price, product quality or sales volume, but by a new, 'powerful lever' behind long-term improvements in output and cheapness: new consumer goods, technologies and methods of production or transport, as well as new markets, sources of supply and forms of organisation. Thus technology formed only a part of Schumpeter's approach to economic development. [41]

We agree that technology is only a part – today, an underrated part – of economic development. Similarly, when Hamel upheld in 2000 both 'non-linear' innovation in *and* the continuous improvement of the basic components of a business, as well as continuous process improvements in its complete system, we believe that he captured three important aspects of innovation. Nor do we fret about his fourth aspect, which

---

Theorist of innovation: the social-democratic economist Joseph Alois Schumpeter. Briefly Austria's minister of finance after the First World War, Schumpeter moved to Harvard in 1932, teaching a generation of students there until he died in 1950

was both radical and systemic: 'business concept' innovation. Within the confines of inter-firm competition, above-average profits may indeed come not from directly *confronting* rivals, but from *avoiding* them – by offering the customer big, innovatory differences from what is already on the market. [42]

But in a world of governments, not just firms, innovation cannot be reduced to inter-firm competition. Worse, in Hamel's framework, variety 'in all components of the business model' tended to relativise the role of technology in innovation. Indeed, while Hamel held that Schumpeter's gale of creative destruction had become a hurricane, [43] he missed the key change in the weather: top managers' enthusiasm for alternative business models of a very financial type. And that was why Hamel eulogised Enron's 'new core competences' in fields 'such as finance, law, insurance, credit analysis, and energy market analysis'. [44]

Hamel also applauded IKEA and Virgin Atlantic as 'innovators' that were 'not technology pioneers'. Innovation in business concepts, he argued, often had:

> '… little to do with new technology… Technology, especially IT, is available to all. The question is whether you can apply that technology in a unique way.' [45]

But technology *isn't* available to all. IT isn't, as Nicholas Carr argues, a commodity like electricity – although it would be good if electricity itself were to remain in cheap and plentiful supply. [46] In both IT and energy technologies, as in others, there is scarcity. *But the scarcity is not of precious natural resources, but of great knowledge, investment and ambition.*

Today's experts in innovation give a nod to R&D; but they add and implicitly prefer just about any other stratagem to it. In 2008, a committee of US academics and top CEOs (Microsoft, IBM, 3M, UPS) defined innovation as:

> 'The design, invention, development and/or implementation of new or altered products, services, processes, systems,

organisational structures, or business models for the purpose of creating new value for customers and financial returns for the firm.' [47]

Here business models have the last word, and innovation can mean whatever you want. The only reference to energy reported on is Wal-Mart's 'stretch' goals of a 25 per cent more efficient trucking fleet in three years and a 20 per cent reduction in energy use in new stores in four years. [48]

To reach these goals, Wal-Mart will no doubt have to make more Green changes than it has already done. But to represent measures to improve energy efficiency and conservation as the same as innovation – that really is a stretch.

A brief history of business models

Too many of today's 'new' business models turn out to be very familiar ways by which firms can enhance revenue streams – without making much technological innovation. The ways are familiar because they derive, ultimately, from the realm of finance.

Type of model	Example	Date
Nearly free media	Ochs, Pulitzer, Hearst	1890
Blades, not razors	King Gillette	1905
Easy credit	Henry Ford	around 1911
Hire purchase	Radio Rentals	1950s, 1960s
Software and consumables	Xerox Corporation Hewlett-Packard, Sony	1960s
Subscriptions	AOL	1980s
Pay As You Go	Mobile phones	1990s
Pre-payment meters	Energy utilities	1899, 2000s [49]
Discounts for Direct Debit	Energy utilities	2000s

These models have in common:

- a limited amount of genuine technological innovation
- a stress on expensive, high-margin consumables and software
- an attempt to drive up switching costs on the part of users ('lock-in' to proprietary systems)
- a reliance on advertising, branding, retailing, franchising and more or less regular 'hits' on users' finances.

Mobile phone companies, digital TV broadcasters and Internet Service Providers have scores of formulae for their subscriptions and call rates. In energy, utilities do something similar. The net result, however, has been consumer annoyance that, for all the babble about payment regimes, basic service can often go wildly wrong.

User-centredness and open innovation: cop-outs from R&D

The revered management consultants Booz Allen likewise downplay R&D. After long experience and exhaustive enquiries, Booz Allen confirmed in 2008 that there is no statistically significant relationship between corporate expenditure on R&D and corporate financial performance. Yet R&D is, by its nature, an unpredictable affair – so why would spending on it tally nicely with a firm's market performance?

We've seen that R&D, and especially energy R&D, has long been stagnant in the West. Exactly at this moment, however, Booz Allen proposes that 'just throwing money' at R&D 'isn't the answer'. [50]

In fact, throwing money at energy R&D would make a change. But a culture of passing the buck to others has made taking responsibility for investments in R&D as exceptional in the energy sector as it is in other sectors.

For more than two decades now, the doctrine has grown

that any source of innovation is valuable so long as it does not emanate from a company's own efforts in technology. Like business models, dogmas of *user-centred, open, outsourced and networked ecosystems of innovation* are always presented as new intellectual breakthroughs. But with the same old ideas, breathlessly expressed, there's also the same old insouciant tone about innovation's technological aspects.

With the rise of the PC and of Apple in the 1980s, both Americans and Europeans have called for products, systems and the innovation process itself to orientate more to 'users'. [51] But while energy suppliers, like other firms, still have plenty to learn about their customers, the logic of user-centred innovation in energy today is for governments to shrug off their duty to provide for the *macrogeneration* of energy, and instead leave households to engage in *microgeneration*.

What user-centredness means in energy, Gordon Brown made clear in his UN speech on stand-by lights (see Chapter 2). Brown boasted that, in the UK,

> '(W)e are pioneering risk based regulation... only on the basis of risk will we demand information, form filling and inspection.
>
> 'We are recognising too that even the most basic addition of information can play a powerful role in making self-driven change happen: providing people with their right to information enables them to meet their responsibilities for environmental change.
>
> 'In Britain... we are now piloting better labelling on electric goods and smart meters in home.' [52]

Here 'responsibilities for environmental change' are to be exercised by consumers. Governments must merely label and meter popular energy use so that people can engage in 'self-driven' change.

In fact, copping out of technological initiative in favour of the consumer conservation of energy exemplifies a wider trend first upheld by America's Henry Chesbrough. In his *Open Innovation* (2003) and *Open Business Models* (2006), Chesbrough argued that large, vertically-integrated firms needn't reinvent the wheel, but should instead rely on others to innovate for them. Innovation must change from 'closed' to 'open', basing itself on a 'landscape of abundant knowledge' that lay well beyond central research labs – knowledge lying with customers, yes, but also with other companies, suppliers, universities, national laboratories, industrial consortia, and start-up firms. [53]

Chesbrough missed the point that the world still confronts not a knowledge economy, but rather a relative scarcity of knowledge. Instead, by praising sources of innovation external to the firm, he upheld the *outsourcing* of innovation.

That was music to the ears of the large, vertically-integrated firms that still dominate the world's energy sector. They do precious little of their own R&D. Among energy firms that in 2006/7 *led the sector* in the R&D, the average ratio of R&D to sales is dismal:

How much R&D groups of energy firms did on average in 2006/7, as a percentage of sales [54]

Sector	Average R&D as a percentage of sales	Firms
Electricity	0.9	AREVA, EdF, Korea Electric Power (EP), Tokyo EP, Kansai EP, Chubu EP, Vattenfall, Hydro-Quebec, Kyushu EP, Tohoku EP, Taiwan Power, EP Development, Chugoku EP, ENDESA, British Nuclear Fuels
Gas, water & multiutilities	0.3	RWE, Suez, GdF, Veolia Environnment, Osaka Gas, Tokyo Gas, Nalco
Oil and gas producers	0.3	Royal Dutch Shell, TOTAL, Exxon Mobil, Petroleo Brasiliero, PetroChina, Gazprom, BP, China Petroleum & Chemical, Eni, Statoil, ConocoPhilips, Norsk Hydro, Repsol YPF, Nippon Oil, SK, CNOOC, Nexen

Networked ecosystems of innovation

Shortly after Chesbrough's works were published, McKinsey agreed with him. Noting how firms worldwide expected big revenues from licensing their intellectual property (IP) out to others, McKinsey argued conversely that it was vital to license in ideas from the outside. [55]

Outside in or inside out, *McKinsey's focus on the market for technology licenses highlighted money-spinning transactions in IP more than the hard job of applying it.* As McKinsey observed, companies licensing in IP from outside 'not only develop better products but also gain a better understanding of IP markets.' [56]

For us, the business drive toward outsourcing grows not just from concerns about costs, but also from a broader sensitivity to the *risk* of doing things in-house. However, outsourcing itself is also felt to carry serious risks. [57] That's why *Business Week*, noting 'the new not invented here syndrome' among US IT firms, soon worried about them perhaps going too far in outsourcing innovations from Asia. [58] Others contended that innovation couldn't all be outsourced. [59] Nevertheless, the general consensus on innovation today is to discover and rediscover the open sort. [60]

Even in insightful works on innovation, business models retain their allure. [61] Meanwhile innovation, when open, is vaunted as happening through *networks*. [62] And though government support for open innovation might differentiate between large and small firms, it should also focus on *ecosystems* of firms. [63] Openness is also favoured inside the firm – between departments and across functions. Lastly, innovation is often presented as a *combination* of existing technologies. [64]

In energy, combination can indeed help innovation. But combining energy technologies demands hard work and resolution. It cannot be a substitute for in-depth, specialised knowledge and investment in particular disciplines.

As with our old friend Alfred Marshall, today's fondness for networked innovation and for ecosystems of innovation betrays an emphasis on wealth created not directly, but through factors external to the firm. The premises are that

- the world has become a knowledge economy
- the knowledge that counts is available, but just needs to be distributed, or networked, more sensibly.

Yet people no more live in a knowledge or network economy than they live in a consumer society. Knowledge is the basis for innovation, but the fact is that the world doesn't have enough of that either. People never lived in the atomic age or the space age. They are not engaged in life after the oil crash; [65] nor, as we have said (see Chapter 5), are they about to move into a hydrogen economy. [66]

Open, networked ecosystems of innovation represent a disingenuous reading of the field through the spectacles of IT and of biology. There's no reason why technological and other kinds of innovation should resemble the flow of electrons around a physical grid, and no reasons why the conscious, human activity of innovation should emulate the unconscious, Darwinian world of random mutation and natural selection.

Analogies and metaphors are all very well. But only in today's dumbed-down society can they successfully pass as theories. As we'll see,

- to present innovation in ecological terms is a new twist on an old myth. Indeed, the illusion that the Earth's ecosystem, rather than human activity, is the key source of wealth is now deeply embedded in mainstream economics
- genuinely innovative physical grids – in IT, but also beyond it – are more important to a humanised planet than rhetoric about networked innovation.

Does today's boom in wind turbines amount to a Schumpeterian gale of creative destruction?

Creative destruction, Schumpeter said, should be assessed over decades or centuries. [67] But in a footnote, he also wrote of 'discrete rushes' of the new, 'separated from each other by spans of comparative quiet'. [68] Either way, do renewable forms of energy today amount to a gale of new consumer goods, technologies and methods of production or transport? Do they also amount to a gale of new markets, sources of supply and forms of organisation? Are we in a gale of the New, except that the new New is Green? In commercial competition, is it a case of who's Greenest wins?

These are difficult questions to answer. However, taking the period until 2050, and given everything we have said about the weakness of energy R&D and the financialisation of energy, some things are clear. The process of investing in Green technologies has already been going for about three decades, and has certainly received a boost in the most recent few years. However, the world is still at the very beginning of what promises, on current form, to be a lengthy process. The conservationist impulses and suspicion of human achievement and technology that characterise many environmentalists mean that the Green revolution in many people's minds is still much bigger than the Green investment on the ground.

Some clear differences have also emerged between the early part of the 21st century and Schumpeter's time.

Schumpeter was pointing to the way in which the destructive side of capitalism, with its periodic recessions, had a positive role to play in clearing away the old as the new took its place. Yet today, the process of creative destruction is broadly not about market forces killing off older businesses to make space for new ones. Instead, it's about regulation.

In energy, on the side of creation, renewable technologies have gained and will continue to gain a foothold, not so much when their established competitors are forced to the wall, but more often when the state mandates their use. Chapter 6 has shown that more than just the market is at play. While the wind sector is now becoming more independent of the state, it gained its start through state subsidies. It still depends a lot on mandatory targets for renewable energy set by the state. Solar has also relied on state subsidies, particularly in Germany and Japan, but it still isn't economic. As we have seen in Chapter 5, bioethanol gained its spurs with the help first of the Brazilian government, and then, in the EU, with measures organised from Brussels.

On the side of destruction, older energy businesses feel pressure not so much through market forces and competition as through a continuous ratcheting up of regulation.

As yet, little destruction of old energy supply industries has taken place. To write off existing functional assets is expensive. If renewable energy was so cheap it simply put coal, gas and oil out of business – that would be real creative destruction. But the world's miners, extractors and operators of conventional power stations are not about to lose all their jobs. The IEA points out that, without early retirement, three quarters of projected electricity output in 2020, and over half in 2030, will come from power stations operating in 2008. [69]

No doubt some coal, gas and oil companies will go under in the current economic downturn. However, the general rule in the past has been long-term underinvestment more than outright closure. It would be easy to pin all the blame for underinvestment on regulatory difficulties. Nevertheless, the rise of regulation does act so as to reinforce today's business models, which see new investment in physical infrastructure as a last resort.

It's also true that shareholders are unlikely to complain if minimal investment results in shortages that force up prices.

In energy, there isn't a real storm of new Green products and processes. Sometimes thoughtful, large-scale developments really do happen – but there are plenty of quack remedies out there too. *Energise!* has no doubts about the potential, in new technologies, for innovation on a grand scale. It has also highlighted the barriers to that kind of innovation. But so long as precaution and environmentalism dominate the thinking of society, today's boom in wind turbines is unlikely to assume the dimensions of Schumpeter's gale.

Energy conservation and IT

The International Telecommunications Union reports that IT 'contributes' about 2.0-2.5 per cent of worldwide GHG emissions. [70] And in and around UK private and public sector IT departments, the priority now given to reducing energy consumption has reached absurd new heights. [71] IT departments put pressure on vendors to build energy-efficient hardware. The facilities managers who run the buildings that house IT are also obsessed with cutting energy use. Meanwhile human resources departments add 'buying Green' to the criteria by which they appraise managers, all the while fervent in their hope that the War Against Climate Terror will prove a cause that every member of staff can rally around.

Much of today's agenda in IT is about energy-efficient hardware, processing-light software, server packaging and server disposal, and turning the lights out on data centres. None of these items, however, exemplifies Schumpeter's gale of creative destruction. Though a whirlwind of uncritical acceptance of energy conservation has certainly overwhelmed IT managers as much as other professionals, saving money on the £3m that a typical data centre runs up in annual energy costs isn't the kind of innovation Schumpeter had in mind.

In California it's true, IT gurus now put money on all-electric cars succeeding petrol or electric hybrids. Google CEO Eric Schmidt has offered a plan – backed up by investment by Google's philanthropic arm – to get all US electric energy from renewables in 20 years and eliminate half of petrol consumption. [72] Meanwhile, Shai Agassi, a former president of the software giant SAP, has raised $200m to build an infrastructure of electric battery-swapping stations. [73]

The need for batteries in mobile phones may contribute

important innovations. But today's parsimony and breast-beating about power for IT herald neither the emergence of a new sector, nor the radical reorganisation of an old one.

What the examples of energy innovation associated with IT show is that it's only when innovators move beyond the narrow realm of consumption and efficiency that big gains are really there to be made.

The pace of Green innovation is grindingly slow

Take, as another example, BP Alternative Energy – BP's non-fossil fuel activities. Founded in 2005, BP Alternative Energy was going to be accompanied by investments of $8bn over the following 10 years. That sounded a lot; but from 2004 to 2007, BP's total revenues and other income grew from just under $200bn to more than $291bn. Purchases, production and manufacturing expenses accounted for most expenditures; profits rose from $25bn to more than $30bn. [74] Whatever it's measured against at BP, the £0.8bn invested each year by the company in Alternative Energy is chickenfeed.

Let's now inspect the record of the US, often a laggard in applying environmentalist policies on energy, and also of Germany, which has long prided itself on its leading role. Is there clear evidence of a Schumpeterian dynamic in either country's energy sector?

In the US, from the end of 2002 to the end of 2007, wind power notched up annual increases in generating capacity that averaged an impressive 29 per cent. But in generating an estimated 48 billion kWh in 2008, wind accounted for barely one per cent of US electricity supply. [75] Though no nuclear power stations are being built in the US right now, even Barack Obama's boost to renewables is unlikely to build capacity the way coal and gas were forecast to before he won the Presidency:

Before Obama: planned capacity additions from new generators, by energy source, 2007 through 2011, GW [76]

Source	2007	2008	2009	2010	2011
Nuclear	0	0	0	0	0
Coal	1,679	920	12,611	6,839	7,649
Natural Gas	9,891	12,896	11,050	7,569	4,622
Petroleum	255	1	835	50	0
Renewables	5,714	2,032	350	217	56
Total	17,552	16,432	25,617	14,675	12,833
Renewables as percentage of total extra generating capacity	32.6	12.4	1.4	1.5	0

Here renewables were scheduled to take an important 33 per cent of planned capacity increases in 2007. But after that, their contribution to increments in capacity was set for relative *decline*.

Looking at renewables in the production of US *primary energy*, the long-term prognosis of the US Energy Information Administration (EIA), again drawn up before Obama's election victory, was pessimistic:

US: EIA 2008 projections for the production of primary energy, reference case, 2006-2030, quadrillion Btu [77]

Source	2006	2010	2020	2030
Nuclear	8.21	8.31	9.05	9.57
Coal	23.79	23.97	25.2	28.63
Dry Gas	19.04	19.85	20.24	20.0
Petroleum	13.16	15.03	15.71	14.15
Renewables	6.71	8.48	11.42	13.57
Total	71.41	76.17	82.21	86.56
Renewables as percentage of total	9.4	11.1	13.9	15.7

Of course, these forecasts are highly debatable. They may reflect the modest environmentalist ambitions of the Bush years; but in the wake of the Energy Independence and Security Act of December 2007, the EIA revised its growth forecasts upward for non-hydroelectric renewable energy.

Obama will ramp up renewable energy more rapidly still. But it will be very hard for him to leave a legacy such that, by 2020, renewables take, say, a 20 per cent market share of US production. Yet only such a pace of development would really qualify for the cachet 'Schumpeterian'.

What about Green cars in the US? Here there is rapid growth – but once again, from a very low base. The market share among new US cars taken by petrol or electric hybrids rose by more than 27 per cent over the year to January 2008. But hybrids still amount to little more than two per cent of new US car sales. [78]

The case of renewables in Germany

In Germany, renewables take more than 10 per cent of electric power generation, and GHG emissions are down by 17 per cent since 1990. Plans for 2020 are for:

- renewables to take 25 per cent of electricity supply – not least, because nuclear power, which supplies more than a quarter of Germany's electricity today, is due to end in 2025
- GHG emissions to drop by as much as 40 per cent from 1990 levels.

It might, then, seem obvious that a Schumpeterian gale is underway. If German nuclear power is indeed wound down, for example, that would amount to creative destruction – if not by the market, then by popular disapproval.

Yet for once, McKinsey's observations may be right here. Authors from three of the company's German offices suggest that the emissions reductions achieved so far follow 'almost entirely' from the restructuring and modernisation of the high-emission power and industrial sectors of the old East Germany. The reunification of Germany in 1989 and the drawn-out changes in the East afterward had what McKinsey describes 'a big one-time effect'. [79]

So, the end of the Cold War *did* bring a kind of Schumpeterian gale of creative destruction to East Germany. Energy and other plants were closed, or re-equipped, and the energy efficiency of industry improved. McKinsey also notes that Germany now has a strong opportunity to export Green technologies abroad – another sign of a gale. But as the consultancy adds, bringing more renewables and fewer emissions to Germany will be harder in the future than it has been in the past. Carbon capture and storage could make a big difference, but popular opposition to the infrastructure of CCS might stop that difference being made. [80]

How Green are the world's banks?

In October 2007, Deutsche Bank's asset management division appointed Mark Fulton to an exciting new role: that of Climate Change Strategist. By October 2008, the enthusiasm of division global head Kevin Parker for investing in Green infrastructure was still more palpable: carbon in the atmosphere had reached an 800,000 year high, and global warming might be 'only a few years away from the point of no return'. [81]

So far, however, funereal banker alarmism about global warming has been matched only by caution about backing renewables at scale. A January 2008 report by Ceres, an alliance between investors and Greens, found that 23 out of 40 US and non-US banks surveyed mentioned climate change in their latest annual shareholder reports, and that the sample as whole published no fewer than 58 research reports on the subject just in 2007. 'What this report can say with certainty', it reassured itself, 'is that climate change has galvanised the attention of the banking community'. [82]

In November 2008, after Obama's election victory,

Deutsche Bank published an update of its thinking. By then, however, Mark Fulton had postponed the tipping point for 'self-sustaining global warming'. It was now 'perhaps only 15 years away'. Perhaps that was why Deutsche's new White Paper, Economic stimulus: the case for "Green infrastructure", Energy Security and "Green" jobs, had kind words for improvements in the US electric power grid and funding for technologically proven renewables – but nothing to say about Deutsche Bank backing such projects. Instead, the bank strongly proposed a radical innovation in the US: the formation of... a Green National Infrastructure Bank. [83]

Assessing the charge of 'Greenwash'

From the practice and theory of innovation, through the examples of IT, BP, America, Germany and the world's banks, there is little evidence that a Green, Schumpeterian gale of creative destruction is truly about to sweep through the world economy. But if big business hasn't really yet gone Green, are all the advertisements and the bankers' research reports on climate change just so much *Greenwash* – a cynical outburst of rhetoric, unmatched by genuine actions? [84]

In fact the accusation of Greenwash, though right about the low level of Green investments in practice, is itself too cynical. When, in 1995, Greenpeace succeeded in whipping up an international campaign against Shell's plan to dispose of its Brent Spar oil storage and tanker-loading buoy in the North Sea, it was a turning point in the evolution of Western elites. While Greenpeace filmed the offending buoy for £300,000 from bright orange helicopters, and went on to spend a further £1m on public relations around the whole affair, the children of Shell executives in Northern Europe were harassed in school playgrounds. Forty-something Shell managers sensed that they were losing touch with society, and in the first place, with youth. From then on, their motives embraced not just profits and the bottom line, but also a desire to obey society's new Green norms.

That desire is real – whatever Green critics say. If Green technologies have yet to amount to a Schumpeterian gale, that isn't because capitalist fat-cats are greedy or silver-haired creeps who are always on the take and always keeping great innovations in the filing cabinet because of a still greater short-term interest in profits. To explain sclerotic energy innovation, it's glib simply to 'follow the money' behind the elite's actions.

A hand in the till, a demanding Wall Street, a hoarding of super new technologies, the credit crunch – none of these things begin to explain what has retarded the drive toward innovation in the West. Rather, *Energise!* has highlighted how a longstanding history of vulgar economics, risk consciousness and financialisation has put the establishment into a straitjacket. It desperately *wants* to go Green, but finds it difficult to do so.

If elites were simply myopic about climate change, and so needed just to 'see sense' in order to be shaken into acting, Green investments and technologies would be everywhere. So why is business so reluctant to do more than pontificate about climate change? Not because it's unconcerned, but rather because it has made weak innovation into a whole way of life.

Business seems unable to accelerate Green innovations, let alone the wider ones that are needed. *That's one more reason why people need to stop worrying about their personal carbon footprint, and start acting as citizens engaged in a political, society-wide mobilisation for a better energy supply, and for broader efforts to transform the planet in a human direction.*

In a moment we'll conclude with some elements of the programme of transformation that we have in mind. For now, however, it's worth recalling that innovation and economics are today interpreted as being to do with *ecosystems*. So let's now briefly trace how a simple idea – that nature knows best – has come to put the dubious concepts of *natural capital* and *ecosystem services* into the mainstream of economics. Once that tracing is done, it will become clear how fundamentally hostile to technological innovation environmentalist theory is.

The doctrine of natural capital

Today's Greens argue that the natural world forms the bulk of the world's wealth. But it was E F Schumacher's *Small is Beautiful* (1973) that first introduced the idea of 'natural capital.'

Schumacher acknowledged the role of human beings in creating a 'large fund' of scientific, technological and other kinds of knowledge, as well as infrastructure and capital equipment. He noted, however,

> '… all this is but a small part of the total capital we are using. Far larger is the capital provided by nature and not by man – and we do not even recognise it as such.' [85]

Schumacher's first example of natural capital was inanimate: fossil fuels. 'If we treated them as capital items, we should be concerned with conservation', he claimed. [86]

While squandering fossil fuels threatened civilisation, squandering 'the capital of living nature' threatened life itself. This second kind of natural capital, for Schumacher, was the *animate* world's capacity to absorb *pollution* – a capacity, he claimed, that had been degraded by man's synthesis of 'compound substances unknown to nature'. Such activity, Schumacher argued, used up 'a certain kind of irreplaceable asset, namely the tolerance margins which benign nature always provides'. [87]

Today, natural capital is about *lifeless resources*, while the living natural world's continuing and beneficial processes are termed 'ecosystem services'.

Schumacher's ideas, which drew upon Catholicism, had humanistic components. His divided loyalties came out in his famous phrase: 'Man is small, and, therefore, small is beautiful'. [88] The phrase elevated man, making him the measure of the good, but at the same time diminished him with the proposition that, because of the size of his body, *not* his mind, smallness must prevail.

Today's successors to Schumacher are not so balanced.

E F Schumacher: more humanistic than his successors

Born in Bonn in 1911, Schumacher went to Oxford in 1930, and then to Columbia University, New York City. Without completing his studies, he returned to Germany to work in banking. Then, with the rise of Nazism, he left for England again in 1936.

Later interned as an enemy alien after war broke out, Schumacher was sent to labour in England's fields, where he wrote on economics. Keynes spotted his work, and incorporated it in his proposal for a new Bank for International Settlements to manage finance between the post-war Western powers.

In 1950, Schumacher began 20 years as chief economic advisor to Britain's National Coal Board. Given that the NCB employed no fewer than 800,000 people, energy soon became central to his thinking.

In 1955, whilst acting as a UN advisor to Burma, he came to question 'development' and the way mainstream economics favoured simply maximising consumption. He felt that what was consumed really mattered.

It was only after *Small is Beautiful*, however, that his ideas gained widespread recognition.

Catholicism made Schumacher suggest that man live peacefully not only with others, 'but also with nature and, above all, with those Higher Powers which have made nature and have made us'. He also insisted that science could *not* produce ideas by which people could live. The methods of physical science, he wrote, 'cannot be applied to the study of politics or economics'. [89]

Schumacher was right there. By contrast, today's Greens deify science – only to go on to invert Darwin by casting humanity as inferior to animals. For Greens, after all, human beings are culpable. And animals? They're not just innocent, but provide ecosystem services.

Schumacher was far-sighted in his designation of living organisms as a capital superior to fossil fuels. At the 1992 Rio summit, the Convention on Biological Diversity (CBD) was signed alongside the better-known – and also legally binding – International Framework Convention on Climate Change. [90] Then, in 1997, Robert Costanza and others famously valued world natural capital and ecosystem services by adding up what people would pay to replace them. Constanza and his team found that the annual value contributed by 17 ecosystem services was $16-54 trillion – much more than the world's GNP of $18 trillion. [91]

For Schumacher, work wasn't to be automated; rather, it was good for body and soul – and production methods and machines should 'leave ample room for human creativity'. [92] But Schumacher's heirs ignored even this Catholic view of labour. With nature now worth trillions, their aim became simply to minimise its use. Thus in 1999, about their 'next industrial revolution', Paul Hawken, Amory Lovins and L Hunter Lovins said: 'It is people who have become an abundant resource while *nature* is becoming disturbingly scarce'. [93]

Labour was to be exploited; nature, left free. Technology had a role not to free mankind from toil and humanise the planet through higher *labour* productivity, but to win *resource* productivity – derive 'four, ten, or even a hundred times as much benefit from each unit of energy, water, materials, or anything else borrowed from the planet and consumed'. [94]

Here, technological innovation was once more constrained. Its purpose was not to serve thoughtful human beings, but to cosset plants, animals and inanimate nature.

Birth control for cars

For all their market-orientated praise for high technology, Hawken and the Lovinses cramped its potential and, in the process, denigrated human beings. Once they vaunted natural capital, their focus for innovation in energy had to be on efficiency and conservation in use – on 'energy productivity instead of energy production'. [95] Thus, out simply to make homes energy efficient, they wanted 'superwindow and efficient-lamp factories instead of power stations and transmission lines'. [96] 'Hypercars' could be ultra-light, ultra-low-drag and hybrid-electric or hydrogen-powered; but the world also needed 'birth control for cars'. Anyway, *land use* had to come before mobility, which was 'a symptom of being in the wrong place'. [97]

To want to travel was to be in the wrong place. Instead, Hawken and the Lovinses favoured a world in which land use was restrained – one which 'put the places people live, work, shop and play all within *five minutes' walk* of one another'. [98]

Though too miniaturised for many people, such a world at least sounds laid-back. But in playing up resource productivity, Hawken and the Lovinses were quite explicit that *labour productivity should go down*, not up. Resource productivity was 'a basis to increase worldwide employment with meaningful jobs' and, indeed, safeguard against the loss of social cohesion. [99] The Hawken/Lovins 'service paradigm' meant that their meaningful factories making superwindows and efficient lamps would be 'considerably more labour-intensive' than power stations – and that this was something to be celebrated.

So you might, if Hawken and the Lovinses had their way, walk to work. But work might be just a little labour-intensive. Employment would be buoyant; but what exactly would you be paid? Hawken and the Lovinses said nothing about that.

Time to start manufacturing houses

The UN says nearly 100,000 new homes are needed *every day* if the world is to house its urban population by 2030, as well as re-house today's 1 billion people living in urban slums.

The UN on world population, cities and homes, 2030 [100]

Urban population, 2003	3,043,934,680
Estimated urban population, 2030	4,944,679,063
So:	
Additional urban population 2003–2030	**1,900,744,383**
Population living in slums, 2001	923,986,000
So:	
Nos in need of homes and urban services by 2030	**2,824,730,383**
Increase in households over 25 years	877,364,000
New housing units needed per year	35,094,000
New housing units needed per day	96,150
New housing units needed per hour	**4,000**

These figures may not be very accurate. [101] But clearly the world needs houses to be *manufactured* on a high-tech, high-productivity, high-volume basis. That way, there could also be continuous competition over the basic energy efficiency of different types of houses. Moreover, once hi-tech, energy-saving technology for the home gets put into mass manufactured houses at the start of the production process, it will come cheap enough to attract even the most recalcitrant 'consumer.'

Boeing performs final assembly on its 777 planes on a line moving at nearly 4cm a minute. Already, too, Toyota manufactures 5000 homes a year at a plant at Kasugai. [102] So what is needed to meet the UN's rough target of an annual 35m new homes for the world? About 7000 Kasugais.

For today's backward housebuilding industry, that really would amount to a gale of creative destruction.

The UN's Millennium Assessment popularises ecosystem services

In 2000, UN Secretary General Kofi Annan proposed that, just as the International Framework Convention on Climate Change took scientific advice from the IPCC, so the Convention on Biological Diversity also needed advice. Accordingly the UN, the World Bank, the Washington environmental think-tank World Resources Institute, together with various wealthy US foundations, set up the Millennium Ecosystem Assessment (MA). In 2005, the MA published thousands of pages of reports. [103] Its board, which included representatives from Unilever, Lucent Technologies and Skanska, issued what it called 'a stark warning':

> 'Human activity is putting such strain on the natural functions of Earth that the ability of the planet's ecosystems to sustain future generations can no longer be taken for granted... Nearly two thirds of the services provided by nature to humankind are... in decline worldwide... the benefits reaped from our engineering of the planet have been achieved by running down natural capital assets.' [104]

The MA broadly defined 'ecosystem services' as 'benefits people obtain from ecosystems', such as

- supporting – nutrient cycling, soil formation, primary production
- provisioning – food, fresh water, wood and fibre, fuel
- regulating – climate, flood, disease, water purification
- cultural – aesthetic, spiritual, educational, recreational.

An astounding diagram (see diagram on page 460) illustrated the 'linkages' between these services and human security, material goods, good health, good social relations and, ultimately, freedom of choice and action. [105]

For the MA, then, humanity's engineering of the planet not only ran down natural capital, but also, through that process, destroyed the source of 'human well-being' itself.

Once nature is described as a capital, and its value added up in the usual Benthamite way, engineering could only be a bad idea – and minimising humanity's imprint on nature became the way to reach 'well-being'.

The MA frames every modern problem as a product of human misdeeds with the environment. Take the MA Synthesis Report on Health. The diagram on the next page is said to describe 'the causal pathway from escalating human pressures on the environment through to ecosystem changes resulting in diverse health consequences'. [106]

As we can see, the environment is held to have 'impacts' on almost every aspect of human health.

Now, around the planet, floods, malnutrition and slums still constitute severe and urgent problems. As the MA also notes, a billion people lack access to safe water supplies, 2.6 billion lack adequate sanitation, and 3.2 million people die each year from water-associated infectious diseases. *But why see access to clean water as a question of ecosystem services?* If that were true, then the best strategy for increasing access to water would indeed be to minimise human impact.

Yet such a strategy would be absurd. In the developed world, access to clean water isn't determined by ecosystem services. To turn on a tap isn't to rely directly on nature, but upon a complex network of pipes, pumps, sewage works, dams and reservoirs.

If people had to rely directly on nature, cities like London, New York and Tokyo simply couldn't exist. What cities demonstrate is that human infrastructure can massively multiply the quantity of water 'provided' by nature.

Human beings have always had to *work up* nature's resources. Nature doesn't, in a conscious or beneficent fashion, service humanity. As we've seen, those who laud ecosystem services don't mind human beings working harder. But they clearly also imagine that nature works too hard for humanity.

The MA rightly argues that ecosystem degradation hits rural populations and the poor especially hard. [107] But this enforced dependence on nature isn't itself a law of nature. Developed-world technologies can work among the poor.

Malnutrition isn't a simple product of degraded ecosystems. Yes, the poor depend heavily on the soil, and on the natural availability of water for irrigation. But again, that isn't a reason why the technologies that have made agriculture so productive in the developed world cannot be applied among the poor.

It's equally wrong to root *war*, whether in Darfur now or in the Middle East later, in climate change. [108] Ever since the Enlightenment of which he was a part, the philosopher Jean-Jacques Rousseau (1712-78) has been accused of idealising nature and condemning civilisation. But at least he saw that human conflicts arise from society, not from nature. In 1754 Rousseau explained:

> 'The first man who, having enclosed a piece of ground, bethought himself of saying *This is mine*, and found people simple enough to believe him, was the real founder of civil society. From how many crimes, wars and murders, from how many horrors and misfortunes might not any one have saved mankind, by pulling up the stakes, or filling up the ditch, and crying to his fellows, "Beware of listening to this impostor; you are undone if you once forget that the fruits of the earth belong to us all, and the earth itself to nobody."' [109]

In 1642, in his *De Cive*, Thomas Hobbes had primitive life as a 'war of all against all'. For Rousseau, by contrast, every feature of human life, good and bad, only emerged after primitive man had left the state of nature and embarked on civilisation.

Making a decent world, to repeat the point, means fixing all aspects of infrastructure, production and environment. So in that respect, at least, the MEA usefully widens the discussion beyond

Only connect: how the UN's Millennium Ecosystem Assessment tries to link everything to ecosystem services

Environmental changes and ecosystem impairment

Examples of health impacts

Climate change

1
Direct health impacts ☉
Floods, heatwaves, water shortage, landslides, increased exposure to ultraviolet radiation, exposure to pollutants

Stratospheric ozone depletion

Forest clearance and land cover change

Land degradation and desertification

2
'Ecosystem-mediated' health impacts
altered infectious diseases risk, reduced food yields (malnutrition, stunting), depletion of natural medicines, mental health (personal, community), impacts of aesthetic / culural impoverishment

Escalating human pressure on global environment

Wetlands loss and damage

Biodiversity loss

Freshwater depletion and contamination

3
Indirect, deferred, and displaced health impacts
Diverse health consequences of livelihood loss, population displacement (including slum dwelling), conflict, inappropriate adaptation and mitigation

Urbanisation and its' impacts

Damage to coastal reefs and ecosystems

energy and climate change. However, we already suggested in Chapter 3 that what we called the *transformation* of the planet is a better route to progress than the mitigation of CO_2 emissions or adaptation to climate change.

To give more detail on transformation, we now return to this triad. By thinking big and having faith in human talents, a programme of transforming the planet in the direction of humanism could do much for energy supply and CO_2 reduction.

'Mitigation' means, literally, to make a problem less bad. Certainly we want to mitigate the problem of climate change, and indeed other problems. But to alleviate is still to leave things as they were before.

Preventing the Earth from warming more than a few degrees is sensible. But why not go further?

Mitigation is based on targets for concentrations of gases and aerosols in the atmosphere and for factors such as land use. So suppose that such a system were fully implemented. Far from bringing an end to 'anthropogenic' climate change, the system would for the first time make such change purposeful. Climate would become human and anthropogenic in its fullest sense. The idea of mitigation, then, implies that human beings have a preference about the kind of climate they want. Yet that is what the concept of *geo-engineering* is all about.

Many environmentalists have come out against engineering the planet in this way on the grounds that humanity is just not up to it. Bill Becker, executive director of the Presidential Climate Action Project at the University of Colorado, explains that:

> 'Geo-engineering is born of the dangerous conceit that human engineering is superior to nature's engineering. In reality, the first wonder of the world is the world itself, a system that has taken billions of years to evolve through endless trial and error – or, depending on your cosmology, that was created by God – and that performs immeasurable and largely unappreciated services to support life as we know it.' [110]

About geo-engineering, climate sceptics often come surprisingly close to agreement with Greens. Philip Stott, Emeritus Professor of Biogeography at the School of Oriental and African Studies at the University of London, writes:

> 'Climate is the most complex, chaotic, non-linear system. The idea that climate can be managed "in a predictable way" by manipulating one factor, carbon dioxide, out of the millions of factors involved is Alice-in-Wonderland science, with the verdict before the trial. This is the ultimate flaw: the sheer hubris of humans maintaining a "sustainable climate" vividly demonstrates the delusions of the sustainability myth.' [111]

We agree that geo-engineering is not easy, cheap, or even attainable in the short term. But we disagree that the aspiration to engineer is a 'dangerous conceit' or shows up the 'sheer hubris' of human beings.

Environmentalists don't want to follow through on the consequences of mitigation. But they are even more nervous about adaptation. To the extent that humanity can adapt to climate change, there will be less of a problem to mitigate through the adoption of more ethical lifestyles. That's anathema to Greens.

The aspiration to go geo-engineering

Geo-engineering, the idea of engineering the earth on a planet-wide scale, is an idea that the Greens love to hate. To them it takes human arrogance and impact on the planet to the highest degree.

When geo-engineering is raised today as a way of coping with climate change, there are two responses. Green opponents denounce it as beyond the pale, suggesting that we have had quite enough human interference with the climate as it is. Its proponents take a surprisingly similar line, arguing that it should not be used as an excuse to do nothing about cutting emissions.

The reluctance to back geo-engineering is very much a product of the nervous noughties. In many respects suspicion of climate engineering parallels that of nuclear power. Like atomic engineering, meteorology received a tremendous boost from the Second World War. During the Cold War there was a great deal of military interest in the possible use of weather modification in war.

James R Fleming points to the attempted use of weather modification by the US military in Vietnam as a turning point in attitudes. By 1977, a UN Convention on the Prohibition of Military or any other Hostile Use of Environmental Modification Techniques was in place. [112]

Of course, modifying climate or the weather is not yet practical in the way that nuclear energy is. But people should be open to the possibility of such technologies being developed in the future. As with nuclear technology, the possibility of abuse is a reason to promote a humanist politics, not put the brakes on the whole idea of development.

In reality, the prospect of geo-engineering is not so scary. After all, even the idea that people prefer a certain global temperature and will cut emissions or sequester carbon to achieve it is a modest step in the direction of geo-engineering. More generally, it's far from surprising that recent interest in climate science has generated plenty of ideas for intervening in the climate.

There's increasing interest in geo-engineering. A recent special issue of the Royal Society's Philosophical Transactions dealt with techniques such as cloud seeding clouds and fertilising and oceans. [113] These investigations are in their early stages. However, about the prospect of humanity reshaping the planet, researchers should be bold, not apologetic.

More than adapting to the world, mankind has gone about transforming it

In the hyperbolic style preferred by many environmentalists, Mark Lynas has described adaptation as *genocide* – since adapting to a changed environment requires resources, and to rely on it means denying the poor the ability to deal with climate change, which leaves them to suffer the full consequences of this. [114] But there can be no doubt that adaptation will play an important part in the future. Around the planet, humans live in a *built* environment, whether in the countryside or the city. The effect of the climate and weather on human beings depends as much on that built environment as it does on climate itself.

Damage from hurricanes in the US, for example, has increased. Yet that isn't because there are more severe storms, but because more people have built expensive properties in the path of these storms. If 2005 levels of development in the US are projected all the way back to 1900, the US has suffered no increasing trend in the damage caused by hurricanes – even including Hurricane Katrina. [115]

Protecting yourself against hurricanes has its place. Yet like mitigation, the way in which adaptation is interpreted today can reveal a sad narrowing of human horizons. Right away, for example, adaptation nowadays tends to connote *biological* adaptation through evolution. But when humans adapt, they do so with foresight. Evolution isn't like that.

More importantly, people don't so much adapt *to* their environment, as adapt *it* to fit *them*. In extreme cases, they've carried a bubble of breathable atmosphere into outer space or to the bottom of the oceans.

Third and most critical, it's easy to overlook how, in the past, human beings haven't so much adapted as *totally transformed* the planet – and how they retain the potential to pursue more transformation in the future.

In a comprehensive assessment of what's known about how humans have interacted the world's forests, the geographer

Michael Williams has shown that little of the world has been untouched over the millennia. After the Americas were discovered, Europeans saw the New World as virgin territory. But in fact indigenous populations had burned, cleared and foraged among forests, only for human diseases to usher in reforestation. In the shape of the land, humanity was already a key factor.

As Williams suggests, there's been a big switch over the decades, from colonial mindsets to a Green myth of natives living in harmony with nature. And that's only made people still more myopic about humanity's role in shaping the American landscape. Yet in the Americas, it's probable that the forest landscape of 1750 was:

> '... less humanised than that of 1492, when Indian numbers and their impact was [sic] at their peak. With such evidence, the terms presettlement and postsettlement should be consigned to the intellectual trash can. States of "natural" and "equilibrium" have probably not existed since the end of the Ice Age.' [116]

Whereas Williams brilliantly evokes the changing relationship between humanity and nature since the Ice Age, Lord Nicholas Stern fails to do that. Stern wrote:

> 'A warming of 5° C on a global scale would be far outside the experience of human civilisation and comparable to the difference between temperatures during the last ice age and today.' [117]

But like landing on the Moon, finding America was 'far outside the experience of human civilisation'. And a temperature rise even of 5°C in the future would take place in a context different from that which has occurred since the Ice Age. The difference is precisely *human civilisation*, and what it can now do – about climate change, and about everything else.

Even in pre-Columbian Latin America, humans transformed

the land. In a famous paper, *The Pristine Myth: The Landscape of the Americas in 1492*, University of Wisconsin geographer William M Denevan wrote:

> 'The tropical rainforest has long had a reputation for being pristine, whether in 1492 or 1992. There is, however, increasing evidence that the forests of Amazonia and elsewhere are largely anthropogenic in form and composition.' [118]

Above the treetops, the famous lost cities of Latin America are sometimes visible today, but only as the peaks of the highest temples. Yet these peaks just conceal the remains of a farming civilisation that first quelled the forests, then was buried under their re-advance. Concurring with Denevan that, in Latin America, there were no virgin tropical forests in 1492, Williams widens his insight about the era before that year:

> 'Whether it was Europe, the Americas, Australia, New Zealand, or Asia, it was a far more altered world than we have ever thought.' [119]

Since civilisation began, humanity has continually changed the Earth. Today, know-alls distort that achievement to draw the most pessimistic conclusions. It's as if mankind can't credit itself with the improvements it has made over the centuries.

The transformation to which *Energise!* looks forward will differ from those of old. Humans now know a lot about the environment, and can be more discriminating and ingenious in their handling of it. But transformation will not be based on the patronising premise that, in tackling a cure for AIDS, for example, 'money is not the answer'. [120] Nor will it be about everyone minimising their *water footprint*. [121]

Transformation, as we've said, will be about major investment in a Schumpeterian, 30-50 year *gale of new-generation technologies*.

Was it really only luck that brought humanity this far?

According to Nassim Nicholas Taleb, best-selling author of *The Black Swan: The Impact of the Highly Improbable* (2007), to argue that past successes result from anything other than luck is like 'someone playing Russian roulette and finding it a good idea because he survived and pocketed the money.' [122]

Taleb was once a trader on financial markets. There luck indeed plays a larger role than skill. But Taleb's book suggests a wider role for luck, holding that it trumps aptitude throughout human affairs. For Taleb, the appearance of a black rather than a white swan is the unforeseen, unknowable event that changes everything – helping some and ruining others. Using this idea, Taleb contends that people today *underestimate* luck, because they are *over-optimistic about humanity*. Worse,

> 'Justification of [such] over-optimism on the grounds that "it brought us here" arises from a far more serious mistake about human nature: the belief that we are built to understand nature and that our decisions are, and have been, the result of our own choices. I beg to disagree. So many instincts drive us.' [123]

In common with today's biological interpretations of everything, Taleb takes human instincts as evidence that humans are simply not able to exercise choice.

In the ancient world, humanity's achievements were thought to be the work of the gods. Pre-modern thinkers couldn't believe that human beings were truly responsible for big breakthroughs. Today, acclaimed by modern critics, Taleb reverts to this pre-modern view.

In Chapter 1 we mentioned the Greek myth of Prometheus stealing fire from the Gods. But that myth never happened in reality. Humans decided to master fire for themselves – whatever Nassim Nicholas Taleb might want his readers to believe.

Smart electrical grids and global ones

Modern electricity grids are perhaps the most remarkable engineering achievement ever. That's not generally appreciated – in part, because grids are too large to see all in one place.

People mostly encounter the grid only as sockets and switches on a wall. But the entire grid is essentially a single construction. The electrons in the wires that stretch across whole nations and continents march in lockstep in every power station and every consumer device attached to the grid. Keeping the system in balance with power delivered just where it needs to be is no mean feat.

Like many others, Barack Obama is keen on a new 'smart grid'. On the campaign trail in 2008, he warned that the Chinese were 'preparing for a very competitive 21st century economy' – especially in infrastructure. One of the most important infrastructure projects America needed, Obama went on, was

> '… a whole new electricity grid. Because if we're going to be serious about renewable energy, I want to be able to get wind power from North Dakota to population centers, like Chicago. And we're going to have to have a smart grid if we want to use plug-in hybrids then we want to be able to have ordinary consumers sell back the electricity that's generated from those car batteries, back into the grid. That can create five million new jobs, just in new energy.' [124]

Obama is right that America's power grid needs updating and expanding. However, while national grids are good, international ones can be even better. They represent a chance to expand economies of scale in energy supply beyond national boundaries.

So when Obama hinted at heightening US rivalry with China around infrastructure, it was a pity. Technology, after all, has no special allegiance to any country. Patriotic feeling should not be allowed to stand in the way of international grids.

Obama is also right that America will have to make new connections to bring renewable energy to where it is consumed. But the new President's suggestion that a smart grid could allow plug-in hybrid owners to sell electricity back to it is the wrong premise on which to promise millions of new jobs.

Here Obama seems carried away with the green ideal of distributed generation. In truth when plug-hybrids draw power from the grid, there's little point in their owners selling electricity back into it. More important, plug-in hybrids will be still very much be powered by carbon-based fuels – but these fuels can never generate electricity in the large, centralised and thus efficient style of, say, gas-fired electricity plants.

New grids will be smarter, no question. They will integrate power systems with the latest IT. But more thoughtful IT should not be used to entangle the consumer in minute to minute bills, selling power back to the grid, or managing a way round power cuts. Smart grids should work behind the scenes – like present-day grids, only better, more stable and on a larger scale.

Transport, water and electronic grids

Chapter 5 looked at oil and gas grids, but also at cars. So what about *roads*? Here, as with the infrastructure of energy, there's a clear story of underinvestment. In the UK, British Chancellor Alistair Darling proclaimed in the 2008 Budget speech: 'We need more capacity on our roads but we cannot build our way out of all the problems we face'. [125] Yet over 1996-2006, Britain built just 50 miles of main roads a year. [126] Meanwhile in the US, ever since a steel-deck truss bridge in Minneapolis collapsed into the Mississippi River in 2007, attention has been focused on the country's 600,000 road bridges. As Barack Obama has clearly registered, more than a quarter of these are either structurally deficient or functionally obsolete. [127]

Alongside roads, *railway lines* deserve a long gale of creative destruction. Russia, for example, has no less than 85,000km of train track: much of which is in need of replacement and extension.

In Europe, plans to rationalise 27 national systems for *air traffic control* into just a few regional grids have the potential to make fuel-intensive holding patterns a thing of the past. [128]

Humans have constructed a variety of interlocking *water grids*. A first grid supplies water to our taps. In the developed world, water, like the energy we seek in the future, is always

Algal bloom off the coast of Cornwall. This kind of problem confirms the need for better water grids

on and always ready to hand. The same grid delivers irrigation for agriculture, as well as water for industrial uses – the largest of which, predictably enough, is in energy supply. Meanwhile a second grid, just as essential as the first, carries away used water and sewage.

Environmentalists worry that humans pollute water with the residues of pesticides and with the hormones contained in contraceptive pills. It's true that traces of many man-made chemicals are distributed in the water supply. But in fact natural pollution can be a bigger problem. [129]

Humans need better water grids so that they can take control over water flows and cleanliness. For example, a more serious problem than either pesticides or hormones in water is the high concentration of nitrate fertiliser in the run-off from agricultural land. Nitrate-enriched water can trigger a bloom of algae and bacterial growth that chokes off other life by consuming available oxygen – a process known as eutrophication. Blooms can also contaminate water supplies and be toxic for humans.

On top of better water grids – for transport, leisure, flood control and sea-borne trade – the world also needs *more water*. At present it relies on capturing fresh water from the rain, or mining water from underground reservoirs. In the future, mankind will both disinfect used water and feed fresh water back into grids through desalination. [130]

The most efficient desalination plants work by forcing the water through a membrane with microscopic pores. The right sort of membrane will keep salt from passing through. Naturally, salt water is not in short supply, and the only other major requirement is energy to power the pumps.

That's yet another reason why the world will definitely need more energy.

The most famous grid, of course, is the Internet, which grew out of – and is now absorbing – the telephone grid. The Internet has taken mankind beyond voice communication and is now rapidly expanding through the Web, audio, moving pictures, email and ecommerce.

At CERN's Large Hadron Collider, the Internet has been beefed up to carry the 15 million Gigabytes of data a year that the experiment will generate.

Grids connect us to one another. They tie us in to civilisation. And they will also equip us with another means to go transforming the planet: bio-engineering.

Transformation vs other approaches (1): the IPCC

So much for geo-engineering, grids and bio-engineering. We now contrast our sketch of transformation to the ideas of the IPCC, the consultants McKinsey, and the Danish statistician Bjørn Lomborg. We begin with the IPCC and McKinsey on energy and climate.

In Chapter 3, we attacked the IPCC's Working Groups II and III for their eclectic computer modelling of adaptation and mitigation. It's also true that, just as Working Group II shows a marked enthusiasm for 'altered food and recreational choices' and more regulation, so Working Group III looks forward to motorists adopting an 'efficient driving style'. [131] But to be fair, Working Group III offers a useful table assessing what it calls 'key mitigation technologies and practices' over six sectors; and, in a technical summary, Working Group II offers another useful table assessing how the world might adapt to droughts, floods, warming and storms over the 'vulnerable sectors' of food, fibre and forestry; water resources; human health, and industry, settlement and society. [132]

In due course, we will condense the IPCC's two tables and add three extra columns for comparison: the first, a distillation of what McKinsey has said might be done around climate change; the second, a summary of the views of Bjørn Lomborg; and the last, our own views.

On the relationship between mitigation and adaptation, the IPCC says this:

'Mitigation will have global benefits but, owing to the lag times in the climate and biophysical systems, these will hardly be noticeable until around the middle of the 21st century... The benefits of adaptation are largely local to regional in scale but they can be immediate, especially if they also address vulnerabilities to current climate conditions... [So] climate policy is not about making a choice between adapting to and mitigating climate change. If key vulnerabilities to climate change are to be addressed, adaptation is necessary because even the most stringent mitigation efforts cannot avoid further climate change in the next few decades. Mitigation is necessary because reliance on adaptation alone could eventually lead to a magnitude of climate change to which effective adaptation is possible only at very high social, environmental and economic costs...' [133]

Sensible stuff. However, it's apparent that the IPCC is unable to move beyond mitigation and adaptation. What's more, the consultants McKinsey also concentrate on mitigation.

Transformation vs other approaches (2): McKinsey

In a Marshallian mood, McKinsey announced in February 2007 that 'market-distorting subsidies, information gaps, misaligned incentives, and other market inefficiencies now undermine energy productivity'. There was a need to apply conventional technologies more than renewables, and by 2020, proven technological opportunities could cut the growth in annual global energy demand from a base case of 2.2 down to a desirable 0.6 per cent. For McKinsey, 'removing policy distortions, making the price and usage of energy more transparent, and selectively deploying demand-side energy policies' could, by 2020, lower world energy consumption by a quarter. The emphasis was on correcting market failures: on conservation, not innovation. [134]

McKinsey did note the scale of the grids of energy that the next decade alone would demand:

> 'In natural gas, the amount of indigenous production
> consumed within countries will continue to decline,
> replaced by cross-border flows delivered by long-distance
> pipelines and by ships carrying liquefied natural gas (LNG).
> Oil production too will increasingly switch to regions that
> are more and more remote...' [135]

Yet the McKinsey downplayed energy innovation. It believed
that energy had the potential for causing a transformation, and
insisted that energy helped firms 'transform capital and labour
into finished goods and services'. [136] But McKinsey didn't want
to do too much transforming.

What detained McKinsey was that, from 2000 to 2005, the
value of the top 25 *private equity* deals in energy and materials
had tripled to $64 billion. It trumpeted that the developing world
could contribute to

- the profitability of energy firms: in Asia, energy firms had
  adapt their 'traditional capital-intensive business models'
  to take better advantage of low labour costs [137]
- the 'abatement' of more than half of the world's GHG
  emissions, given a 'low cost' scenario for trading them –
  that is, a price up to a cool €40/tonne. [138]

Yes, it's cheaper to apply clean technologies to a new power plant,
house, or car than it is to retrofit an old one. Thus McKinsey
enthused that it's cheaper to abate new emissions growth in the
developing world than to cut existing emissions in the West. But
the firm said nothing about the developing world's potential role
in R&D in energy supply. Instead, it highlighted the potential for

- conservation among the Third World's large populations
- fixing the Third World's *forests*, so as to allow emissions to
  be avoided quite cheaply.

In short, McKinsey wanted... 'low-tech abatement'. [139]

The question of forests

As we pointed out in Chapter 3, climate change is not all about energy. Changes in land use also have a significant effect. The amount of carbon locked up in plant life and soil, and particularly in forests, is vast.

The clearing of the world's forests is taking place most rapidly in Brazil and Indonesia, but also in Myanmar, sub-Saharan Africa and elsewhere. According to the IPCC, global forest cover in 2005 was nearly four *billion* hectares, about a third of the earth's surface. Gross deforestation came to 13.1 *million* hectares per year over 1990-2000, falling slightly to 12.9m hectares per year over 2000-2005. Bringing into account land reconverted back to forest, net losses slowed more dramatically – from 8.9m hectares per year in 1990-2000 to 7.3m hectares per year in over 2000-2005. [140]

As land is cleared, much of carbon stored by the forests is released into the atmosphere. Even at relatively slow rates of clearing, the amounts are large. Most controversial has been burning of Indonesian peat bogs to make way for palm oil plantations. In the exceptionally warm year of 1998, burning of these Indonesian forests alone was estimated to equate to 13-40 per cent of annual CO_2 emissions from fossil fuels. [141]

Greens complain that biofuels accelerate deforestation by pushing up demand for land; they cannot be regarded as carbon neutral. But to see the arrest of deforestation as a cheap-and-cheerful resolution of climate change ignores the full relationship between development and land clearing.

In some respects deforestation is a symptom of underdevelopment. As urbanisation continues and agriculture becomes more industrialised, the rate of deforestation ought to fall. A big slowing up of deforestation tends to come when the rural poor of an underdeveloped nation gain access to modern energy. In China forests have *spread* as consumption of wood fuel has fallen. But in Africa, consumption of wood fuel is growing very rapidly, and significantly adds to deforestation. [142]

Another positive contribution to the world forests will come when a better capitalised forestry sector is able to invest in long-term plantations rather than the simple harvesting of wild trees.

Altogether, deforestation is likely to continue for several more decades as developing countries industrialise. Unlike in the developed world, however, in the tropics the process is likely to come to a natural halt before all the forests are gone. Those who wish to see forests preserved should seek to accelerate the industrialisation of the tropics, rather than halt that process.

In 2007, McKinsey favoured not innovation in energy supply, so much as efficiency and conservation in transport, buildings, forestry, and agriculture. Even though this potential, it conceded, was 'difficult to capture, as it involves billions of small emitters – often consumers'. [143] Indeed, McKinsey found that 70 per cent of its possible GHG abatements

> '... would not depend on any major technological developments. These measures either involve very little technology (for example, those in forestry or agriculture) or rely primarily on mature technologies, such as nuclear power, small-scale hydropower, and energy-efficient lighting. The remaining 30 per cent of abatements depend on new technologies or significantly lower costs for existing ones, such as carbon capture and storage, biofuels, wind power, and solar panels. The point is not that technological R&D has no importance for abatement but rather that low-tech abatement is important in a 2030 perspective.' [144]

In April 2008, however, looking at the US, Germany, the UK and Australia, McKinsey changed its line on technology. 'Many opportunities' now involved low-carbon energy technologies. At low cost and without changing popular lifestyles, the US, Germany and the UK could by 2030 cut emissions by 25, and

Australia by 70 per cent. [145]

To cut emissions by 80 per cent for 2050, forests would once more need fixing – yet now some new technologies, recorded in the tables that follow, were felt worthy. However, McKinsey also argued:

> 'Countries could choose to influence the consumer's behavior through regulations, financial incentives, or both: citizens might be motivated to travel less, to use public transportation more, or to buy smaller cars. Countries could also motivate their people to consume less water, use less floor space and fewer appliances, and unplug idle appliances.' [146]

What an adventurous perspective!

Water: dams do more than just provide energy

As we saw in Chapter 6, large dams have faced heavy criticism for causing environmental and social disruption. There can be no doubt that dam construction on the largest scale poses challenges that go far beyond getting the engineering right. In China, the Three Gorges dam displaced about a million people. The benefits of dams make it essential to find ways to meet such challenges.

It isn't just their enormous energy output and social significance that gives giant dams their importance. As Chapter 6 also suggested, dams play a key role in the smooth running of an electricity grid, through their ability to hold on to electricity – 'pumped storage'. At times of low demand for energy, dam water is pumped back uphill using energy generated elsewhere. At times of high demand, dam water is released and extra capacity rapidly brought on line.

The flexibility of hydroelectric turbines, which can be turned on or off far more rapidly than the high temperature sort, is also often vital to the smooth running of an electricity grid. But the greatest benefits of dams emerge from taking a still broader view. It's in application rather than in construction that dams will truly astonish in years to come.

Unlike more rapidly developing technologies, the basic techniques of constructing dams and hydroelectric turbines are mature. These fields will see incremental as much as revolutionary development. There is, however, big scope for the development of wider water management systems, of which dams will an integral part.

Water management is essential to a wide range of human activities: the supply of drinking water; agricultural irrigation; industry (not least, managing the water that cools power stations is important); sewage management; flood control, and, increasingly, recreation. The future may also likely see the development of new activities such as aquaculture (fish-farming), or unforeseen applications such as a revival of water transport.

The trend will be toward management of entire drainage basins consisting of river systems and coastal regions. No doubt this will be driven in part by concern that climate change will raise sea levels, increase storminess and increase rainfall.

Yet as an increasing part of the world's population builds on flood plains along its coastal regions, the world will anyway need to take control of water flows, global warming or no global warming. By doing so, it can do much better than merely adapt to a changing climate, but rather open up whole new possibilities in land, settlement and agriculture, as the Dutch have been finding out for centuries.

The future, then, will see not just more dams, but the coordination of dams through IT. Radar tracking and the direct chemical and electronic monitoring of water will more and more happen in real time.

Taken together, these developments will allow water to be put to new uses – and on a greater scale than ever before.

Transformation vs other approaches (3): Bjørn Lomborg

Bjørn Lomborg shot to notoriety in 2001 with his publication of *The Skeptical Environmentalist: Measuring the Real State of the World*. He took aim at what he called the 'litany' of environmental disaster stories with which people are nowadays assailed, and set out to present a corrective. His work covered energy, global warming, chemicals, water, forests and more. Overall, he claimed that while environmental problems remain, most are getting resolved rather than getting worse.

Environmentalists panned the book. [147] Mark Lynas personally slapped Lomborg with a cream pie during a book launch. 'I wanted to put a Baked Alaska in his smug face,' explained Lynas, 'in solidarity with the native Indian and Eskimo people in Alaska who are reporting rising temperatures, shrinking sea ice and worsening effects on animal and bird life'.[148] Meanwhile, a review in *Nature* compared Lomborg's treatment of the extinction of animal species with the views of those unprepared to accept that the Nazis had tried to extinguish the Jews. [149]

As a climate scientist might put it, this kind of mudslinging didn't create a very calm atmosphere.

For ourselves, we find Lomborg's approach wanting – but not as wanting as that of environmentalism. While critics attacked many of his specific scientific claims, they missed his key political argument. First, Lomborg claimed that one of the most serious consequences of the litany was that it undermined

> '... our confidence in our ability to solve our remaining problems. It gives us a feeling of being under siege, constantly having to act with our backs to the wall, and this means that we will often implement unwise decisions based on emotional gut reactions. The Litany gets to modern man and impacts us directly: the Litany frightens us.' [150]

The one critic who did reply to this argument was Kathryn Schulz, editor at large of the environmentalist magazine *Grist*. While conceding that media coverage could influence people, Schulz averred that

> 'the Litany may be compelling because of people's lived experiences. Lomborg… could not be less interested in the community concerned about the cyanide in its water supply, the neighbourhood battling the smelter in its backyard, or any of the millions of people all over the world who confront the consequences of environmental degradation every day.' [151]

But while the impressionistic 'lived experiences' of communities may simply suggest environmental degradation, the larger reality may be *not enough economic development*. To take 'lived experience' as the basis of decision-making is a profoundly anti-scientific approach – and one that can only be hostile to innovation.

For all their moralising, environmentalists are simply blind to the idea that anybody might disagree with them on social or political grounds. For them, disagreement can only issue from Stupid People Who Do Not Agree With The Science.

In *Cool it* (2007), Lomborg's argument is as follows:

1. Global warming is real and man-made
2. Statements about the strong, ominous, and immediate consequences of global warming are often wildly exaggerated, and this is unlikely to make for good policy
3. We need smarter solutions for global warming
4. Many other issues are much more important than global warming. [152]

Grinning and bearing it: Bjørn Lomborg. Staying though he does within the technocratic framework of cost-benefit analysis, the Danish statistician has scored some palpable hits against climate orthodoxy – only to be vilified for his pains

We completely agree with this perspective. But what we propose around the transformation of the planet leads us into some significant differences with Lomborg.

First of all, while Lomborg's call for cool tempers is coupled with a devastating critique of many Green dogmas, he doesn't look too closely at technological innovation – and he doesn't mount any kind of critique of ecosystem services.

For Lomborg, the irrationality of environmentalism today is to do with the media's commercially-orientated search for scare stories, stories about conflict, and stories about guilt. It is bound up with politicians' desire to capture the moral high ground, to distance themselves from the usual squabbles, and to make some taxes popular. [153]

These comments are true, but don't get to grips with the detailed, if faulty, political economy of environmentalism. Despite useful historical excavations of past doom-mongers on climate, Lomborg rather de-historicises the culture surrounding climate change today, speaking of 'a deep-seated human tendency to believe that things were better in the old days', and of 'the age-old media focus on bad news about the natural world'. [154]

In *Cool It,* global warming boils down to the costs and benefits of taking different courses of action. In this scheme, Kyoto comes out badly, adaptation to climate change in the medium-term future is sensible, and geo-engineering might at least be a pragmatic course. Separate measures are necessary to beat malaria, poverty, starvation, water stress, HIV/AIDS, damage to eyesight through malnutrition, poor drinking water and sanitation. [155]

Lomborg's programme of what we might call 'adaptation plus' may well be more efficacious than environmentalist policies. But by failing to take up Green theory enough, he is reduced to saying that Greens show 'a neglect of thinking out priorities'. Everything is a question of *trade-offs* between what you would like to happen and what it will cost. [156]

That won't cut it. Environmentalism would not be wrong to question the whole idea of cost/benefit analysis, for the benefits sought by society are never a technocratic, and always a political question. Environmentalists deify nature; *Energise!* puts human needs first. Natural science cannot decide that.

Charmingly but naïvely, Lomborg says that the debate on global warming 'has often become so fixated on CO_2 cuts that it neglects what presumably is our primary objective – to improve the quality of life and the environment'. [157] But he presumes too much of his opponents. This chapter's discussion of ecosystem services has shown that environmentalists do not make a 'trade-off' between quality of life on the one hand and environment on the other. At bottom – and the Millenium Ecosystems Assessment was very clear on this – they attribute all aspects of human life, including quality of life, to nature.

A kind of global 'business case', complete with a simple defence of free trade, will do little to lessen Green influence. Nor will Lomborg's advocacy of a tax of $2-14 per tonne on CO_2. [158] But perhaps the most disappointing aspect of Lomborg's programme is its limited ambitions in technological innovation and R&D.

Lomborg rightly notes that energy R&D has fallen sharply since the early 1980s, whether private or public, and over nuclear, fossil fuels and renewables alike. He wants nations to commit themselves to spending not the current 0.006 per cent of GDP on renewables R&D, which is what OECD countries do, but 0.05 per cent, or about $25bn a year. [159]

Perhaps this is enough. It certainly needs to be accompanied by big expenditures on R&D in non-renewable energy, and outside the field of energy altogether. That will require, politically, a sterling defence of technological innovation.

Lomborg doesn't mount such a defence – perhaps because his cost/benefit approach isn't so different from that advocated by Jeremy Bentham and his modern interpreters.

Which way to beat malaria?

Very properly, Bjørn Lomborg emphasises that malaria cannot be understood as a function of temperature alone. Up to the 19th century, malaria was endemic throughout Europe, even as far north and Finland and Siberia. In 1933, no less than 30 per cent of the population of the Tennessee River valley was infected, while an epidemic took place in the Netherlands in 1943-46, during and after the war years.

The lesson Lomborg draws is that in Europe and the US:

'... we eliminated malaria while the world warmed over the past century and a half. While temperature does impact malaria, it is clearly not destiny. What probably matters much more is a wide array of factors, from nutrition and health care, through draining and mosquito eradication, to income and availability of quinine or newer treatments.' [160]

That's right. But how sad that Lomborg feels called upon to disparage the 70 million cases of malaria putatively saved by Kyoto till the year 2085, and instead talk up what he calls the 'simple and cheap' United Nations Millennium Development Goal of *halving* the incidence of malaria by 2015! [161] By 2085, achievement of the UN's goal early on in the century might ensure that more than 28 billion cases of malaria had been prevented. But what about also *increasing R&D into malaria* dramatically, so that *a total cure* for it might be found by 2015 – if not before?

Lomborg calls for 'targeted policies, mosquito nets, medicine and mosquito eradication'. [162] Yet it is not far-fetched to suggest that drugs and vaccines should be developed so that *malaria itself* can be *eradicated*. That, after all, is one of the goals of the Global Malaria Action Plan, announced at the UN Millennium Development Goals malaria summit in New York in 2008. The Plan points out that the Maldives, Tunisia, and most recently the United Arab Emirates have eliminated malaria from within their borders. [163]

Much has been learned about the sequencing of the genome of three parasites that carry malaria. Much remains to be learned. [164] But on the whole, research in genetics and pharmaceuticals has a lot more going for it than the distribution of mosquito nets.

Summing up this book (1): Obama's collective solutions and ours

We started this book by suggesting that man-made climate change can and should be solved as part of a wider effort – an effort to address society's need for a lot more cheap energy. In the process, we upheld the energised citizen, informed and able to win arguments about the kind of energy supply people and organisations can and should have. We made a trenchant attack on those who believe that changes in their habits as individual consumers, once added up over millions, are the right way to save the planet.

In an unguarded moment, Barack Obama was on the same wavelength. Facing himself with an imagined question, on a televised debate about what he *personally* had done that was Green, Obama revealed a forthright, if not coarse attitude. He told *Newsweek* magazine that he'd have said:

> 'Well, the truth is, Brian, we can't solve global warming because I f***ing changed light bulbs in my house. It's because of something collective.' [165]

In looking at climate change, energy supply and, in this chapter, at broader and more longstanding problems to do with the physical world, *Energise!* has put forward solutions that are indeed 'something collective'. Obama's remarks should serve as a rebuke to those in Britain who make conspicuous their highly personal non-consumption.

Obama's collective solutions, however, are not our collective solutions. He wants society to auction permits to emit

CO_2; we look to technical fixes, global grids and an international division of labour to get things done. He wants individuals to engage in informal but politically binding partnerships with the state – partnerships that are designed to save on the personal consumption of energy. By contrast, we favour a citizenry that is thoughtful and alert about every aspect of energy supply, and that is ready to participate in the *execution* of a much bigger and better supply.

Obama has said that Americans 'can lead the world, secure our nation, and meet our moral obligations to future generations'. [166] The generations we are interested in, however, are rather different from his. This book has dwelt not on the resentment that future generations are thought bound, in due course, to bear toward folks today, but rather on future generations of technology that need to be nurtured with great researchers and proper budgets: 4G nuclear reactors, 3G biofuels, 3G kites to harness the wind, and 3G photovoltaic panels,

There will come a time when millions of citizens are moved to stand up for new generations of technology, and for the R&D that will make such developments possible. People won't easily be fooled again by the state's lack of resolve in energy supply. Nor will they be sympathetic to the hesitations of private energy suppliers.

Summing up (2): for elites, security has more or less become energy security

Throughout this book, we've tried to show that the blinkered perspective of the independent consumer and his decentralised microgeneration of energy is, like the doctrine of the carbon footprint, the wrong way to think about energy and climate change. However, once society's focus shifts from hidden energy supply to tangible energy consumption, energy comes to be perceived as a winnable squabble over resources that are finite and scarce.

That tendency must be resisted. Energy and climate change now mediate East-West relations as never before: they are, perhaps, the main mediator nowadays, more important than arsenals, trade or even – dare we say it? – Olympic Games.

In fact energy is only a mediator in tension, and rarely an originator of it. Yet whether it's to do with fissile nuclear materials, dirty coal, Russian gas, Asian demand for oil, or Obama's confidence that 'green collar' American workers are better than any other kind of workers, energy, and especially oil, has now become, in the public mind, a cipher for tense manoeuvres between West and East, and between North and South. In that sense, every informed and active citizen needs to know about and take an independent position on energy.

Late in November 2008, the US National Intelligence Council published one of its most path-breaking reports. *Global trends 2025: a transformed world* was widely commented upon for its forecast of an end to US global leadership. [167] What was also striking about the report, however, was its mechanistic assertion that, 'under any scenario', what it called 'energy dynamics' could produce 'a number of new alignments or groupings with geopolitical significance'. [168] What might those alignments be? They could be Russia dominating Central Asia, Beijing hooking up with Riyadh, Beijing getting friendly with Tehran, or India making overtures to Burma, Iran and Central Asia. But whatever they might be, the whole discussion only goes to show one thing: that energy and oil are, like climate change, meant to explain everything nowadays.

In the minds of many, security has more or less become energy security. Independence from oil was a key plank in Obama's election manifesto. Like few other issues, energy has come to concentrate national and international fears.

Summing up (3): until 2030, handwringing and hysteria about GHGs – unless energised citizens make a good political riposte

What, though, about choice of technique? Summing up chapters 4-6, we can forecast an overall future for energy with respect to climate and GHG emissions.

Al Gore has put forward a goal of making the US electricity supply carbon-neutral by 2018. [169] There's a difference between ambition and foolishness. Gore's target is foolish for two reasons.

First, even the most ambitious clean energy program will not reduce emissions unless and until the existing stock of power stations are closed down. That puts one limit on how fast emissions can be reduced.

A second reason why electricity cannot be decarbonised inside just one decade is that energy technologies take time to development. Almost every week brings news of incremental advances in the methods needed to turn solar power into a practical source of energy. Yet a long period of incremental advance is likely to be required for solar to come of age.

We have predicted that, though it's on the verge of becoming economic in the sunniest areas, solar will take a decade to take off the way that wind has taken off today.

Solar may eventually become the world's largest single source of energy. Other technologies lie in a more distant future. Innovative varieties of geothermal and high altitude wind have huge potential, but at present exist only in demonstration form. They will likely take at least two decades to be implemented on a scale that makes a difference.

While Gore was referring to electricity, similar points apply to transport: efficient hybrid cars will take at least a decade to become commonplace, even after the technology has advanced enough to make them competitive on price. Although biofuels could replace oil relatively quickly, the advanced variety is still in development.

For these reasons, and even with all-out investment in clean energy, we don't expect that emissions of CO_2 will begin consistently to fall much before 2030. After that, we do expect that investment in clean energy will begin to pay off. Emissions will begin to fall, slowly at first and more rapidly after 2040.

Unless the terms of debate on energy and climate are altered soon, then, hand-wringing and hysteria about emissions could well dominate much of political and international life in the next 20 years.

It's a dismal prospect. The carbon footprint blame game will do nothing to spur the investment in energy necessary to transform the world. It will do nothing to prepare people for the rest of what could still be a great century.

We hope that *Energise!* will contribute to an alternative approach. There is a need to challenge the precautionary principle, whether applied to governments' choice of technique in energy supply, or to the decisions made by energy companies.

Summing up (4): precaution means no real state commitment to nuclear, and little risk-taking private investment in energy

The impact of the precautionary approach to the environment taken by the British *state* is seen most clearly in relation to nuclear energy. Few British politicians are prepared to go beyond the idea that nuclear must be an 'option'. In practice, elected representatives fall back on ageing reactors, and postpone taking real action.

We argue, by contrast, that nuclear should be a mainstay of electricity supply. Even in 2050, when solar might account for a quarter of electricity, nuclear should supply another quarter. If the nuclear revival doesn't reach that extent, there will certainly be problems meeting the world's energy needs.

In Britain successive governments have equivocated over nuclear for more than a decade. As a result there is an annual discussion over whether winter gas demand can be met. If decisive action had been taken, the problem would now be solved. Of course new nuclear will not come on line instantly, but that's all the more reason to start soon.

While precaution is publicly celebrated around new nuclear and coal-fired plants, it's less visible in the low-carbon domain. Nevertheless, we have pointed to many examples. The future of genetically modified biofuels may also prove particularly contentious.

The impact of *corporate* precaution can be seen in the

failure to invest. This is likely to be accentuated in the wake of the Crash of 2008. In its *World Energy Outlook 2008*, the IEA projected that, even without the expense of investing in clean energy, investment of more than $26 trillion will be needed over the period 2007-2030 – 52 per cent of the sum going on power generation, and most of the rest going on oil and gas. [170] In particular, the IEA worried about *oil*. With oil, it said, 'the immediate risk to supply is not one of lack of global resources, but rather a lack of investment where it is needed'. It went on that there was a 'real risk' that under-investment would cause an 'oil-supply crunch' before 2015. [171]

The IEA was right to conclude that, even giving geology its due, economics remains the determining factor in oil supply – as it does in energy generally. Yet if there's now a real dearth of investment in oil, that cannot be explained simply as an inevitable downswing in the business cycle – one in which oil prices are so low, investment in oil production no longer makes sense.

No doubt that's part of the story. But even before the Crash of 2008, investment in oil, as in all parts of the international energy industry, was far from impressive. As this chapter has shown, energy companies have for some time been more interested in financial juggling and exotic business models than they have in actually generating higher levels of energy.

The IEA says that while the current financial crisis is 'not expected to affect long-term investment', it could lead to 'delays in bringing current projects to completion'. [172] That verdict might prove too sanguine. For years industry has been looking for any excuse not to invest. Failure to invest in the oil sector in the late 1990s was blamed on low oil prices, despite a growing economy. Today the world is still feeling the consequences of that period of low investment. After enjoying a ramp up to record high oil prices, oil firms have found themselves a new, up-to-date version of an old excuse for not investing – low oil prices.

They have found it just in the nick of time. Indeed, low oil prices have also been invoked by the renewables industry as a means of explaining that sector's unwillingness to invest.

It's reasonable to ask why companies as large as the oil majors, at least, now feel unable to take the long-term view of investment. There was a time when they were able to look past the unavoidable swings in prices or growth that take place in the first few years of a new project's life.

Late in 2008, BP withdrew from a CCS project in the UK and Shell pulled back from investing in Canadian oil sands. We believe that the reluctance to invest is a form of corporate risk aversion, one that is just as unhealthy as the green sort.

How can this reluctance to invest be dealt with? It's no good just exhorting people or firms to innovate. What is at issue is a struggle not to improve personal behaviour, but to change the whole way society thinks and acts about producing energy. It's a struggle to preserve science from manipulation, individuals from regulation, and society from power cuts.

We've no doubt that citizens can rise to that challenge.

How our perspective of transformation stacks up against other perspectives

In the tables below, we show how our programme of transforming the planet in line with mankind's needs and talents goes beyond the measures that the IPCC and McKinsey propose in relation to climate change. We show, too, how our proposals are more ambitious than those proposed by Lomborg.

For each sector, beginning with energy supply but going well beyond it, we describe innovations and investment strategies whose benefits should exceed those that are required purely to deal with climate change.

Why do the benefits of our approach go further than simply dealing with climate change? Because our strategies were in the first place designed to deal with the roots of longstanding problems other than climate change.

The world needs a great deal of inexpensive energy. In building a new round of supply, it can start to minimise climate change – and start, too, to maximise the kind of achievements it wants to make.

Comparing strategies for the world (1): energy supply, transport, buildings

Sector	IPCC's key mitigation technologies on the market today [173]	IPCC's key mitigation technologies to be on the market by 2030 [174]	IPCC's examples of adaptation for vulnerable sectors [175]
Energy supply	Improved supply and distribution efficiency; switch from coal to gas; nuclear; renewable heat and power (hydropower, solar, wind, geothermal, bioenergy); CHP; early applications of CCS (eg store removed CO_2 from natural gas)	CCS for gas, biomass and coal-fired electricity generation; advanced nuclear and renewables, including tidal, wave, concentrating solar, solar PV	
Transport	Fuel efficiency; hybrids; cleaner diesel vehicles; biofuels; shift from road to rail/public transport; cycling, walking; land-use; transport planning	2G biofuels; more efficient aircraft; advanced electrics and hybrids with more powerful and reliable batteries	
Buildings and settlement	Efficient lighting, use of daylight; efficient electrical appliances, heating and cooling devices; improved cook stoves, insulation; passive and active solar design for heating and cooling; alternative refrigeration fluids, recover/recycle fluorinated gases	Integrated design of commercial buildings including technologies such as intelligent meters that provide feedback and control; integrated solar PV	FLOODS: Improve adaptation capacities, especially for livelihoods; incorporate climate change in development programmes; improve water supply systems, co-ordinate jurisdictions. DROUGHT: Improve flood protection infrastructure; flood-proof buildings; change land use in high-risk areas; flood hazard mapping, warnings; empower community institutions. WARMING: Assistance programmes for especially vulnerable groups; improve adaptive capacities; technological change. STORMS: Emergency preparedness, including Emergency Warning Systems (EWS); more resilient infrastructure; financial risk management

McKinsey's GHG abatement measures	Lomborg	Energise!
	Increase energy R&D by a factor of ten to 0.05 per cent of GDP, or $25bn a year: pilot programmes, public-private partnerships on high risk projects, training scientists and engineers, prizes for crossing technological thresholds, international collaboration and research centres. policy encouraging adoption of new and existing technologies to speed learning [176]	Increase all forms of energy R&D, especially in 4G nuclear, fusion; burning coal better, CCS; 3G biofuels, New Carbon Infrastructure; floating wind turbines, kites; 3G PV, CSP; tidal, engineered geothermal. To speed growth in developing world, ramp up investment in clean energy as it becomes competitive with fossil fuels. Political leadership to overcome precautionary opposition to investment, especially in nuclear. Private sector leadership to increase R&D and investment long term
Efficiency measures, mainly in transport and buildings, cut demand for energy, and at no net cost, cut GHGs by 6 Gt. [177] Biofuels replace 20-30 per cent of current transport fuels by 2030, cut emissions by 80 per cent below the level they would reach with fossil fuels [178]	Biofuels won't cut CO_2 emissions, but double them. Stopping flying would 'quite plausibly be one of the worst ways' to help starving Ethiopians. [179]	Improve road, air and public transport to expand capacity, make travel easier and cheapen the costs of world trade. Hybrids to contribute to efficiency. 2G and 3G biofuels to reduce emissions. Mag-lev trains; supersonic, then hypersonic passenger planes (London to Sydney in 5 hours)
With 25 per cent of global energy demand, housing is the largest energy-use segment. Fit out new homes with tight building shells, including chemically treated windows to cut ingress of winter cold and summer heat; high-grade insulation; compact fluorescent lighting; solar water heaters. Higher efficiency standards for appliances; cut standby power requirements. Measures in lighting, heating, and cooling could slow annual growth in residential energy demand from 1.4 to 0.5 per cent, and reduce 2020 total world energy demand by three per cent. [180] Biomass, geo-thermal, district heating [181]	FLOODS: barriers, dikes, levees, coastal barriers and – rarely – giving up land. [182] WARMING: Green spaces and water features can produce local cooling, while also making more beautiful cities'. Painting surfaces white can avoid heat build up. Air conditioning [183]	Expand production of clean energy to allow buildings to do what they're meant to: provide comfort and convenience, including generous space requirements, good insulation and air conditioning, solid flood- and storm- resistant design. Mass manufacture of housing would help reach these goals. FLOODS, WARMING: full planning of infrastructure around homes to allow house manufacturers maximum flexibility of construction and design

Comparing strategies for the world (2): industry, agriculture, waste and water

Sector	IPCC's key mitigation technologies on the market today	IPCC's key mitigation technologies to be on the market by 2030	IPCC's examples of adaptation for vulnerable sectors
Industry	More efficient end-use electrical equipment; heat and power recovery; material recycling and substitution; control of non-CO_2 emissions; process-specific technologies	Advanced energy efficiency; CCS for cement, ammonia, iron manufacture; inert electrodes for making aluminium	
Agriculture	Improved crop and grazing land management to increase soil carbon storage; restoration of cultivated peaty soils and degraded lands; improve rice cultivation and livestock/manure management to cut CH_4 emissions; improve application of N_2 fertiliser to cut N_2O emissions; dedicated energy crops to replace fossil fuel use; improved energy efficiency	Improvements of crop yields	DROUGHT: Crops: development of new drought-resistant varieties; intercropping; crop residue retention; weed management; irrigation and hydroponic farming; water harvesting. Livestock: supplementary feeding; change stocking rate; alter grazing and rotation of pasture. Social: Improve extension services; debt relief; diversification of income. FLOODS: Crops: polders and improved drainage; development and promotion of alternative crops; adjustment of plantation and harvesting schedule; floating agricultural systems. Social: Improved extension services. WARMING: Crops: development of new heat-resistant varieties; altered timing of cropping; pest control and surveillance of crops. Livestock: housing and shade provision; change to heat-tolerant breeds. Social: diversification of income. STORMS: Crops: development of wind-resistant crops (eg vanilla)
Waste management and water resources	WASTE: Landfill CH_4 recovery; waste incineration with energy recovery; composting of organic waste; controlled waste water treatment; recycling and waste minimisation	WASTE: Biocovers and biofilters to optimise CH_4 oxidation	DROUGHT: Leak reduction. Water demand management through metering and pricing. Soil moisture conservation eg through mulching. Desalination of seawater. Conservation of groundwater through artificial recharge. Education for sustainable water use. FLOODS: Enhanced implementation of protection measures including flood forecasting and warning, regulation through planning legislation and zoning; promotion of insurance; and relocation of vulnerable assets. WARMING: Water demand management through metering and pricing. Education for sustainable water use. STORMS: Coastal defence design and implementation to protect water supply against contamination

McKinsey's GHG abatement measures	Lomborg	Energise!
General: replace halogen lamps with Light Emitting Diodes.[184] Steel: in US, expand co-generation, improve recuperative burners to cut energy demand by 30 per cent. Bigger opportunity in developing world mills – efficiencies and human maintenance costs are lower. Paper: extended nip presses extract 5-7 per cent more water from intermediate products, so cutting the need for dryers. Cement: fit out traditional ball mills with high-pressure roller presses, or replace with horizontal roller mills. [185] German industry could cut 30m tonnes/year by 2020 by adopting more energy-efficient motor systems with variable-speed drives [186]	Many mitigation options today too expensive to pursue. Trade liberalisation and lowering the cost of starting a business will increase industrial growth [187]	Expand production of clean energy to allow the rest of industry to restructure around saving human labour, rather than saving energy
In agriculture and waste disposal, developing economies represent more than half of the 1.5 Gt of possible abatements [188]	Improved soil health, water management, research, school meals, nutrient fortification. Increasing income above subsistence agricultural levels to allow food imports from more areas more suited to agriculture [189]	Increase mechanisation and use of IT, especially in the developing world, so as to end malnutrition. Increase properly managed application of fertilisers, pesticides and herbicides. New crops, including genetically modified ones. Improve transport and storage infrastructure in developing world, to connect agricultural products to markets. Integrate biofuels and other bioproducts into agriculture
	WASTE: Bring basic water and sanitation services to 3bn people over 8 years. WATER: Trade between water rich and water scarce regions. Drip irrigation can cut water usage 30-70 per cent while increasing yields. Industry could cut usage by 30-90 per cent at a low cost. Water in the developing world should be priced to reflect true cost [190]	WASTE: Where recycling makes sense, do it industrially, in a mechanised and professional manner – not at the level of individual households. WATER: Abundant cheap and clean energy supply will make desalination practical. Increase supply through dams, grids. FLOODS, STORMS: use water grids to control effects of bad weather across entire water basins

Comparing strategies for the world (3): forestry and health

Sector	IPCC's key mitigation technologies on the market today	IPCC's key mitigation technologies to be on the market by 2030	IPCC's examples of adaptation for vulnerable sectors
Forestry	Afforestation; reforestation; forest management; reduced deforestation; harvested wood product management; use of forestry products for bioenergy to replace fossil fuel use	Improve tree species to up biomass productivity and carbon sequestration. Improve remote sensing technologies to analyse vegetation/soil carbon sequestration potential and to map land use change	WARMING: Fire management through altered stand layout, landscape planning, dead timber salvaging, clearing undergrowth. Insect control through prescribed burning, non-chemical pest control. Social: Diversification of income
Health			DROUGHT: Grain storage; emergency feeding stations; safe drinking water/ sanitation; strengthen public institutions and health systems; access to international food markets. FLOODS: EWS; disaster preparedness; post-event emergency relief. WARMING: World surveillance for disease emergence; strengthen public institutions and health systems; national and regional heat warning systems; cut urban heat islands through green spaces; adjust clothing and activity levels; up fluid intake. STORMS: EWS; disaster preparedness planning; effective post-event emergency relief

McKinsey's GHG abatement measures	Lomborg	*Energise!*
Forestry measures – protecting, planting, and replanting forests – save 6.7 Gt. Halving deforestation rates in Africa + cutting them by 75 per cent in Latin America saves nearly 3 Gt. Big abatements in Asia's forests cost more: land is scarce, commercial logging has a higher opportunity cost than subsistence farming in Africa and commercial agriculture in Latin America [191]	Debt-for-nature swaps; labelling of sustainable timber; 'higher economic growth and a better economic foundation so as to secure the countries concerned the resources to think long term' [192]	Replace slash-and-burn agriculture with industrial agriculture. Provide modern energy to those relying on woodfuel. Invest capital and R&D in timber production. Plan landscapes to control fires, insects
	HIV/AIDS, malaria and malnutrition are higher priorities than reducing the impact of climate change [193]	Increase R&D in genetics, stem cells, emerging diseases and pharmaceuticals. Eliminate malaria through vaccines and drugs; treat HIV, cancer and heart disease better. DROUGHTS, FLOODS: Accelerate economic development, including infrastructure such as dams. Aim to eliminate food and water shortages resulting from drought, and minimise the consequences of floods. WARMING: Improved nutrition and overall health should minimise effects of stress from extreme events.

REFERENCES

PREFACE

1 See the *Energy Improvement and Extension Act of 2008 (Engrossed Amendment as Agreed to by Senate)*, H.R.6049, 23 September 2008, and *Energy Improvement and Extension Act of 2008 (Introduced in House)*, H.R.7201, 28 September 2008, available through http://thomas. loc.gov. For an overview, see Christopher Helman, 'Green Energy Boom In Bailout Bill', *Forbes*, 2 October 2008, on www.forbes.com/business/ energy/2008/10/02/ green-energy-taxes-biz-energy- cx_ch_1002energy08_taxes. html. On the tax breaks for plug-in hybrid vehicles, see John O'Dell, 'Plug-In Tax Credits Hitching Ride On Wall Street Bail-Out Bill', *Green Car Advisor*, 3 October 2008, on http://blogs. edmunds.com/greencaradvisor/ FuelsTechnologies/Plug- insandElectric/.

2 Vaclav Smil, *Energy at the Crossroads*, Chapter 3, 'Against forecasting', MIT Press, 2003, p121.

3 Man-made greenhouse gases (GHGs) are 85 per cent carbon dioxide (CO_2). The other 15 per cent is composed of methane/ natural gas (CH_4), nitrous oxide, (N_2O) and three rather hairy kinds of fluorinated gases, (F-gases): sulphur hexafluoride, hydrofluorocarbons and perfluorocarbons. The last two kinds of emissions have come to prominence because they are substitutes for chlorofluorocarbons (CFCs), which have been phased out for damaging the Earth's ozone layer.

4 Nigel Lawson, *An Appeal to Reason: A Cool Look at Global Warming*, Gerald Duckworth & Co, 2008.

5 EdF, 'The Energy Deal', on www.savetodaysavetomorrow. com.

6 Jonathan Freedland, 'When old dogmas die, there is room for all kinds of radical new thinking', *The Guardian*, 15 October 2008, on www.guardian.co.uk/ commentisfree/2008/oct/15/ economic-policy-banking.

7 See for example Jonathan Porritt, *Capitalism as if the World Matters*, Earthscan, 2005.

8 'Obama On Responding To 9/11', *Wall Street Journal* Video, 9 September 2008, on http:// online.wsj.com/video/obama-on- responding-to-911/1CE9C161- C4E3-4F9C-AF11- 7CC8BA712F93.html.

9 Ibid.

10 Department of Business, Enterprise & Regulatory Reform (BERR), *The UK Fuel Poverty Strategy*, November 2001, on www.berr.gov.uk/whatwedo/ energy/fuel-poverty/strategy/ index.html.

11 'Government facing fuel court case', *BBC News online*, 6 October 2008, on http://news. bbc.co.uk/1/hi/uk/7653939.stm.

12 Patrick Hennessy, 'Alistair Darling turns to Keynes as he looks to spend his way out of recession', *The Sunday Telegraph*, 19 October 2008, on www.

telegraph.co.uk/finance/
economics/article3223224.ece.

13 See John Maynard Keynes,
Preface to the German edition of
*The General Theory of
Employment, Interest and Money*,
7 September 1936, on http://
etext.library.adelaide.edu.au/k/
keynes/john_maynard/k44g/
k44g.html#preface2.

CHAPTER 1

1 See for example Hamish McRae,
'$130 a barrel and rising: it's a
Seventies-style shock but this
time we won't be held to ransom',
The Independent on Sunday,
25 May 2008, on www.
independent.co.uk/news/
business/comment/hamish-
mcrae/hamish-mcrae-130-a-
barrel-and-rising-its-a-
seventiesstyle-shock-but-this-
time-we-wont-be-held-to-
ransom-833815.html.

2 'Turmoil In Financial Markets
Overspills To Energy', *Energy
Insider*, October 2008, p3, see
http://members.aol.com/
eeinformer/.

3 Jamal Saghir, World Bank
Director of Energy, Transport and
Water, quoted in 'Inside Africa:
Africa's Energy Crisis', *CNN.com*,
aired on 12 May 2007, on http://
transcripts.cnn.com/
TRANSCRIPTS/0705/12/i_if.01.
html.

4 Charts from H-Holger Rogner,
Dadi Zhou and others,
'Introduction' in B Metz and
others, eds, *Climate Change
2007: Mitigation. Contribution of
Working Group III to the Fourth
Assessment Report of the
Intergovernmental Panel on
Climate Change (IPCC)*, Chapter
1, p104, on www.ipcc.ch/
ipccreports/ar4-wg3.htm. In the
first chart, 'Industry' emissions
arise mostly from energy-
intensive sectors such as steel
production. 'Deforestation'
includes use of fuel wood. 'Other'
includes non-road transport such
as trains; cement production, and
flaring of gas during oil
production. 'International
transport' includes aviation and
marine transport.

5 Ibid., p104.

6 Chart from ibid., p105. 'Energy
supply' excludes refineries and
coke ovens, 'Transport' includes
international transport, 'Buildings'
includes the traditional use of
biomass, 'Industry' includes
refineries and coke ovens, and
'Agriculture' includes the burning
of waste and savannah. 'LULUCF'
means land use, land-use change
and forestry: it includes
deforestation and decay of
biomass following deforestation
or logging. 'Waste' includes
methane from landfill and CO_2
from incineration.

7 Ibid., p104.

8 Vaclav Smil, *Energy at the
Crossroads*, MIT Press, 2003,
p6.

9 IEA, *Energy Technology
Perspectives 2008*, 6 June
2008, pp58, 75, 570. Executive
Summary available on www.iea.
org/Textbase/techno/etp/
ETP_2008_Exec_Sum_English.
pdf. The $227 trillion figure for

2050 is in 2005 dollars.

10   On cleaning and maintenance of solar panels, see for example Queensland Government *Environmental Protection Agency, Solar Power Systems: how they work and ways to keep them working*, on www.epa.qld.gov.au/publications/p01732aa.pdf/Solar_power_systems_how_they_work_and_ways_to_keep_them_working.pdf.

11   Melanie Reid, 'A world of hemp lingerie? No thanks', *The Times*, 21 April 2008.

12   John Guillebaud, *Youthquake: Population, Fertility and Environment in the 21st Century*, Optimum Population Trust, 11 July 2007, on www.optimumpopulation.org/Youthquake.pdf.

13   'Please please me – go veggie and save the world', interview, *The Sunday Times*, 23 December 2007.

14   Austin Williams, *The Enemies of Progress: the Dangers of Sustainability*, Societas, 2008, Chapter 4, 'The indoctrinators'. See also Geraldine Hackett, 'Ditch lessons, schools are told', *The Sunday Times*, 24 June 2007.

15   Paul Ehrlich, 'An ecologist's perspective on nuclear power', in Federation of American Scientists, *Public Interest Report*, Vol. 28, No. 506, May-June 1975, p5, on www.fas.org/faspir/archive/1970-1981/May-June1975.pdf.

16   Amory Lovins, 'Energy Strategy: The Road Not Taken?', *Foreign Affairs*, October 1976, pp68, 69, 74, 76-78, 81. Available on www.rmi.org/images/PDFs/Energy/E77-01_TheRoadNotTaken.pdf.

17   The only extra detail to remember here is that Brazil, with its strong biofuels sector, also makes the West nervous. In energy matters, the West views Brazil almost as part of the East.

18   Elisabeth Rosenthal, 'China Increases Lead as Biggest Carbon Dioxide Emitter', *The New York Times*, 14 June 2008, on www.nytimes.com/2008/06/14/world/asia/14china.html?_r=2&th&emc=th&oref=slogin&oref=slogin.

19   Jonathan Amos, 'Brown urged to resist coal rush', *BBC News*, 14 December 2007, on www.news.bbc.co.uk/1/hi/sci/tech/7143567.stm.

20   Lester B Brown, *Plan B 2.0: Rescuing a Planet Under Stress and a Civilisation in Trouble*, Norton, 2006, p197.

21   Valerie J Karplus, *Innovation in China's Energy Sector*, Stanford University Program on Energy and Sustainable Development, Working Paper #61, March 2007, p18, on http://iis-db.stanford.edu/pubs/21519/WP61__Karplus_China__Innovations.pdf.

22   Bill Emmott, *Rivals: How the Power Struggle Between China, India and Japan Will Shape Our Next Decade*, Allen Lane, 2008, pp180-181. For another assessment, see David Smith, *The Dragon and the Elephant: China, India and the New World Order*, Profile Books, 2007, pp214-219.

23   Alan Beattie, 'Green barricade' *Financial Times*, 24 January 2008. The compulsory purchase of permits to emit is also a tactic that the West may use to frustrate Eastern exporters. Ibid.

24   WWF biannual Living Planet Report, cited in 'Global ecosystems "face collapse"', *BBC News*, 24 October 2006, on http://news.bbc.co.uk/1/hi/sci/tech/6077798.stm.

25   IEA, op. cit., and BP, *BP Statistical Review of World Energy* 11 June 2008, on www.bp.com/liveassets/bp_internet/globalbp/globalbp_uk_english/reports_and_publications/statistical_energy_review_2008/STAGING/local_assets/downloads/pdf/statistical_review_of_world_energy_full_review_2008.pdf.

26   Barbara Ward and René Dubos, *Only One Earth: The Care And Maintenance Of A Small Planet*, WW Norton, 1972.

27   Paul Scherrer Institut, 'Neutrons for research and nuclear waste disposal', 31 January 2007, on www.psi.ch/medien/Medienmitteilungen/mm_megapie_jan07/mm_megapie_jan07_bild_E.html, and 'The Megapie Experiment – Facts & Figures', 31 January 2007, on www.psi.ch/medien/Medienmitteilungen/mm_megapie_jan07/PSI_MM_Megapie_Jan07_BgInfo_E.pdf.

28   Willie D Jones, 'Synthetic Fuel From a Solar Collector', *IEEE Spectrum*, 7 January 2008, on http://spectrum.ieee.org/

jan08/5866.

29   See 'Avoiding dangerous climate change', International Symposium on the stabilisation of greenhouse gas concentrations, Hadley Centre, Meteorological Office, Exeter, UK, 1-3 February 2005, Report of the International Scientific Steering Committee, May 2005, pp5, 6 and 10, on www.stabilisation2005.com/Steering_Commitee_Report.pdf.

30   Naama Goren-Inbar and others, 'Evidence of Hominin Control of Fire at Gesher Benot Ya`aqov, Israel', *Science*, Vol. 304, No. 5671, 30 April 2004, pp725-727, on www.sciencemag.org/cgi/content/full/304/5671/725?maxtoshow=&HITS=20&hits=20&RESULTFORMAT=&andorexacttitle=or&andorexacttitleabs=or&fulltext=Goren-Inbar&andorexactfulltext=phrase&searchid=1&FIRSTINDEX=0&sortspec=relevance&fdate=4/1/2004&tdate=5/31/2004&resourcetype=HWCIT.

31   Michael Balter, 'Earliest Signs of Human-Controlled Fire Uncovered in Israel', *Science*, Vol. 304, No. 5671, 30 April 2004, pp663-665, on www.sciencemag.org/cgi/content/full/304/5671/663a.

32   Jonathan Leake, 'A lungful of carbon delusion', *The Sunday Times*, 16 December 2007.

33   OECD, *Main Science and Technology Indicators (MSTI): 2008/1*, 26 May 2008, sample charts, p3, on www.oecd.org/dataoecd/49/45/24236156.pdf.

34   Richard Doornbosch and Simon Upton, *Do we have the right R&D*

priorities and programmes to support the energy technologies of the future?*, OECD, Round Table on Sustainable Development, SG/SD/RT(2006)1, 30 June 2006, pp31, 32, 36, 37, on www.oecd.org/document/31/0,3343,en_39315735_39312980_39398815_1_1_1_1,00.html

35   John Krenickie, quoted in Ed Crooks, 'Twin threats and a lack of leadership', *Financial Times Special Report: Energy*, 9 November 2007, on www.ft.com/cms/s/1/79e56ee8-8d15-11dc-a398-0000779fd2ac,dwp_uuid=4bf66816-8c39-11dc-b887-0000779fd2ac.html.

36   See for example Greenpeace USA, 'Alaska Oil Pipeline Cracks, BP Shuts Down Prudhoe Bay', 8 August 2006, final paragraph, on www.greenpeace.org/usa/news/alaska-oil-pipeline-cracks-bp.

37   For an entire website hostile to biofuels, see www.biofuelwatch.org.uk.

38   The idea that biofuels would lead to deforestation leapt to high profile after the publication of Timothy Searchinger and others, 'Use of US Croplands for Biofuels Increases Greenhouse Gases Through Emissions from Land-Use Change', *Science*, Vol. 319, No. 5867, 29 February 2008, pp1238-1240

39   Merlin Hyman, director of the UK's Environmental Industries Commission, quoted in Andrew Bounds, 'Industry warns on crop fuel targets', *Financial Times*, 15 January 2008.

40   Andrea Rossi and Yianna Lambro,

Gender and equity issues in liquid biofuels production: minimizing the risks to maximize the opportunities, Food and Agriculture Organization of the United Nations, May 2008, on ftp.fao.org/docrep/fao/010/ai503e/ai503e00.pdf.

41   See the furore around an April 2008 World Bank internal working paper, later published as a formal policy research working paper, criticising US biofuels for their inflationary impact. Donald Mitchell, *A Note on Rising Food Prices*, The World Bank Development Prospects Group, July 2008, on www-wds.worldbank.org/external/default/WDSContentServer/IW3P/IB/2008/07/28/000020439_20080728103002/Rendered/PDF/WP4682.pdf. www.ethanolrfa.org/objects/documents/1812/lecg_work_bank_critique.pdf.

42   'Tide power plan is "wrong option"', *BBC News*, 1 October 2007, on http://news.bbc.co.uk/1/hi/uk/7021355.stm.

43   David W Keith and others, 'The influence of large-scale wind power on global climate', *Proceedings of the National Academy of Science*, Vol. 101, No. 46, 16 November 2004, on www.pnas.org/cgi/reprint/101/46/16115.

44   Energy Saving Trust, *Rise of the Machines*, June 2006, p10, on www.energysavingtrust.org.uk/uploads/documents/aboutest/Riseofthemachines.pdf.

45   Ibid., p35.

46   Office for National Statistics,

Labour Market Statistics, September 2008. First release, p3, on www.statistics.gov.uk/pdfdir/lmsuk0908.pdf.

47 Peter W Huber and Mark P Mills, *The Bottomless Well: The Twilight of Fuel, the Virtue of Waste, and Why We Will Never Run Out of Energy*, Basic Books, 2005, pp xxiii, 144 and Chapter 9, 'Insatiable demand'.

48 The somewhat sinister-sounding International Tanker Owners Pollution Federation publishes figures which suggest that the quantities of oil, in tonnes, spilt annually are as follows: 1970s –314,200; 1980s – 117,600; 1990s –113,800 2000-2007 – 24,000. See ITOPF, 'Statistics', on www.itopf.com/information%2Dservices/data%2Dand%2Dstatistics/statistics.

49 'Getting there? Oil-rig safety', *The Economist*, 10 August 1985; James Woudhuysen, 'Naval supremacy still rules the world', *The Listener*, 12 June 1986.

50 Health and Safety Executive, 'Offshore industry has more to do says HSE', press release E045:07, 21 November 2007, on www.hse.gov.uk/press/2007/e07045.htm.

51 Lord Nicholas Stern, *Stern Review on the Economics of Climate Change*, HM Treasury, 2006, on www.hm-treasury.gov.uk./independent_reviews/stern_review_economics_climate_change/stern_review_report.cfm.

52 Ibid., p1.

53 Paul Hawken, Amory B Lovins and L Hunter Lovins, *Natural Capitalism: The Next Industrial Revolution*, Earthscan, 1999, and Jeffrey Sachs (ed), *Investing in Development: A Practical Plan to Achieve the Millennium Development Goals*, Earthscan, 2005, on www.unmillenniumproject.org/reports/fullreport.htm.

54 Alfred Marshall, *The Principles of Economics*, Macmillan, 1890. Chapter 2 – Wealth, on http://socserv.mcmaster.ca/econ/ugcm/3ll3/marshall/prin/prinbk2.

55 Arthur Cecil Pigou, *The Economics of Welfare*, 1920, Part II, Chapter IX – 'Divergences between marginal social net product and marginal private net product', on www.econlib.org/library/NPDBooks/Pigou/pgEW20.html#Part%20II,%20Chapter%209.

56 Ronald H Coase, 'The Problem of Social Cost', *Journal of Law and Economics*, October 1960, p9, on www.sfu.ca/~allen/CoaseJLE1960.pdf.

57 US Environmental Protection Agency, *Cap and Trade: Acid Rain Program Basics*, on http://epa.gov/airmarkets/cap-trade/docs/arbasics.pdf. The other 97 per cent of permits are distributed free.

58 US Environmental Protection Agency, *Clean Air Mercury Rule*, on www.epa.gov/oar/mercuryrule/basic.htm.

59 Figure derived from Energy Information Administration,

International Energy Annual 2005, 'World Primary Energy Consumption (Btu), 1980-2005', Table E1, on www.eia.doe.gov/pub/international/iealf/tablee1.xls.

60   International Energy Agency, *Key World Energy Statistics 2007*, pp6, 7, on www.iea.org/textbase/nppdf/free/2007/key_stats_2007.pdf. The OECD is the Paris-based Organisation of Economic Cooperation and Development, a research body whose 30 members, most of which are in the developed world, are listed on www.oecd.org.

61   Ibid., pp28-9. Here, fossil fuels burnt to generate electricity figure under 'Electricity', and 'other' includes combustible renewables, waste, geothermal, solar and wind; it thus includes some electricity.

62   In the first place, the world needs more nuclear engineers. We take up the UK's predicament in Chapter 4; as for the US, the Nuclear Energy Institute notes that nearly half the US nuclear workforce is aged 47 or older. See 'Nuclear Renaissance Presents Significant Employment Opportunities, NEI Tells Senate Panel', Nuclear Energy Institute press release, 6 November 2007, on www.nei.org/newsandevents/newsreleases/berrigansenaterelease/.

63   IEA, *Energy Technology Perspectives 2008*, op. cit., pp3, 56.

64   One way of stating the Second Law of Thermodynamics is that heat can never be converted into useful work with 100 per cent efficiency. If, say, 99.9 per cent efficiency could be achieved, that would only be an unimportant detail. In fact, in combination with other laws of thermodynamics, it turns out that maximum efficiency is much more limited. For example, a power station boiler at 540°C dumping waste heat into a cooling tower at 20°C cannot be more than 65 per cent efficient at converting heat to work.

65   IEA, *Energy Technology Perspectives 2008*, op. cit., p43.

66   Ibid., p40.

67   Matt Power, 'Don't Buy That New Prius! Test-Drive a Used Car Instead', *Wired*, 16 June 2008, on www.wired.com/science/planetearth/magazine/16-06/ff_heresies_09usedcars.

68   The White House, 'Global Climate Change Policy Book', Executive Summary, February 2002, on www.whitehouse.gov/news/releases/2002/02/climatechange.html.

69   Going back millions of years, there is evidence that CO_2 levels, and temperatures, have been much higher than those of the present. But overall conditions on the planet were very different from those that obtain today.

70   Myles Allen and David Frame, 'Call off the Quest', *Science*, Vol. 318, No. 5850, 26 October 2007, p583.

71   Ibid.

72   All figures are for ppm CO_2eq unless otherwise stated.

73 Richard B Alley and others, 'Summary for Policymakers', in Susan Solomon and others, editors, *Climate Change 2007: The Physical Science Basis. Contribution of Working Group I to the Fourth Assessment Report of the IPCC*, Cambridge University Press, 2007, p12, on www.ipcc.ch/pdf/assessment-report/ar4/wg1/ar4-wg1-spm.pdf.

74 IEA, *Energy Technology Perspectives 2008: Scenarios and Strategies to 2050*, 2008, p51.

75 Lord Nicholas Stern, op. cit., Chapter 7, p117, on www.hm-treasury.gov.uk/d/Chapter_7_Projecting_the_Growth_of_Greenhouse-Gas_Emissions.pdf.

76 Lord Turner of Ecchinswell, *Interim Advice by the Committee on Climate Change*, 7 October 2008, on www.theccc.org.uk/downloads/Interim%20report%20letter%20to%20DECC%20SofS.pdf.

77 Barack Obama and Joe Biden, *New Energy for America*, fact sheet, no date, p2, on www.barackobama.com/pdf/factsheet_energy_speech_080308.pdf.

78 IEA, *Energy Technology Perspectives 2008*, op. cit., p51.

79 Lord Nicholas Stern, op. cit., p12, Chapter 13, p299, on www.hm-treasury.gov.uk/d/Chapter_13_Towards_a_Goal_for_Climate-Change_Policy.pdf.

80 James Hansen, 'Statement of James E Hansen', to a Maidstone criminal court in defence of the Kingsnorth Six, 10 September 2008, p10, on www.columbia.edu/~jeh1/mailings/20080910_Kingsnorth.pdf.

81 Kyoto Protocol to the United Nations Framework Convention on Climate Change, available from http://unfccc.int/resource//docs/convkp/kpeng.pdf. For individual country targets, see Annex B, p20.

82 Council of the European Union, *Presidency Conclusions of the Brussels European Council (8/9 March 2007)*, 2 May 2007, p12, on http://register.consilium.europa.eu/pdf/en/07/st07/st07224-re01.en07.pdf.

83 Lord Turner of Ecchinswell, *Interim advice*, op. cit., and Department of Energy and Climate Change, 'UK leads world with commitment to cut emissions by 80 per cent by 2050', 16 October 2008, on http://nds.coi.gov.uk/environment/ fullDetail.asp?ReleaseID=381477&NewsAreaID=2.

84 Barack Obama and Joe Biden, *New Energy for America*, fact sheet, op. cit., p2.

85 IEA, *Energy Technology Perspectives 2008*, op. cit., p55.

86 Lord Nicholas Stern, op. cit., Chapter 8, p201, on www.hm-treasury.gov.uk/d/ Chapter_8_The_Challenge_of_Stabilisation.pdf.

87 James Hansen, 'Statement of James E Hansen', op. cit., p10.

88 Alley and others, op. cit., Table SPM3, p13.

89 IEA, op. cit., p37. The IEA here relies on the authority of the IPCC.

90 Lord Nicholas Stern, op. cit., p12.

91 IEA, op. cit., p58.

92 James Hansen, 'A Slippery Slope: How much global warming constitutes "Dangerous Anthropogenic Interference"?' *Climatic Change*, Vol. 68, pp269-276, 2005, p278.

93 Richard B Alley and others, op. cit., Table SPM3, pp13, 14.

94 IEA, op. cit., p37.

95 Lord Nicholas Stern, op. cit., 'Part II: The impacts of climate change on growth and development', far right column in Table 3.1, p57, on www.hm-treasury.gov.uk./media/D/8/Part_II_Introduction_group.pdf.

96 Al Gore, *An Inconvenient Truth*, 2006, transcript on http://forumpolitics.com/blogs/2007/03/17/an-inconvient-truth-transcript/.

97 James Hansen, 'Statement of James E. Hansen', to a UK criminal court in defence of the Kingsnorth Six, 10 September 2008, p8, on www.columbia.edu/~jeh1/mailings/20080910_Kingsnorth.pdf.

98 IPCC, *Climate Change 2007: Synthesis Report,* 'Global anthropogenic GHG emissions', Figure 2.1, 2007, p36, on www.ipcc.ch/pdf/assessment-report/ar4/syr/ar4_syr.pdf.

99 G A Meehl and others, 'Global Climate Projections', Chapter 10 in Susan Solomon and others, op cit, p798, on www.ipcc.ch/pdf/assessment-report/ar4/wg1ar4-wg1-chapter10.pdf.

100 J D Annan and J C Hargreaves. 'Using multiple observationally-based constraints to estimate climate sensitivity', *Geophysical Research Letters*, Vol. 33, L06704, 18 March 2006. Abstract on www.agu.org/pubs/crossref/2006/ 2005GL025259.shtml.

101 Barack Obama, *New Energy for America*, Speech in Lansing, Michigan, 4 August 2008, transcript on www.pbs.org/newshour/ bb/politics/july-dec08/obamaenergy_08-04.html.

102 'PM Speech on climate change', *No10.gov.uk*, 14 September 2004, on www.number10.gov.uk/, p6333.

103 H J Schellnhuber and others, Preface to *Avoiding Dangerous Climate Change*, Cambridge University Press, 2005. pxi, on www.defra.gov.uk/environment/climatechange/research/dangerous-cc/pdf/avoid-dangercc.pdf.

104 In 2005, total distribution losses were eight per cent in the UK grid and six per cent in the US grid. Data from IEA, 'Electricity/Heat in United Kingdom in 2005', on www.iea.org/Textbase/stats/electricitydata.asp?COUNTRYCODE=GB&Submit=Submit.

CHAPTER 2

1 'The Kitchen Debate' was held at the American National Exhibition in Moscow. It is on www.cnn.com/SPECIALS/cold.war/episodes/14/documents/debate/.

2 The famous and often misquoted statement on society is in *Woman's Own*, 31 October 1987, on www.MargaretThatcher.org/speeches/displaydocument.asp?docid=106689. The clarification, which was carried out in a later Downing Street statement, was published in the *Sunday Times* 'Atticus' column, 10 July 1988.

3 HM Treasury, Speech by the Rt Hon Gordon Brown MP, Chancellor of the Exchequer, to United Nations Ambassadors, New York, 20 April 2006, on http://www.hm-treasury.gov.uk/1888.htm.

4 House of Lords Science and Technology Committee, *2nd Report of Session 2005-06: Energy Efficiency, Volume I*, 15 July 2005, Annex 2, Table 6.1, p126, using DTI 2004 figures and on www.parliament.uk/parliamentary_committees/lords_s_t_select/efficiency.cfm. Research by Dr Phil Sinclair, of the University of Surrey, may give better guidance than government accounts: Ibid., sections 2.36-2.43 and Appendix 4. We have used Sinclair's path 1, which takes a pessimistic view of UK carbon intensity, 1990-2003, and uses DEFRA, *Review of the UK Climate Change Programme*, 2004, Table 6.

5 The Institution of Electrical Engineers, *Energy Efficiency in the Home: A Guide to Effective Energy Conservation in the Home, Revised Edition*, June 2004, p4, gives a figure of 11.9 per cent for the CO_2 contribution made by lighting back in 1995/96. Given the growth of dishwashers, consumer electronics and IT in the home, since that time, a figure of 10 per cent now seems reasonable.

6 DEFRA, *Climate Change: The UK Programme 2006*, March 2006, p75, on www.defra.gov.uk/environment/climatechange/uk/ukccp/index.htm.

7 James C Franklin, 'Employment outlook: 2006-16 – an overview of BLS projections to 2016', *Monthly Labor Review*, November 2007, pp3-4, on www.bls.gov/opub/mlr/2007/11/art1full.pdf.

8 Betty W Su, 'The US economy to 2016: slower growth as boomers begin to retire', in 'Employment outlook: 2006-16', *Monthly Labor Review*, November 2007, pp17-18, on www.bls.gov/opub/mlr/2007/11/art2full.pdf.

9 DEFRA, 'Estimated emissions of carbon dioxide (CO_2 expressed as Carbon) by National Communication Source Category: 1970-2006', Table 5, 25 September 2008, on www.defra.gov.uk/environment/statistics/globatmos/alltables.htm. The total includes other small items; percentages are rounded.

10 Thomas Malthus, *An Essay on the Principle of Population*, 1798, first edition, Ch. 1, paragraph 15; Ch. 3, paragraph 14; and Ch. 5, paragraph 14, on www.econlib.

org/library/Malthus/malPop.html. Letter to David Ricardo, 16 July 1822, quoted in John Maynard Keynes, *The General Theory of Employment, Interest and Money*, 1936, Chapter 23, Section VII, on http://etext.library.adelaide. edu.au/k/keynes/john_maynard/ k44g/k44g.html.

11   Thomas Malthus, *An Essay*, op. cit., Ch. 11, paragraph 6.

12   Jared Diamond, 'What's Your Consumption Factor?', *The New York Times*, 2 January 2008, on www.nytimes.com/2008/01/02/ opinion/02diamond.html?_ r=2&pagewanted=all&oref =slogin.

13   Andy Jones, *Eating Oil: Food Supply in a Changing Climate*, Sustain and Elm Farm Research Centre, November 2001, available price £10 on www. sustainweb.org/publications/ order/98/.

14   Northern Ireland Electricity, 'Are You Energy Obese?', press release, 24 January 2006 www. nie.co.uk/media/newsstory. asp?idcode=900. Note, too, that the Energy Saving Trust (EST) asks: 'Do you know where your home sits on the energy obesity scales?' See EST, 'The energy obesity quiz', on www.est.org.uk/ myhome/whatcan/quiz.

15   Centre for Alternative Technology, *Zero Carbon Britain: An Alternative Energy Strategy*, 2007, p7, available price £15 on www2.cat.org.uk/ shopping/product_info. php?products_ id=759&osCsid=64.

16   See Anson Rabinbach, *The Human Motor: Energy, Fatigue, and the Origins of Modernity*, University of California Press, 1992.

17   See Rob Lyons, 'The dangers of fried food and a fried planet', *spiked*, 17 October 2007, on www.spiked-online.com/index. php?/site/article/3975/, and James Woudhuysen, 'Brown's "get fit" towns: Kim Jong-il would be proud', *spiked*, 5 November 2007, on www.spiked-online. com/index.php?/site/ article/4038/.

18   Thomas Malthus, *An Essay*, op. cit., but sixth edition, Book IV, Ch. XIV, paragraph 14, on www. econlib.org/library/Malthus/ malPop.html.

19   Thorstein Veblen, *The Theory of the Leisure Class*, 1899, Chapter Four, on http://socserv2. mcmaster.ca/%7Eecon/ ugcm/3ll3/veblen/leisure/chap04. txt.

20   Tim Jackson, 'Sustainable consumption: the psychology and practicalities of changing the way we shop', lecture to the Royal Society for the encouragement of Arts, Manufactures and Commerce (RSA), London, 19 October 2005, on www.thersa. net/acrobat/ forumforfuture_191005.pdf.

21   Camilla Toulmin, 7 November 2005, quoted in 'So how green is my carbon?', *RSA Carbon Limited Project*, 8 November 2006, on www.rsacarbonlimited.org/ viewarticle.aspa?pageid=569

22   Quoted in Richard Gray, 'Chief

scientist in sports cars warning to women', *The Daily Telegraph*, 17 December 2007, on www.telegraph.co.uk/news/main.jhtml?xml=/news/2007/12/16/ncar116.xml.

23   Thorstein Veblen, 'The preconceptions of economic science, part III', *The Quarterly Journal of Economics*, Volume 14, 1900, on http://socserv2.mcmaster.ca/%7Eecon/ugcm/3ll3/veblen/prec3.txt.

24   Thorstein Veblen, *The Engineers and the Price System,* on http://socserv2.mcmaster.ca/%7Eecon/ugcm/3ll3/veblen/prec3.txt.

25   J A Hobson and A F Mummery, *The Physiology of Industry: Being an Exposure of Certain Fallacies in Existing Theories of Economics* (1889), Kessinger Publishing, 2008.

26   J A Hobson, *Imperialism, A Study*, Part II, Chapter IV, Part I, on www.marxists.org/archive/hobson/1902/imperialism/pt2ch4.htm, and Part I Chapter V, on www.marxists.org/archive/hobson/1902/imperialism/pt1ch6.htm.

27   J M Keynes, *The General Theory*, op. cit., Chapter 23, 'Notes on mercantilism, the usury laws, stamped money and theories of under-consumption'.

28   J M Keynes, *Economic Possibilities for our Grandchildren*, 1930, on www.econ.yale.edu/smith/econ116a/keynes1.pdf.

29   Abraham Maslow, 'A theory of human motivation', *Psychological Review*, No. 50, September 1943, pp370-396, on http://psychclassics.yorku.ca/Maslow/motivation.htm.

30   Ibid.

31   Peter Tertzakian, *A Thousand Barrels a Second: The Coming Oil Break Point and the Challenges Facing an Energy Dependent World*, McGraw-Hill, 2006, pp19-20.

32   John Kenneth Galbraith, *The Affluent Society* (1958), Penguin Books, second edition, 1970, pp147, 149, 152-154.

33   Daniel Ben-Ami, 'The midwife of miserabilism', *spiked review of books*, No. 9, January 2008, on www.spikedonline.com/index.php?/site/reviewofbooks_article/4363/.

34   Karl Marx, *Introduction to a Contribution to the Critique of Political Economy*, 1857, on www.marxists.org/archive/marx/works/1859/critique-pol-economy/appx1.htm.

35   Vance Packard, *The Hidden Persuaders*, David McKay Company, 1957.

36   Vance Packard, *The Wastemakers* (1960), Pocket Books, 1963, p7.

37   Ibid., pp171, 175.

38   *The Economist*, 13 December 2001; Richard Heinberg, *PowerDown: Options And Actions For A Post-Carbon World*, New Society Publishers, 2004, p110; Thomas Friedman, 'The Oil-Addicted Ayatollahs', *The New York Times*, 2 February 2007.

39   George W Bush, State of the Union address, 31 January 2006,

on www.whitehouse.gov/
stateoftheunion/2006/index.html;
Gordon Brown, press conference,
12 June 2008, on www.
number10.gov.uk/output/
Page15739.asp; Barack Obama,
New Energy for America, Speech
in Lansing, Michigan, 4 August
2008, transcript on www.pbs.org/
newshour/bb/politics/july-dec08/
obamaenergy_08-04.html.

40　Frank Furedi, 'Environmentalism',
spiked, 12 September 2007, on
www.spiked-online.com/index.
php?/site/article/3817/.

41　John Elkington and Mark Lee,
'Catch a Wave', *Grist*, 15
November 2005, on www.grist.
org/biz/fd/2005/11/15/rosa/.

42　*Records of the Cabinet Office*,
minutes of 16 October 1973
meeting, The National Archives,
on www.nationalarchives.gov.uk/
documentsonline/details-result.
asp?Edoc_Id=1070738&queryTy
pe=1&resultcount=282.

43　Richard Wightman Fox and T J
Jackson Lears, *The Culture of
Consumption: Critical Essays in
American History, 1880-1980*,
Pantheon Books, 1983, px.

44　Donella Meadows and others,
*The Limits to Growth: A Report to
the Club of Rome's Project on the
Predicament of Mankind*,
Universe, 1972; Victor Papanek,
*Design For the Real World:
Human Ecology and Social
Change*, Pantheon Books, 1971;
Ernst Schumacher, *Small is
Beautiful: Economics As If People
Mattered* (1973),
HarperPerennial, 1989.
Schumacher is treated in more
detail in Chapter 7.

45　Ivan Illich, *Energy and Equity*,
Marion Boyars Publishers, 1974,
Chapter I, p18. Chapter 1 was
first published in *Le Monde* in
early 1973. The whole book is on
www.cogsci.ed.ac.uk/~ira/illich/
texts/energy_and_equity/
energy_and_equity.txt.

46　Ivan Illich, *Energy and Equity*,
Marion Boyars Publishers, 1974,
pp23, 28, 36, 37, 41, 45.

47　Ibid., pp16, 20, 29.

48　Ivan Illich, *Tools for Conviviality*,
1973, Chapter III, on http://
opencollector.org/history/
homebrew/tools.html, and *Energy
and Equity*, op. cit., p22.

49　Robin McKie and Juliette Jowit,
'Can science really save the
world?', *The Observer*, 7 October
2007, on http://observer.
guardian.co.uk/focus/
story/0,,2185343,00.html.

50　'Is this what it takes to save the
world?' *Nature*, No. 7141, 10
May 2007, p132.

51　Lisa Rosner, 'Introduction', in
Rosner, editor, *The Technological
Fix: How People Use Technology
to Create and Solve Problems*,
Routledge, 2004, pp1, 3.

52　Michael Gibbons, review of René
Dubos, *Reason Awake: Science
for Man*, *Nature*, Vol. 228. No.
5269, p387, 24 October 1970,
available for $30 on www.nature.
com/nature/journal/v228/n5269/
pdf/228387a0.pdf.

53　Shelley McKellar, 'Artificial
hearts: a technological fix more
monstrous than miraculous?', in
Rosner, op. cit., p24.

54 Thomas P Hughes, 'Afterword', in Rosner, op. cit., p241.

55 Lisa Rosner, 'Introduction', and Thomas P Hughes, 'Afterword', in Rosner, op. cit., pp2, 242.

56 Joe Sharkey, 'The Nation: "Enabling" Is Now a Political Disease', *The New York Times*, 27 September 1998, on http://query. nytimes.com/gst/fullpage.html?re s=9C04E1D71539F934A1575A C0A96E958260&sec=&spon=& pagewanted=all.

57 Margaret Shotton, *Computer Addiction? A Study of Computer Dependency*, Taylor & Francis, 1989; Julia Scheeres, 'The quest to end game addiction', *Wired*, 5 December 2001, on www.wired. com/gaming/gamingreviews/ news/2001/12/48479; Andy McCue, 'Employers crack down on Facebook "addicts"', *silicon. com*, 31 August 2007, on www. silicon.com/ciojury/ 0,3800003161,39168320,00. htm; Richard George, roads campaigner at the Campaign for Better Transport, quoted in Dan Milmo, 'Local authorities face £671m bill after road schemes go over budget', *The Guardian*, 7 July 2008, on www.guardian.co. uk/politics/2008/jul/07/transport. localgovernment.

58 Tim Jackson, op. cit.

59 Michael Common and Sigrid Stagl, *Ecological Economics*, Cambridge University Press, 2005, 'Economic performance and the environment – energy use', Supplement to Chapter Seven, pp1-4, on www. cambridge.org/ resources/0521816459/3283_

Supplement%20ch.7%20 Energy%20use.pdf.

60 Katrin Rehfanz and David Maddison, 'Climate and Happiness', *Ecological Economics*, Vol. 52, Issue 1, January 2005, on www.disefin. unige.it/finanza/Ipses/vol%20 52%20issue%201%20n.10.pdf.

61 See James Woudhuysen and Ian Abley, *Why is Construction so Backward?,* Wiley, 2004, Chapter 2, Section 1, 'The reduction of strategy to measurement'.

62 Jeremy Bentham, *An Introduction to the Principles of Morals and Legislation*, Chapters I and IV, 1781, on www.utilitarianism.com/ jeremy-bentham/index.html#five.

63 William Allan, *Studies in African Land Usage in Northern Rhodesia*, 1949, cited in James Heartfield, 'Celebrating the "human footprint"', *spiked*, Thursday 26 April 2007, on www. spiked-online.com/index.php?/ site/article/3201/.

64 Martin Wackernagel and William Rees, *Our Ecological Footprint: Reducing Human Impact on the Earth*, New Society Publishers, 1996; Global Footprint Network, 'Ecological Footprint: Overview', on www.footprintnetwork.org/ gfn_sub.php?content=footprint_ overview.

65 Global Footprint Network, 'October 6 is Ecological Debt Day', on www.footprintnetwork. org/gfn_sub.php?content =overshoot.

66 Thomas Wiedmann and Jan Minx, *A Definition of 'Carbon Footprint'*, ISAUK Research & Consulting,

June 2007, pp1, 2, on www.isa-research.co.uk/docs/ISA-UK_Report_07-01_carbon_footprint.pdf.

67   Josie Appleton, 'Let us bin the moral fable of climate change', *spiked*, 20 August 2007, on www.spiked-online.com/index.php?/site/boxarticle/3738/.

68   J M Keynes, *The General Theory*, op. cit., Chapter 24, 'Concluding notes on the social philosophy towards which the general theory might lead'.

69   James Bellini, *High-Tech Holocaust*, David & Charles, 1986; Laurie Garrett, *The Coming Plague: Newly Emerging Diseases in a World out of Balance*, Atlantic Books, 1994.

70   Sustainable Development Commission, *Shows promise. But must try harder – an assessment by the SDC of the Government's reported progress on sustainable development over the past five years*, 13 April 2004, paragraphs 35-37, p16, on www.sd-commission.org.uk/publications/downloads/040413-Shows-promise.But%20must-try-harder.pdf.

CHAPTER 3

1   Meteorological Office, *England S 1971-2000 averages*, on www.metoffice.gov.uk/climate/uk/averages/19712000/areal/england_s.html.

2   Meteorological Office, *November 2007*, on www.metoffice.gov.uk/climate/uk/2007/november.html.

3   Piers Forster, Venkatachalam Ramaswamy and others, 'Changes in Atmospheric Constituents and in Radiative Forcing', Chapter 2 in Susan Solomon and others, editors, Climate Change 2007: *The Physical Science Basis. Contribution of Working Group I to the Fourth Assessment Report of the IPCC*, Cambridge University Press, 2007, p136, on www.ipcc.ch/pdf/assessment-report/ar4/wg1/ar4-wg1-chapter2.pdf, and IPCC, *Summary for Policymakers*, figure SPM 2, in ibid, on www.ipcc.ch/ipccreports/ar4-wg1.htm.

4   V Ramanathan has pioneered the use of autonomous unmanned aerial vehicles – 'drones' – to study clouds. A recent paper from his group motivates its effort with this observation: 'For the first time, the IPCC 2007 assessment provides the best estimate for the aerosol direct and indirect effects... however, the magnitude of the uncertainty is still more than a factor 2 greater than those estimates and is still greater than the sum of uncertainties of all other components contributing to climate forcing'. See GC Roberts and others, 'Simultaneous observations of aerosol-cloud-albedo interactions with three stacked unmanned aerial vehicles', *Proceedings of the National Academy of Sciences of the United States*, 27 May 2008, Vol. 105, No. 21, p7370.

5   Ibid.

6   Gabrielle Walker and Sir David King, *The Hot Topic: How to Tackle Global Warming and Still Keep the Lights On,* Bloomsbury

Publishing, 2008.

7 See for example DEFRA, *Climate Change: The UK Programme*, March 2006, 2006, para. 8, p10, on www.defra.gov.uk/ environment/climatechange/uk/ ukccp/pdf/ukccp06-all.pdf.

8 On climate change as a cause of wars in Ethiopia and Darfur, see for example Jeffrey Sachs, *Climate Change and War*, Global Policy Forum, 1 March 2005, on www.globalpolicy.org/socecon/ develop/africa/2005/0301sachs. htm. On climate as a source of conflict in China, 1400-1900 AD, see David D Zhang and others, 'Global climate change, war, and population decline in recent human history', *Proceedings of the National Academy of Sciences of the United States of America*, Vol. 104, No. 49, 4 December 2007, pp19214-19219, on www.pnas. org/content/104/49/19214.full. pdf+html?sid=af32f787-268b-4a57-8170-625d40d67fc0.

9 See for example James Heartfield, 'Celebrating the human footprint', *spiked*, 26 April 2007, on www.spiked-online. com/index.php?/site/ article/3201/.

10 Frank Furedi, *Invitation to Terror: the Expanding Empire of the Unknown*, Continuum, 2007.

11 Sonja Boehmer-Christiansen, 'The Precautionary Principle in Germany – enabling Government', in Tim O'Riordan and James Cameron, editors, *Interpreting the Precautionary Principle*, Earthscan, 1994, pp35-36, 38-39. Interestingly, others claim

that the Principle's roots lie in 1930s German social democracy and its politics of good household management – and that the Principle embraces the cost-effective prevention of risk, and the protection of natural systems, in the face of human fallibility. Tim O'Riordan and James Cameron, 'The History and Contemporary Significance of the Precautionary Principle', Chapter 1 in ibid., p16.

12 Philippe Sands, *Principles of International Environmental Law, Volume 1: Frameworks, standards and implementation*, Manchester University Press, 1996, pp209-211.

13 For an insider's account, see the autobiography of Maurice Strong, who was Secretary General of the Rio Summit. Maurice Strong, *Where on Earth Are We Going?*, Texere, 2001.

14 United Nations Environment Programme, *Rio Declaration on Environment and Development*, 1992, on www.unep.org/ Documents.Multilingual/Default.a sp?DocumentID=78&ArticleID=1 163.

15 The American economist Frank Knight first introduced this distinction. Frank Knight, *Risk, Uncertainty and Profit*, Houghton Mifflin Company, 1921. Available online at www.econlib.org/ LIBRARY/Knight/knRUP.html.

16 Jeroen van der Sluijs and Wim Turkenburg, 'Climate change and the precautionary principle', in Elizabeth Fisher and others, *Implementing the Precautionary Principle: Perspectives and*

Prospects, Edward Elgar Publishing, 2006, pp264, 265.

17   David J Ball and Sonja Boehmer-Christiansen, 'Societal concerns and risk decisions', *Journal of Hazardous Materials*, No. 144, 2007, p560.

18   Ibid., p558.

19   See DoD News Briefing, 'Secretary Rumsfeld and Gen. Myers', US Department of Defense News Transcript, 12 February 2002, on www.defenselink.mil/transcripts/transcript.aspx?transcriptid=2636, and Frank Furedi, Ibid., pp55-60.

20   See Lord Nicholas Stern's Launch Presentation, 30 October 2006, on www.hm-treasury.gov.uk./media/0/3/Slides_for_Launch.pdf.

21   Partha Dasgupta, 'A challenge to Kyoto', review of Bjørn Lomborg, *Cool It: The Skeptical Environmentalist's Guide to Global Warming*, Cyan and Marshall Cavendish, 2007; in *Nature*, Vol. 449, No. 7159, 13 September 2007, pp143-144, on www.nature.com/nature/journal/v449/n7159/full/449143a.html.

22   Thomas Malthus, *An Essay on the Principle of Population*, op. cit., sixth edition, Book I, Chapter I, paragraphs 16, 25, on www.econlib.org/library/Malthus/malPop.html.

23   Stanislaw Ulam, quoted in David K Campbell, 'Nonlinear physics: Fresh breather', *Nature*, Vol. 432, No. 7016, p455, 25 November 2004, on www.nature.com/nature/journal/v432/n7016/full/432455a.html.

24   Ivars Peterson, *Newton's Clock: Chaos in the Solar System*, W H Freeman & Co, 1993.

25   Timothy M Lenton and others, 'Tipping elements in the Earth's climate system', *Proceedings of the National Academy of Sciences*, Vol. 105, No. 6, 7 February 2008.

26   Ibid., p1790.

27   Ibid., p1790.

28   Ibid., pp1790-1791.

29   Ibid., p1791.

30   See Rob Lyons, 'Bali: no more jaw-jaw, this is climate war', *spiked*, 6 December 2007, on www.spiked-online.com/index.php?/site/article/4161/.

31   Gordon Brown, Speech on climate change, 19 November 2007, on www.number10.gov.uk/output/Page13791.asp.

32   Reported in 'Defeating climate change by air', *Financial Times*, 13 February 2008.

33   See for example Meyer Hillman, *How We Can Save the Planet*, Penguin, 2004, pp130-131.

34   Mark Lynas, 'Why we must ration the future', *New Statesman*, 23 October 2006.

35   David Roberts, 'Global warming and the Holocaust', *Grist*, 26 November 2007, on http://gristmill.grist.org/story/2007/11/26/152631/38; Peter Christoff, 'Climate change is another grim tale to be treated with respect', *The Age*, 9 July 2007, on www.theage.com.au/

news/opinion/climate-change-is-another-grim-tale-to-be-treated-with-respect/2007/07/08/1183833338608.html; George Monbiot, *Heat: How to Stop the Planet Burning*, Allen Lane, 2006, p41.

36   'President Bush Addresses Members of the Knesset', 15 May 2008, White House press release, on www.whitehouse.gov/news/releases/2008/05/20080515-1.html.

37   The *Bhagavad Gita* text, probably recalled accurately by Oppenheimer if not the Carbon Trust, is 'I have become *Death*, the destroyer of worlds', our emphasis.

38   Dean Acheson, *Present at the Creation: My Years in the State Department* (1969), WW Norton, 1987, p219.

39   Eisenhower, President's news conference, 7 April 1954, on www.historytools.org/sources/domino.html. NB

40   Paul N Edwards, *The Closed World: Computers and the Politics of Discourse in the Cold War*, The MIT Press, 1996, and Philip Mirowski, *Machine Dreams: Economics Becomes a Cyborg Science*, Cambridge University Press, 2001. On cybernetics, the pioneering work is Norbert Wiener, *Cybernetics, or Control and Communication in the Animal and the Machine*, John Wiley & Sons, 1948. Cybernetics emerged from wartime research on the automatic aiming and firing of guns.

41   Herman Kahn, *On Thermonuclear War* [1965], Transaction Publishers, 2007.

42   Herman Kahn, *On Escalation: Metaphors and Scenarios*, Pall Mall Press, 1965, pp37, 38-40.

43   James Woudhuysen, 'Interim report', interview with Herman Kahn, *Design*, July 1982, p31, on www.woudhuysen.com/index.php/main/article/226.

44   John Gillott and Manjit Kumar, *Science and the Retreat from Reason*, Merlin, 1995, pp.174-77.

45   Gabrielle Walker, 'The tipping point of the iceberg', *Nature*, Vol. 441, No. 7095, 15 June 2006.

46   Remarks in reply to one of the authors, DEMOS conference to launch *The Atlas of Ideas: Mapping the New Geography of Science*, the Institution of Engineering and Technology, Savoy Place, London, 17 January 2007.

47   Chris Freeman, 'Introduction: Malthus with a computer', in HSD Cole and others, editors, *Thinking About the Future: a Critique of the Limits to Growth*, Chatto & Windus for Sussex University Press, 1973.

48   Paragraph 1.1 in Lord Nicholas Stern, op. cit., in Stern, Part 1, 'Approach', Chapter 1, p2, on www.hm-treasury.gov.uk./media/3/6/Chapter_1_The_Science_of_Climate_Change.pdf.

49   Martin Parry, Osvaldo Canziani and others, editors, *Climate Change 2007: Impacts, Adaptation and Vulnerability*.

Contribution of Working Group II to the Fourth Assessment Report of the Intergovernmental Panel on Climate Change, IPCC, 2007, on www.ipcc.ch/ipccreports/ar4-wg2.htm.

50 Bert Metz, Ogunlade Davidson and others, editors, *Climate Change 2007: Mitigation. Contribution of Working Group III to the Fourth Assessment Report of the Inter- governmental Panel on Climate Change*, IPCC, 2007, on www.ipcc.ch/ipccreports/ar4-wg3.htm. For a critique of Working Groups II and III, see James Woudhuysen and Joe Kaplinsky, 'Let's fight back against the new Model Army', *spiked*, 12 July 2007, on www.spiked-online.com/index.php?/site/article/3598.

51 The IPCC defines mitigation as 'Technological change and substitution that reduce resource inputs and emissions per unit of output. Although several social, economic and technological policies would produce an emission reduction, with respect to climate change, mitigation means implementing policies to reduce GHG emissions and enhance sinks'. See IPCC, Annex 1 in *Mitigation*, op. cit., p818.

52 Ibid., p809. The IPCC cites as examples of adaptation the raising of river or coastal dikes, and the replacing of sensitive plants by ones more able to resist shock changes in temperature.

53 Wallace S Broecker and Robert Kunzig, *Fixing Climate: What Past Climate Changes Reveal About the Current Threat – and How to Counter It*, Hill and Wang, 2008.

54 Peter M Vitousek and others, 'Human Domination of Earth's Ecosystems', *Science*, Vol. 277, No. 5325, 26 July 1997, pp494-499.

55 Will Steffen, Paul J Crutzen and John R McNeill, 'The Anthropocene: Are Humans Now Overwhelming the Great Forces of Nature?', *Ambio*, Vol. 36, No. 8, December 2007, pp614-621.

56 Jacob Darwin Hamblin, *Oceanographers and the Cold War: Disciples of Marine Science*, University of Washington Press, 2005. On Cold War science more generally, see Daniel S Greenberg, *The Politics of Pure Science*, University of Chicago Press, 1999.

57 Roger Revelle and Hans E Suess, 'Carbon Dioxide Exchange between Atmosphere and Ocean and the Question of an Increase of Atmospheric CO_2 During the Past Decades', *Tellus*, No. 9, 1957, pp18-27.

58 Spencer R Weart, 'Roger Revelle's Discovery', July 2007, pp 4, 5, 7, 8, on www.aip.org/history/climate/Revelle.htm.

59 Ibid., footnote 3, p5, and Weart, 'The public and climate change', July 2007, on www.aip.org/history/climate/public.htm.

60 Weart, 'Spencer R Weart, 'Roger Revelle's Discovery', op. cit.,

61 Spencer R Weart, *Chaos in the atmosphere*, July 2007, www.aip.org/history/climate/chaos.htm.

62 Spencer R Weart, *Rapid Climate Change*, June 2007, on www.aip.org/history/climate/rapid.

htm#M_60.

63 Rachel Carson, *Silent Spring*, Houghton Mifflin, 1962, p8.

64 Thomas S Kuhn, *The Structure of Scientific Revolutions* (1962), third edition, University of Chicago Press, 1996.

65 Tony Gilland, 'Digging up the roots of the IPCC', *spiked*, 28 June 2007, on http://www.spiked-online.com/index.php?/site/article/3540/.

66 Hervé Le Treut, Richard Somerville and others, 'Historical Overview of Climate Change Science', Chapter 1 in Susan Solomon and others, editors, *Climate Change 2007: The Physical Science Basis. Contribution of Working Group I to the Fourth Assessment Report of the IPCC*, Cambridge University Press, 2007, p103.

67 Spencer R Weart, *The Discovery of Global Warming*, Harvard University Press, 2003, p40.

68 John Elkington and Mark Lee, op. cit.

69 Quoted in Tony Elland, op. cit.

70 The seminal study of the idea that a nuclear conflict could be followed by a long period of deep cold, or 'nuclear winter', is by Carl Sagan and others, 'Nuclear Winter: Global Consequences of Multiple Nuclear Explosions', *Science*, Vol. 23, No. 4630, December 1983. This suggested that an exchange of 100 megatonne nuclear weapons could result in months of subzero temperatures.

71 Margaret Thatcher, speech to the Royal Society, 27 September 1988, on www.margaretthatcher.org/speeches/displaydocument.asp?docid=107346.

72 Shardul Agrawala, 'Context and early origins of the Intergovernmental Panel on Climate Change', *Climatic Change*, Vol. 39, No. 4, August 1998. Available for $32 on www.springerlink.com/content/w28x724593566g1t/.

73 Richard B Alley and others, in Susan Solomon, op. cit., on www.ipcc.ch/pdf/assessment-report/ar4/wg1/ar4-wg1-spm.pdfIPCC.

74 IPCC, 'WG1 "The Physical Science Basis" report: The webcast of the press conference (audio)', Information for the press, 33mins 28 seconds, 53.08 and 53.45, on www.ipcc.ch/press/index.htm.

75 UNEP, 'Meet the Executive Director', no date, on www.unep.org/Documents.Multilingual/Default.asp?DocumentID=43&ArticleID=5252&l=en.

76 Quoted in UNEP, 'Missed Opportunity for G8 Leaders on Climate Change', 9 July 2008, on www.unep.org/Documents.Multilingual/Default.asp?DocumentID= 540&ArticleID=5864&l=en.

77 'People's car plan is an eco-disaster: Pachauri', *Deccan Herald*, 16 December 2007, on www.deccanherald.com/Content/Dec162007/business2007121541414.asp.

78 Juliette Jowit, 'UN says eat less

meat to curb global warming',
The Observer, 7 September
2008, on www.guardian.co.uk/
environment/2008/sep/07/food.
foodanddrink.

79 John Lewis Gaddis, *The Long
Peace: Inquiries Into the History
of the Cold War*, Oxford University
Press, 1987.

80 See Office of Science and
Technology, *Tackling Obesities:
Future Choices – Project Report*,
October 2007, especially the
ridiculous models displayed in
Figures 5.2, 5.4 and 5.5, on
www.foresight.gov.uk/Obesity/17.
pdf.

81 Neil Adger and others, 'Summary
for Policymakers', in Martin Parry,
Osvaldo Canziani and others,
editors, *Climate Change 2007:
Impacts, Adaptation and
Vulnerability*, op. cit., p19, on
www.ipcc.ch/pdf/assessment-
report/ar4/wg2/ar4-wg2-spm.pdf.

82 Terry Barker and others,
'Summary for Policymakers', in
Bert Metz, Ogunlade Davidson
and others, editors, *Climate
Change 2007: Mitigation,* op. cit.,
p12, on www.ipcc.ch/pdf/
assessment-report/ar4/wg3/
ar4-wg3-spm.pdf.

83 Malcolm Gladwell, *The Tipping
Point: How Little Things Can
Make a Big Difference* (2000),
Back Bay Books, 2002.

84 Ibid., pp7, 9, 11.

85 Ibid., pp22, 273.

86 Spencer R Weart, *Rapid Climate
Change*, June 2007, on www.aip.
org/history/climate/rapid.
htm#M_60_.

87 John Gillott and Manjit Kumar,
*Science and the Retreat from
Reason*, op. cit., p94.

88 James Hansen and others,
'Earth's Energy Imbalance:
Confirmation and Implications',
Science, Vol. 308, No. 5727, 3
June 2005, pp.1431-1435; P
Lemke, Jiawen Ren and others,
'Observations: Changes in Snow,
Ice and Frozen Ground', Chapter
4 in Susan Solomon and others,
editors, *Climate Change 2007:
The Physical Science Basis.
Contribution of Working Group I
to the Fourth Assessment Report
of the IPCC*, Cambridge University
Press, 2007, p364, on www.ipcc.
ch/pdf/assessment-report/ar4/
wg1/ar4-wg1-chapter4.pdf.

89 James Hansen, 'Scientific
reticence and sea level rise',
Environmental Research Letters,
Vol. 2, No. 2, April-June 2007,
p024002.

90 Amy Craford, 'Al Gore discusses
"An Inconvenient Truth"',
Smithsonian Magazine, July
2006, on www.smithsonianmag.
com/people-places/interviewgore.
html.

91 P Lemke, Jiawen Ren and others,
op. cit., p364.

92 Ibid. See also Andrew Shepherd
and Duncan Wingham, 'Recent
Sea-Level Contributions of the
Antarctic and Greenland Ice
Sheets', *Science*, 16 March
2007, Vol. 310, No. 5818,
pp1529-1532. For comparison,
the East Antarctic ice sheet is
gaining about 25 billion tonnes a
year and the West Antarctic Ice
Sheet is losing about 50 billion
tonnes. The basic issues raised

by the Antarctic are similar to those in Greenland. Shepherd and Wingham, op. cit.

93 Bjørn Lomborg, *Cool It: The Skeptical Environmentalist's Guide to Global Warming*, Cyan, 2007, p77.

94 For both the AD 3000 and AD 5000 model results, see Richard B Alley and others, 'Ice-Sheet and Sea-Level Changes', *Science,* Vol. 310, No. 5747, 21 October 2005.

95 See for example David Biello, 'Conservative Climate', *Scientific American*, March 2007, on www. sciam.com/article. cfm?id=conservative-climate.

96 H Jay Zwally and others, 'Surface Melt-induced Acceleration of Greenland Ice-Sheet Flow', *Science*, Vol. 297, No. 5579, pp218-222; Sarah B Das and others, 'Fracture Propagation to the Base of the Greenland Ice Sheet During Supraglacial Lake Drainage', *Science*, 9 May 2008, Vol. 320, No. 5877, pp778-781.

97 Amy Craford, op. cit.

98 R S W Van de Wal and others, 'Large and Rapid Melt-Induced Velocity Changes in the Ablation Zone of the Greenland Ice Sheet', *Science*, 4 July 2008, Vol. 321, No. 5885, p111.

99 Ibid., p148.

100 Ibid., p144.

101 Ibid., p150.

102 Ibid., p153.

103 Paul Krugman, 'Can this Planet be Saved?' *The New York Times*, 1 August 2008, on www.nytimes.com/2008/08/01/opinion/01krugman.html.

104 Martin Weitzman, 'On Modeling and Interpreting the Economics of Catastrophic Climate Change', February 2008, on www.economics.harvard.edu/faculty/weitzman/files/modeling.pdf.

105 Martin Weitzman, op. cit., p31.

106 William Nordhaus, 'Critical Assumptions in the Stern Review on Climate Change', *Science*, 13 July 2007, Vol. 317, No. 5835, p202.

107 See Scott C Beardsley, Denis Bugrov, and Luis Enriquez, 'The role of regulation in strategy', *McKinsey Quarterly*, Number 4, 2005. Emphasis added.

108 Ibid.

109 ADL, *The Carbon Margin: Translating Carbon Exposure into Competitive Advantage*, December 2007, pp1, 4, 8, on www.adlittle.cn/reports.html?&no_cache=1&view=72.

110 John Thornhill, 'The world should look to Europe as capitalisms clash', *Financial Times*, 25 July 2008.

111 'Member States endorse Commission proposal to reduce standby electricity consumption', press release, Brussels, 8 July 2008, on http://europa.eu/rapid/pressReleasesAction.do?reference=IP/08/1117&format=HTML&aged=0&language=EN&guiLanguage=en.

112 Guido Sacconi, rapporteur, *Draft Report on the proposal for a* *regulation of the European Parliament and of the Council setting emission performance standards for new passenger cars as part of the Community's integrated approach to reduce CO2 emissions from light-duty vehicles*, European Parliament, Committee on the Environment, Public Health and Food Safety, 8 May 2008, p7, on www.europarl.europa.eu/sides/getDoc.do?pubRef=-//EP//NONSGML+COMPARL+PE-406.014+01+DOC+PDF+V0//EN.

113   For a discussion of this concept, see 'What future for the mobile footprint', *Spiked*, no date, on www.spiked-online.com/index.php?/debates/C118.

114   Guido Sacconi, *Draft Report*, op. cit., p6.

115   See James Woudhuysen, 'The Electric Car Conspiracy... that never was', *The Register*, 1 January 2008, on www.theregister.co.uk/2008/01/01/woudhuysen_electric_car.

116   European Union, 'Questions & Answers on Emissions Trading and National Allocation Plans', press release, 8 March 2005, updated 20 June 2005, on http://europa.eu/rapid/pressReleasesAction.do?reference=MEMO/05/84&format=HTML&aged=1&language=EN&guiLanguage=en, and European Union, 'Questions & Answers on the Commission's proposal to revise the EU Emissions Trading System', press release, 23 January 2008, on http://europa.eu/rapid/pressReleasesAction.do?reference=MEMO/08/35&format=HTML&aged=0&language=EN&guiLanguage=en.

117   World Bank, *State and Trends of the Carbon Market*, May 2007, quoted in Commission of the European Communities, *Proposal for a Directive of the European Parliament and of the Council amending Directive 2003/87/EC so as to improve and extend the greenhouse gas emission allowance trading system of the Community*, 23 January 2008, on http://eur-lex.europa.eu/LexUriServ/LexUriServ.do?uri=COM:2008:0016:FIN:EN:PDF.

118   European Union, 'Questions & Answers on Emissions Trading and National Allocation Plans', op. cit.

119   European Union, 'Questions & Answers on the Commission's proposal to revise the EU Emissions Trading System', op. cit.

120   Carl Mortished, 'Policy leap vital for any serious cut in carbon emissions', *The Times*, 5 November 2008, on http://business.timesonline.co.uk/tol/business/columnists/article5083880.ece.

121   European Union, 'Questions & Answers on the Commission's proposal to revise the EU Emissions Trading System', op. cit.

122   McKinsey, op. cit.

123   UN, 'National greenhouse gas inventory data for the period 1990-2006', FCCC/SBI/2008/12, 17 November 2008, p10.

124   Ibid., p8.

125 For example, Nick Davies, 'Truth about Kyoto: huge profits, little carbon saved', *The Guardian*, 2 June 2007, on www.guardian.co.uk/environment/2007/jun/02/india.greenpolitics.

126 Geoff Dyer and Fiona Harvey, 'China toughens stance in emissions debate', *Financial Times*, 28 October 2008, on http://us.ft.com/ftgateway/superpage.ft?news_id=fto102820081452138837.

127 Member States endorse Commission proposal to reduce standby electricity consumption, press release, Brussels, 8 July 2008, on http://europa.eu/rapid/pressReleasesAction.do?reference=IP/08/1117&format=HTML&aged=0&language=EN&guiLanguage=en.

128 Guido Sacconi, rapporteur, *Draft Report on the proposal for a regulation of the European Parliament and of the Council setting emission performance standards for new passenger cars as part of the Community's integrated approach to reduce CO2 emissions from light-duty vehicles*, European Parliament, Committee on the Environment, Public Health and Food Safety, 8 May 2008, p7, on www.europarl.europa.eu/sides/getDoc.do?pubRef=-//EP//NONSGML+COMPARL+PE-406.014+01+DOC+PDF+V0//EN. Emphasis in the original.

129 Ibid., p6.

130 See James Woudhuysen, 'The Electric Car Conspiracy... that never was', *The Register*, 1 January 2008, on www.theregister.co.uk/2008/01/01/woudhuysen_electric_car/.

131 Emission Trading Scheme (EU ETS), 13 June 2008, on http://ec.europa.eu/environment/climat/emission.htm.

132 McKinsey, op. cit.

133 Bob Carter, 'There is a problem with global warming... it stopped in 1998', *The Daily Telegraph*, 9 April 2006, on http://www.telegraph.co.uk/opinion/main.html?xml=/opinion/2006/04/09/do0907.xm; David Whitehouse, 'Has global warming stopped?', *New Statesman*, 19 December 2007, on www.newstatesman.com/scitech/2007/12/global-warming-temperature.

134 Bjørn Lomborg, 'Let the data speak for itself', *The Guardian*, 14 October 2008, on www.guardian.co.uk/commentisfree/2008/oct/14/climatechange-scienceofclimatechange.

135 NASA, Goddard Institute for Space Studies, 'Global-mean monthly, annual and seasonal dT_s based on met.station data', on http://data.giss.nasa.gov/gistemp/tabledata/GLB.Ts.txt.

136 Ibid.

137 Catia M Domingues and others, 'Improved estimates of upper-ocean warming and multi-decadal sea-level rise', *Nature*, Vol. 453, No. 7198, 19 June 2008, pp1090-1093.

138 The assessment report of Working Group I of the IPCC explores all these lines of evidence and more, although of course it cannot be comprehensive. See Susan Solomon and others, editors, *Climate Change 2007: The Physical Science Basis*, op. cit.

139   David Drew, 'Climate Camp for Action Protest', *JPG magazine*, 8 March 2008, on www.jpgmag. com/stories/3189.

140   Martin Visbeck, 'Concept Oceanography: Power of pull', *Nature*, 24 May 2007, Vol. 447, No. 7143, p383.

141   Mark P Baldwin and others, 'How Will the Stratosphere Affect Climate Change?', *Science*, 15 June 2007, Vol. 316, No. 5831, p1577, on www.nwra.com/ resumes/baldwin/pubs/Baldwin_ Dameris_Shepherd_2007.pdf.

142   James C McWilliams, 'Irreducible imprecision in atmospheric and oceanic simulations', *Proceedings of the National Academy of Science of the United States*, 22 May 2007, Vol. 104, No. 21, pp8709-8713, on www.pnas.org/ cgi/content/full/104/21/8709.

143   Gordon B Bonan, 'Forests in Flux', *Science*, 13 June 2008, Vol. 320, No. 5882.

144   Richard Wood, 'Climate change: Natural ups and downs', *Nature*, 1 May 2008, Vol. 453, No. 7191, pp43-45.

CHAPTER 4

1   Alvin Weinberg, 'Social Institutions and nuclear energy', *Science*, Vol. 177, Issue 4043, 7 July 1972, pp27-34.

2   The best treatment of the Faust legend remains Marshall Berman, *All That is Solid Melts Into Air: The Experience of Modernity* (1982), Verso, 1983.

3   For more on technological determinism, see James Woudhuysen, 'Before we rush to a new era', in Geoff Mulgan, editor, *Life After Politics: New Thinking for the Twenty-First Century*, Fontana Press, 1997.

4   *Nucleonics Week* reports on the nuclear industry. It is available through www.platts.com/Nuclear/ Newsletters%20&%20Reports/ Nucleonics%20Week/.

5   A typical atomic size is 10^{-10}m, while a nuclear size is 10^{-15}m.

6   Figures from International Atomic Energy Authority, 'Nuclear power advantages', in *Sustainable Development and Nuclear Power*, no date, on www.iaea.org/ Publications/Booklets/ Development/devnine.html.

7   Chart based on DTI, *The Energy Challenge*, July 2006, Annex 1, Chart A1, p175, on www.berr.gov. uk/files/file31890.pdf. We say 'indicative' because price variations will change the precise make-up of costs. By contrast with uranium ready and waiting to be used as a fuel, uranium ore accounts for only about 1.5 per cent of total nuclear electricity generation costs. The equivalent figure for gas in gas-fired plants is 70 per cent. BERR, *Meeting the Energy Challenge: A White Paper on Nuclear Power*, January 2008, p56, paragraph 2.35, on www. berr.gov.uk/files/file43006.pdf. 'Back end costs' relate to waste recycling, plus an annual reserve to pay for eventual decommissioning. For a treatment of cost compositions in relation to other forms of energy supply, see Nuclear Energy Agency/OECD, *Nuclear Energy Today*, 2005, pp60-62, available through http://www.nea.fr/html/pub/ nuclearenergytoday/welcome.html.

8 James Lovelock, 'Go nuclear, save the planet', *The Sunday Times*, 18 February 2007, on www.timesonline.co.uk/tol/news/uk/article1400073.ece. Another Green turned pro-nuclear apostate is Mark Lynas. See 'Mark Lynas: the green heretic persecuted for his nuclear conversion', *The Sunday Times*, 28 September 2008.

9 To our main three arguments, we would also add that *nuclear reactors beat gas-fired power plants in their ability to scale up well*. In gas-fired electricity production, combustion takes place inside the turbine – as in a jet engine. But it's very costly to scale up turbines, which, as they grow in size, need to increase their strength a great deal if they are not to fly apart. In a nuclear power station, by contrast, heating takes place in a boiler linked to but separate from a turbine (the same is true of a coal-fired power station). With boilers, unlike turbines, there's no problem in scaling up; and as size increases, volumes rise faster than surface areas, making big chambers leak proportionately less energy than small ones. Result: large nuclear reactors tend to achieve higher energy efficiencies than small ones.

10 See World Nuclear Association, on www.world-nuclear.org/.

11 IEA, *Energy Technology Perspectives 2008*, June 2008, pp42, 300. Executive Summary on www.iea.org/textbase/npsum// ETP2008SUM.pdf.

12 Ibid., pp299-300.

13 'Westinghouse AP1000, China', *Power Technology*, on www.power-technology.com/projects/westinghouseap100/.

14 PBMR, 'Project Status', 5 February 2008, on www.pbmr.com/index.asp?content=175.

15 See World Nuclear Association, 'Thorium', March 2008, on www.world-nuclear.org/info/inf62.html.

16 World Nuclear Association, 'World Nuclear Power Reactors 2007-2008 and Uranium Requirements', 9 June 2008, on www.world-nuclear.org/info/reactors.html. Figures are accurate as of May 2008. 'Operating' means connected to the grid. 'Construction' means the first concrete has been poured. 'Planned' are mostly expected in operation within eight years. 'Proposed' means that a specific programme or proposal for a site exists, but expected operation within 20 years. GW plans from IEA, op. cit., p299.

17 DTI, *Our Energy Future – Creating a Low Carbon Economy*, February 2003, p12 paragraph 1.24, and p44, paragraph 4.3, on www.berr.gov.uk/files/file10719.pdf.

18 Gordon Brown, press conference, 12 June 2008, op. cit.

19 John Hutton, statement on energy policy to the House of Commons, 10 January 2008, Commons *Hansard*, Column 517, on www.publications.parliament.uk/ pa/cm200708/cmhansrd/cm080110/debtext/80110-0003.htm#08011057000005.

20 BERR, *Meeting the Energy Challenge*, op. cit.

21 Ibid., p11, paragraph 5.

22 'Whether energy companies choose to invest in new nuclear power stations is, ultimately, a matter for them.' Ibid., p62, paragraph 2.52.

23 Ibid., p17, paragraph 25.

24 Ibid., p46, paragraph 2.9.

25 Ibid., p53.

26 Ibid., p18, paragraph 27.

27 Ibid., p29, paragraph 60, and p105, paragraph 2.183. Foolishly, the White Paper insists on bracketing security of oil supplies, which have little bearing on the generation of electricity nowadays, with security in gas, which is much more relevant to that. Ibid., p16, paragraph 21.

28 Ibid., p57, paragraph 2.38.

29 Ibid., p69, paragraph 2.74.

30 Greenpeace UK, advertisement, *The Times*, 20 June 2008, on www.enoughsenough.org/nuclear.pdf.

31 Pöyry (Oxford) Ltd, *Securing Power: Potential for CCGT CHP Generation at Industrial Sites in the UK: A Report to Greenpeace*, June 2008, on www.greenpeace.org.uk/files/pdfs/climate/securing-power.pdf.

32 4G designs for Gas Cooled Fast Reactors (GFRs) contain the option of generating heat on top of electricity. 4G designs for Very High Temperature Reactors (VTRs) are anyway primarily designed to yield heat – though electricity is also an option. See Idaho National Laboratory, 'Gas Cooled Fast Reactor (GFR)', on http://nuclear.inl.gov/gen4/gfr.shtml, and INL, 'Very High Temperature Reactor (VTR)', on http://nuclear.inl.gov/gen4/vhtr.shtml.

33 World Nuclear Association, *World Nuclear Power Reactors 2007-2008*, op. cit.

34 All the more striking is the fact that Greenpeace fears that nuclear power stations located by the sea face a grave risk of flooding, on account of climate change. Coastal CHP installations, it would seem, do not suffer from such a danger. See Greenpeace, *The impacts of climate change on nuclear power stations sites* (sic)*: a review of four proposed new-build sites on the UK coastline*, a report written by the Middlesex University Flood Hazard Research Centre. March 2007, on www.greenpeace.org.uk/files/pdfs/nuclear/8179.pdf.

35 BERR, *Meeting the Energy Challenge*, op. cit., p13, paragraphs 13, 14, and p16, paragraph 21.

36 Ibid., p14, paragraph 15.

37 Nuclear Energy Agency/OECD, *Nuclear Energy Today*, op. cit p62.

38 Ibid., p137, paragraph 3.11.

39 Ibid., p137, paragraph 3.10, and p138, paragraphs 3.12, 3.13.

40 Ibid., p138, paragraphs 3.13-3.15.

41 Ibid., p40.

42 Ibid., p138, paragraph 2.12, and Communities and Local Government, *Planning for a Sustainable Future*, May 2007, p34, on www.communities.gov.uk/documents/planningandbuilding/pdf/planningsustainablefuture.pdf.

43 BERR, *Meeting the Energy Challenge*, op. cit., p36.

44 Ibid., p142, paragraph 3.26 and p143, paragraph 3.33. The GDA is led by the Nuclear Installations Inspectorate (NII) and the Office of Civil Nuclear Security (OCNS) of the Health and Safety Executive (HSE), as well as by the Environment Agency (EA).

45 Ed Crooks and others, 'UK sees EDF as preferred British Energy buyer', *Financial Times*, 24 August 2008, on www.ft.com/cms/s/0/0b043a6a-7219-11dd-a44a-0000779fd18c.html.

46 BERR, *Meeting the Energy Challenge*, op. cit., p108, paragraph 2.196.

47 Ibid., p112, paragraph 2.213, and Nestoria, 'Buy prices in Mayfair', on www.nestoria.co.uk/mayfair/property/buy. The figure is for July 2008.

48 LB Lave, M Ashworth and C Gellings, 'The Ageing Workforce: Electricity Industry Challenges and Solutions', *The Electricity Journal*, March 2007, Vol. 20, Issue 2, p75.

49 See for example the campaign organised by Earth Hour, on www.earthhour.org/.

50 It is defined as absorption of one joule per kilogram, with an additional weighting depending on biological effectiveness.

51 Figures for artificial civilian sources from BERR, *Meeting the Energy Challenge*, op. cit., Box 2, p80, on www.berr.gov.uk/files/file43006.pdf. With marine radioactivity, 10 per cent is nuclear; the rest comes from the phosphate, oil, gas and other industries. Figures for post-war test and natural background radiation from M C Thorne, 'Background Radiation: Natural and Man-made', *Journal of Radiological Protection*, Vol. 23, 2003, p39. For Three Mile Island, see E O Talbot and others, 'Long-Term Follow-Up of the Residents of the Three Mile Island Accident Area: 1979-1998', *Environmental Health Perspectives*, Vol. 111, No. 3, March 2003.

52 In 1998, 10 years after Saddam Hussein's bombing of Halabja with a chemical cocktail of mustard gas, Sarin, Tabun and VX, the US Senate received evidence that occurrences of genetic mutations among local Kurds appeared comparable with those who were one to two kilometres from ground zero in Hiroshima and Nagasaki. '1998 Congressional Hearings on Intelligence and Security

Testimony of Dr Christine M Gosden, Professor of Medical Genetics in the University of Liverpool, Before the Senate Judiciary Subcommittee on Technology, Terrorism and Government and the Senate Select Committee on Intelligence on Chemical and Biological Weapons' Threats to America: Are We Prepared?', 22 April 1998, on www.fas.org/irp/congress/1998_hr/s980422-cg.htm.

53 See James Woudhuysen, 'Clausewitz after 9/11', *spiked review of books*, October 2007, on www.spiked-online.com/index.php?/site/reviewofbooks_article/4009/.

54 Thomas B Allen and Norman Polmar, *Rickover: Father of the Nuclear Navy*, Potomac Books, 2007.

55 Robert Pool, *Beyond Engineering: How Society Shapes Technology*, Oxford University Press, 1997.

56 'The Man in Tempo 3', *Time*, 11 January 1954, on www.time.com/time/printout/0,8816,819338,00.html.

57 Robert Pool, op. cit., p50.

58 The literature is immense, but of particular use is Lawrence Freedman and Saki Dockrill, 'Hiroshima: A Strategy of Shock', in Dockrill, editor, *From Pearl Harbor to Hiroshima: The Second World War in Asia and the Pacific, 1941-45*, Palgrave Macmillan 1993. See also John Hersey, Hiroshima, *Knopf*, 1946, first published as a special issue *The New Yorker*, 31 August 1946.

59 See for example Robert Jungk, *Brighter Than A Thousand Suns: A Personal History of the Atomic Scientists* (1956), Penguin, 1960; Richard Rhodes, *The Making of the Atomic Bomb*, Simon & Schuster, 1986; Kai Bird and Martin J Sherwin, *American Prometheus: The Triumph and Tragedy of J Robert Oppenheimer*, Vintage, 2006; Cynthia Kelly and Richard Rhodes, editors, *The Manhattan Project: The Birth of the Atomic Bomb in the Words of Its Creators, Eyewitnesses and Historians*, Black Dog & Leventhal Publishers, 2007.

60 Lawrence Freedman and Saki Dockrill, op. cit., p193.

61 Lawrence Freedman, 'The Strategy of Hiroshima', *Journal of Strategic Studies*, Vol. 1, Issue 1, May 1978.

62 Lawrence Freedman and Saki Dockrill, op. cit., p199.

63 The classic but now discredited work suggesting that the Bomb was dropped as a warning to Russia is Gar Alperovitz, *Atomic Diplomacy: Hiroshima and Potsdam: The Use of the Atomic Bomb and the American Confrontation with Soviet Power*, Simon & Schuster, 1965.

64 See the classic works by John Dower, *War Without Mercy: Race and Power in the Pacific War*, Pantheon Books, 1986, and Christopher Thorne, *The Issue of War: States, Societies, and the Far Eastern Conflict of 1941-1945*, Hamish Hamilton, 1985.

65 Catherine Caufield, *Multiple Exposures: Chronicles of the Radiation Age* (1989), Penguin, 1990.

66 Quoted in ibid., p50.

67 Ibid., p53.

68 The Franck report, as it is known, is on http://fas.org/sgp//eprint/franck.html. *The Bulletin of Atomic Scientists* is on www.thebulletin.org.

69 Robert Oppenheimer in a 1965 NBC television documentary, *The decision to drop the Bomb*, cited in Gregg Herken, 'American scientists and US nuclear weapons policy', *Peace & Change*, Spring 1985, Vol. 11 Issue 1, p19.

70 Gregg Herken, op. cit., pp19, 20.

71 Catherine Caufield, op. cit., pp52-53.

72 Sean Collins, 'The dogma of "transparency"', *spiked review of books*, 7 November 2007, on www.spiked-online.com/index.php?/site/reviewofbooks_article/4135/.

73 In a famous scoop conducted in the immediate ruins of Hiroshima, a left-leaning Australian journalist headlined the *Daily Express* with 'THE ATOMIC PLAGUE – I write this as a warning to the world', on 5 September 1945. Catherine Caufield, op. cit., p63; Wilfred Burchett, *The Daily Express*, 5 September 1945. On 13 September, however, the *New York Times* headlined with 'NO RADIOACTIVITY IN HIROSHIMA RUIN; WHAT OUR SUPERFORTRESSES DID TO A

JAPANESE PLANE PRODUCTION CENTER'. See W H Lawrence, *The New York Times*, 13 September 1945, abstract on http://select.nytimes.com/gst/abstract.html?res=F00D12FC3F5D177A93C1A81782D85F418485F9&scp=1&sq=No%20radioactivity%20in%20Hiroshima%20Ruin&st=cse.

74 The NCRP was previously the US Advisory Committee on X-Ray and Radium Protection.

75 Catherine Caufield, op. cit., pp73, 120. The NCRP's weekly limit for nuclear workers under 35 was 3 mSv, but that for those over 45 was double that. All workers could expect to receive a one-off emergency exposure of no more than 250 mSv, though military personnel would be subject to no such restrictions in an emergency. Ibid.

76 Archives of the ABCC are on www7.nationalacademies.org/archives/ABCC_1945-1982.html. The ABCC's General Report of January 1947 is on www7.nationalacademies.org/archives/ABCC_GeneralReport1947.html.

77 Sue Rabbit Roff, *Hotspots: The Legacy of Hiroshima and Nagasaki*, Cassell, 1995.

78 Catherine Caufield, op. cit., pp163-164.

79 Sue Rabbit Roff, op. cit., pp5, 7.

80 David Bradley's *No Place to Hide* (1948), a bestselling account of Crossroads, raised doubts. No fewer than 11 tests performed in Nevada in 1953 also succeeded in bringing fallout to upstate New York. Catherine Caufield, op. cit.,

Chapter 10, 'Operation Crossroads', pp91-98, and pp100-103.

81 Catherine Caulfield, op. cit., pp112-115.

82 A recent treatment of the reception Japan gave the tests is Toshihiro Higuchi, 'An Environmental Origin of Antinuclear Activism in Japan, 1954–1963: The Government, the Grassroots Movement, and the Politics of Risk', *Peace & Change*, July 2008, Vol. 33 Issue 3, pp333-367.

83 The AEC suffered revelations about its attempts to censor a paper on radiation, the meltdown of a research reactor near Detroit, and congressional hearings in which it was reported that the Bravo test had indeed scattered fallout – over an area equivalent to 18,000 km^2. Catherine Caulfield, op. cit., pp 127-128, 151, and Kenneth Rose, *One Nation Underground: The Fallout Shelter in American Culture,* p26, footnote 50, citing Thomas J Kerr, *Civil Defense in the US: Bandaid for a Holocaust?* Westview Press, 1983, p75.

84 In 1957 the AEC had to concede that radiation could cause genetic defects, even though it only published information about this in 1963. In 1958 it belatedly began systematically to monitor exposure to radiation inside the human body, even though it had long been aware of the dangers of food and drink containing isotopes such as carbon-14, iodine-131 and strontium-90. In 1959 it was wrong-footed by the release of a Department of Defense letter stating that fallout in the stratosphere returned to earth not in seven years, as the AEC contended, but in two. Catherine Caufield, op. cit., pp126, 128-130.

85 Gregg Herken, op. cit.. In 1958, the AEC also tried to present a softer face to nuclear weapons. It announced Plowshare, a dubious programme to show that nuclear explosions had a civilian role in major tunnelling and digging projects, and the extraction of natural gas.

86 Alice Stewart and others, 'A survey of childhood malignancies', *British Medical Journal*, Vol. 1, No. 5086, 28 June 1958, p1500, on www. pubmedcentral.nih.gov/picrender. fcgi?artid=2029590&blobtype= pdf.

87 Catherine Caulfield, op. cit., pp151-152.

88 Ibid., pp155-157.

89 *Environmental Effects Of Producing Electric Power – Hearings Before The Joint Committee On Atomic Energy of the United States, Ninety-First Congress, Second Session on Environmental Effects Of Producing Electric Power*, Part 2, Vol. I, 27, 28, 29, 30 January and 24, 25, 26 February 1970, on http://sul-derivatives.stanford. edu/derivative?CSNID=0000221 8&mediaType=application/pdf.

90 The left both anticipated and popularised a phrase used by General Electric chairman Charles Edward Wilson in a speech to the Army Ordnance Association in

1944: the permanent war economy. See Walter Oakes, 'Towards a permanent war economy?', (February 1944) in Seymour Melman, *The war economy of the United States: Readings in Military, Industry and Economy*, St Martin's Press, 1971, and T N Vance and others, *The Permanent War Economy*, *Independent Socialist Press*, 1951. Oakes and Vance were pseudonyms of the same professional economist.

91 Max Schactman, 'Nightfall of capitalism', *The New International*, February 1948, in Vance, op. cit.; Hal Draper, 'The economic state of the union', *Labour Action*, 17 January 1949, in Vance, op. cit.; Paul Sweezy, 'Recent developments in American capitalism', *Monthly Review*, May 1949, in Paul Sweezy, *The Present as History,* Monthly Review Press, 1953.

92 See T N Vance, 'Economic prospects for 1956', *The New International*, Winter 1955/6.

93 C Wright Mills, *The Power Elite*, Oxford University Press, 1956, pp215, 223. There was grist to mill of anti-Vietnam war radicals here. Indeed in 1965, nearly 10 years after *The Power Elite*, the American activist Tom Hayden mounted an important academic defence of it against the 'pluralist' school of Cold War US sociology led by Daniel Bell. See Marc Pilsiuk and Tom Hayden, 'Is there a military industrial complex?', *American Sociological Review*, July 1965.

94 Gabriel Kolko, *Main Currents in Modern American History*, Harper & Row, 1976, pp264-269.

95 Thus, two years after his famous book, Mills announced: 'The immediate cause of WW III is the preparation for it'. See C Wright Mills, *The Causes of World War III*, Simon & Schuster, 1958, p47.

96 Eisenhower, Farewell address, 17 January 1961, on www.eisenhower.archives.gov/speeches/farewell_address.html. Eisenhower's speech is attributed to his chief speechwriter Malcolm Moos, a Johns Hopkins University political scientist, and Ralph Williams, a Navy captain and also a speechwriter for Eisenhower.

97 The observation is in Carroll Pursell, *The Military Industrial Complex*, Harper & Row, 1972.

98 Fred J Cook, 'Juggernaut: the warfare state', *The Nation*, 28 October 1961; abstract on www.thenation.com/archive/detail/13339509. See also Fred J Cook, *The Warfare State: Is the Juggernaut Out Of Control?*, Macmillan, 1962. Those who used Cook's work to further the radical cause included Paul Baran and Paul Sweezy, *Monopoly Capital*, Penguin, 1966, p210 and Sidney Lens, *The Military Industrial Complex*, Kahn & Averill, 1970.

99 Fred J Cook, op. cit., p282.

100 Cook made more than a nod toward the kind of psychobabble we have already encountered among Greens today. 'A Johns Hopkins University psychiatrist', he wrote, 'has noted that many persons are "innately vicious" and might trigger total destruction out of pure malice.' Ibid., p282. Years later, one economist specialising in nuclear matters gave a somewhat more sober account of the dynamics of nuclear war: 'What really traps us… is not the impersonal force of technology but the attitudes of large bureaucracies committed to certain weapons systems and ways of thinking.' See Daniel Ford, *The Button: The Nuclear Trigger – Does it Work?*, Unwin Paperbacks, 1986, p54.

101 He pontificated: 'Most of the prevailing needs to relax, to have fun, to behave and consume in accordance with the advertisements… belong to this category of false needs.' Herbert Marcuse, *One-Dimensional Man: Studies in the Ideology of Advanced Industrial Society* (1964), Routledge, 2002, p7. On the common ground between Mills and Marcuse, see James Panton, 'Intellectual Influences on the New Left in America: C Wright Mills and Herbert Marcuse', *Reconstruction*, Vol. 8 No. 1, 2008, on http://reconstruction.eserver.org/081/panton.shtml.

102 Herbert Marcuse, op. cit., p5.

103 Ibid., pp156-158.

104 See Kennedy, 'Radio and Television Report to the American People on the Berlin Crisis', 25 July 1961, on www.jfklibrary.org/Asset+Tree/Asset+Viewers/Audio+Video+Asset+Viewer.htm?guid=%7B2C529501-7B4E-4E12-8C1D-78F9C53F2BDC%7D&type=Audio.

105 Herbert Marcuse, op. cit., pp83, 93, 252.

106 Ibid., pxxxl.

107 Rachel Carson, *Silent Spring*, Houghton Mifflin, 1962, p7.

108 JK Galbraith, *The New Industrial State* (1967), Mentor Books, fourth edition, Chapter II, 'The imperatives of technology', pp10-18.

109 Bob Hunter, quoted in Rex Weyler, 'Waves of Compassion: The Founding of Greenpeace. Where Are They Now?' *Understanding the Next Evolution*, no date, on www.utne.com/web_special/web_specials_archives/articles/2246-1.html.

110 Nuclear disarmament remains a major concern of Greenpeace today. One thing that has changed, however, is that the organisation boasts an annual income of more than €173m. See Greenpeace, *Annual Report 2006*, p26, on www.greenpeace.org/raw/content/international/press/reports/annual-report-2006.pdf.

111 Rex Weyler, *Greenpeace: How a Group of Ecologists, Journalists, and Visionaries Changed the World*, Rodale Books, 2004, pp391, 392.

112 Peter Pringle and James Jacob Spigelman, *The Nuclear Barons: The Inside Story of How They Created Our Nuclear Nightmare*, Michael Joseph, 1981.

113 Colin Sweet, *The Price of Nuclear Power*, Heinemann, 1983, p22.

114 Roger S Carlsmith, 'Future nuclear energy capacity

requirements', in Pierre Zaleski and others, *Nuclear Energy Maturity: Proceedings of the European Nuclear Conference, Paris, 21-25 April 1975, Plenary Sessions*, Pergamon Press, 1975, pp318-319.

115 IEA, *Energy Technology Perspectives 2008*, June 2008, p300.

116 Ibid., Figure 8.8.

117 Duncan Burn, *The Political Economy of Nuclear Energy: An Economic Study of Contrasting Organisations in the UK and USA*, Institute of Economic Affairs, 1967, p110.

118 Margaret Gowing and Lorna Arnold, *Independence and Deterrence: Britain and Atomic Energy, 1945-52*, Vol. 2, Policy Making, Chapter 17, Palgrave Macmillan, 1974.

119 For the full argument, see Irvin C Bupp and Jean-Claude Derian, *Light Water: How the Nuclear Dream Dissolved*, Basic Books, 1978.

120 Walter Patterson, *Nuclear Power*, Penguin Books, 1976.

121 The text of the Treaty is available at http://disarmament.un.org/wmd/npt/npt%20authenticated%20text-English.pdf.

122 'Nuclear Power Policy Statement on Decisions Reached Following a Review', 7 April 1977, in John T Woolley and Gerhard Peters, *The American Presidency Project*, on www.presidency.ucsb.edu/ws/index.php?pid=7316.

123 Ibid., and also *Presidential Directive/NSC-8*, 24 March 1977, on www.nci.org/new/pu-repro/carter77/index.htm.

124 E O Talbott and others, 'Long-term follow-up of the residents of the Three Mile Island accident area: 1979-1998', *Environmental Health Perspectives*, Vol. 111, No. 3, March 2003, pp341-348.

125 Catherine Caufield, op. cit., p136. We have converted 340,000 and 15 curies to becquerels, the modern measure of radioactivity.

126 S Wing, 'Objectivity and Ethics in Environmental Health Science', *Environmental Health Perspectives*, Vol. 111, No. 14, November 2003.

127 Hans Jonas, *The Imperative of Responsibility: in Search of an Ethics for the Technological Age (1979 and 1981)*, University of Chicago Press, 1984, pp190-191.

128 'By risks I mean above all radioactivity, which completely evades human perceptive abilities, but also toxins and pollutants in the air, the water and foodstuffs, together with the accompanying short- and long-term effects on plants, animals and people.' See Ulrich Beck and others, *Risk Society: Towards a New Modernity* (1986), Sage, 1992, p22.

129 John Gray, 'Nature Bites Back', in Jane Franklin, editor, *The Politics of Risk Society*, Polity Press, 1998, p46. For the UK Government, of course, the problem with Chernobyl was that regulatory scrutiny of reactor operations in the former USSR was 'far less rigorous than it is in the UK today' and that many past accidents with nuclear occurred in power stations 'with designs that would not be acceptable to regulators in the UK'. Whatever the problem, more British regulation seems to be the answer. See BERR, *Meeting the Energy Challenge*, op. cit., Box 2, p75, paragraph 2.89.

130 Greenpeace, *The Chernobyl Catastrophe: Consequences on Human Health*, 2006, pp8, 10, on www.greenpeace.org/raw/content/international/press/reports/chernobylhealthreport.pdf.

131 Chernobyl Forum, *Chernobyl's Legacy: Health, Environmental and Socio-Economic Impacts and Recommendations to the Governments of Belarus, the Russian Federation and Ukraine*, Second Revised version, 2003-5. Available from www.iaea.org/Publications/Booklets/Chernobyl/chernobyl.pdf.

132 Richard Stone, 'The long shadow of Chernobyl', *National Geographic*, April 2006, p44.

133 E Broughton, 'The Bhopal disaster and its aftermath: a review', *Environmental Health: A Global Access Science Source*, Vol. 4, No. 6, 2005.

134 Xinhua, 'Coal mine accidents kill 6,027 in China', *China Daily*, 17 January 2005, on www.chinadaily.com.cn/english/doc/2005-01/17/content_409640.htm; Zhao

Xiaohui and Jiang Xueli, 'Coal mining: Most deadly job in China', *China Daily*, 13 November 2004, on www.chinadaily.com.cn/english/doc/2004-11/13/content_391242.htm. In 2004, the death rate in Chinese mining was about 100 times that in the US, where in 2003 each miner produced more than 45 times as much coal. The Chinese government has rapidly closed thousands of small scale, inefficient, unsafe mines and, as a consequence, the death rate is falling fast. Xinhua, 'China to close 4,861 small collieries by end of 2007', *China Daily*, 4 September 2006, on www.chinadaily.com.cn/china/2006-09/04/content_681187.htm.

135 For Greenpeace UK the opposite is the case. Nuclear power is 'inherently dangerous', and 'scientists agree' that another catastrophe 'on the scale of Chernobyl' could still happen 'any time, anywhere.' See Greenpeace UK, 'Nuclear power – the problems', on www.greenpeace.org.uk/nuclear/problems.

136 Richard Stone, op. cit., p44.

137 Bernard L Cohen, *The Nuclear Energy Option*, Plenum Press, 1990. Available on www.phyast.pitt.edu/~blc/book/chapter7.html.

138 Robert F Kennedy Jr, *Crimes Against Nature: Standing up to Bush and the Kyoto Killers who are Cashing in on our World*, Penguin Books, 2005, p166.

139 Helen Caldicott, *Nuclear Power is Not the Answer*, The New Press, 2006, p62.

140   Douglas M Chapin and others, 'Nuclear Power Plants and Their Fuel as Terrorist Targets', *Science*, Vol. 297, No. 5589, 20 September 2002, pp1997-1999.

141   For more on the AP1000, see Westinghouse, 'AP1000', on www.ap1000.westinghousenuclear.com/.

142   Westinghouse, 'AP1000 at a Glance', on www.ap1000.westinghousenuclear.com/ap1000_glance.html.

143   US Nuclear Regulatory Commission, 'Safety culture', on www.nrc.gov/about-nrc/regulatory/enforcement/safety-culture.html.

144   James Baker III, 'From points to pathways of mutual advantage: next steps in Soviet-American relations', address to American Committee on US-Soviet Relations, US Department of State Dispatch, 22 October 1990, on http://findarticles.com/p/articles/mi_m1584/is_n8_v1/ai_9141030/pg_5?tag=artBody;col1.

145   Ronald Reagan, *Address to the Annual Convention of the American Bar Association*, 8 July 1985, quoted in James Schlesinger, 'Reykjavik and Revelations: A Turn of the Tide', *Foreign Affairs, America and the World 1986*, Vol. 65, No. 3, p440.

146   See the informative article by Petra Minnerop, 'Rogue States – State Sponsors of Terrorism?', *German Law Journal*, No. 9, 1 September 2002, on www.germanlawjournal.com/article.php?id=188#fuss3.

147   On Bull, see Global Security, 'Weapons of Mass Destruction (WMD): Project Babylon Supergun / PC-2', on www.globalsecurity.org/wmd/world/iraq/supergun.htm.

148   Petra Minnerop, op. cit.

149   See www.gnep.energy.gov.

150   Global Nuclear Energy Partnership, 'Statement of Principles', on http://gneppartnership.org/docs/GNEP_SOP.pdf.

151   Leonor Tomero, 'The future of GNEP: The international partners', *Bulletin of Atomic Scientists*, 31 July 2008, on www.thebulletin.org/web-edition/reports/the-future-of-gnep/the-future-of-gnep-the-international-partners.

152   Paul Gordon Lauren and others, *Force and Statecraft: Diplomatic Challenges of our Time*, fourth edition, 2007, p277.

153   See the useful treatment of sanctions by Jay Gordon, 'When Economic Sanctions Become Weapons of Mass Destruction', *Contemporary Conflicts*, 26 March 2004, on http://conconflicts.ssrc.org/archives/iraq/gordon/.

154   'Application for Yucca Mountain Store goes in', *World Nuclear News*, 3 June 2008, on www.world-nuclear-news.org/WR_DOE_submits_Yucca_Mountain_application_0306087.html. Yucca Mountain has been mired in controversy. Lawsuits have been filed both by opponents (including the State of Nevada that has sued the Federal government), and by utility companies angry at the delay.

155 Paul Slovic and others, 'Perceived Risk, Trust, and the Politics of Nuclear Waste', *Science*, Vol. 254, No. 5038, 13 December 1991, pp1603-1607.

156 Bernard L Cohen, *The Nuclear Energy Option*, Plenum Press, 1990, Figure 1 of Chapter 11. Available on www.phyast.pitt. edu/~blc/book/chapter11.html.

157 The half-life is the time taken for half of a sample to decay. After two half-lives, a quarter will remain; after three half-lives, an eighth, and so on.

158 For cadmium, mercury and lead, the tonnages given are for 2007. US Geological Survey Commodity Statistics and Information, on http://minerals.usgs.gov/minerals/ pubs/commodity/. For uranium and plutonium, the tonnages given are the totals for all isotopes, not just those entered on the left-hand column. World Nuclear Association, www. world-nuclear.org/info/inf15.html and www.world-nuclear.org/info/ inf15.html.

159 Barack Obama and Joe Biden, *New Energy for America*, fact sheet, no date, pp6-7, on

160 Barack Obama, interview on energy with the *San Francisco Chronicle*, 17 January 2008, YouTube video, on http://uk. youtube.com/ watch?v=SMwBbl6RoIs &feature=related.

161 Ibid.

162 Noel Annan, *Our Age: the Generation That Made Post-war Britain* (1990), Fontana, 1991, p385.

163 Terry Macalister, 'Report reveals chaos at the heart of Nuclear Decommissioning Authority', *The Guardian*, 23 July 2008, on www. guardian.co.uk/ environment/2008/jul/23/ nuclearpower.energy.

164 DEFRA, 'UK Radioactive Waste Inventory', Table 1, 'The baseline inventory as at 1 April 2007 comprising of the following radioactive wastes and materials', on www.defra.gov.uk/ environment/radioactivity/mrws/ waste/index.htm.

165 DTI, *The Energy Challenge: Energy Review Report 2006*, July 2006, p118, paragraph 5.116, on www.berr.gov.uk/files/ file31890.pdf.

166 DEFRA, Ibid.

167 Apart from BEIR, the United Nations Scientific Committee on the Effects of Atomic Radiation (UNSCEAR), founded in 1955, oversees safety standards. See www.unscear.org/unscear/index. html.

168 George Kneale, Thomas Mancuso and Alice Stewart, 'Radiation exposures of Hanford workers dying from cancer and other causes', *Health Physics*, Vol. 3 No 5, November 1977.

CHAPTER 5

1 In the combustion of fossil fuels, the chemistry of photosynthesis is reversed. When fossil fuels are burnt in machines, stored sunlight is transformed into heat. That heat might be used to raise steam for turning an electrical generator in a coal-fired power station, to cook food using natural gas, or to move the pistons in an internal combustion engine. Combustion reverses the chemical reaction that originally formed the fuel, so that carbon and hydrogen are combined with oxygen to form CO_2, which returns to the atmosphere, and water.

2   Matthew Lockwood, *After the Coal Rush: Assessing policy options for coal-fired electricity generation*, Institute of Public Policy Research, 2 July 2008, on www.ippr.org/ publicationsandreports/ publication.asp?id=617. Registration required.

3   David Wheeler, 'Tata Ultra Mega Mistake: The IFC Should Not Get Burned by Coal', Center for Global Development, 12 March 2008, on http://blogs.cgdev.org/ globaldevelopment/2008/03/ tata_ultra_mega_mistake_the_ if.php.

4   Tim Flannery, *The Weather Makers*, Allen Lane, 2006, p304.

5   Jeffrey Ball, 'Wall Street Shows Skepticism Over Coal', *Wall Street Journal*, 4 February 2008, on http://online.wsj.com/article/ SB120209079624339759.html.

6   IEA Coal Industry Advisory Board, *Meeting Our Energy Needs – driving forward coal's role in a clean, clever and competitive energy future*, 9 November 2005, p7, on www.iea.org/Textbase/ work/workshopdetail.asp?WS_ ID=228

7   See World Coal Institute, 'Coal Facts 2007', October 2007, on www.worldcoal.org/pages/ content/index.asp?PageID=188. Figures for South Africa, Israel, India and Morocco are estimates for 2005.

8   Kebin He, Hong Huo, and Qiang Zhang, 'Urban air pollution in China: Current Status, Characteristics, and Progress', *Annual Review of Energy and the Environment*, Vol. 27, November 2002, pp397-431.

9   Yu Dawei, 'China Holds Its Breath for Clean-Coal Power', *Caijing Magazine*, 30 April 2008, on http://english.caijing.com. cn/2008-04-30/100058988. html.

10   Mark Jaccard, *Sustainable Fossil Fuels*, Cambridge University Press, 2005, p195.

11   MIT, *The Future of Coal: Options for a Carbon-Constrained World*, Interdisciplinary MIT study, 2007, p17, Table 3.1, 'Representative Performance And Economics For Air-Blown PC Generating Technologies', p19, and Table 3.5, 'Representative Performance and Economics for Oxy-Fuel Pulverised Coal and IGCC Power Generation Technologies, Compared with Supercritical Pulverised Coal', p30, on http:// web.mit.edu/coal/The_Future_of_ Coal.pdf.

12   Jon Gibbins and Hannah Chalmers, 'Preparing for global rollout: A "developed country first" demonstration programme for rapid CCS deployment',

Energy Policy, Vol. 36, Issue 2, February 2008.

13 Simon Shackley and Jon Gibbins, 'The case for carbon capture', *New Statesman*, 15 May 2006, on www.newstatesman. com/200605150062.

14 Joseph R McConnell and Ross Edwards, 'Coal burning leaves toxic heavy metal legacy in the Arctic', *Proceedings of the National Academy of Sciences of the United States of America*, 26 August 2008, Vol. 105, No. 34, available for $10 on www.pnas. org/content/105/34/12140.full.

15 John Ashton, interviewed in Roger Harrabin, 'China building more power plants', *BBC News online*, 19 June 2007, on http://news. bbc.co.uk/1/hi/world/asia-pacific/6769743.stm.

16 Jeff Tollefson, 'China: Stoking the fire', *Nature*, Vol. 454, No. 7203, 23 July 2008, pp388-393.

17 John Sauven, quoted in Roger Harrabin, op. cit.

18 Quoted in James Woudhuysen, 'Like it or not, coal is vital to Asia's growth', *spiked*, 12 September 2007, on www. spiked-online.com/index.php?/ site/article/3808.

19 See WWF, *Annual Review 2007*, p19, on www.panda.org/ news_facts/publications/key_ publications/index.cfm.

20 John Vidal, 'Climb every chimney...', *The Guardian*, 12 September 2008, on www. guardian.co.uk/ environment/2008/sep/12/ activists.kingsnorth.

21 Martyn Day, a partner with Leigh Day solicitors, quoted in ibid.

22 For an excellent account of the significance of the trial, see Brendan O'Neill, 'State sanctioned radicalism', *spiked*, 15 September 2008, on www. spiked-online.com/index.php?/ site/article/5721.

23 'Ask the Experts', *AmericasPower. org*, no date, on www. americaspower.org/Ask-The-Experts.

24 Arthur Scargill, 'Coal isn't the climate enemy, Mr Monbiot. It's the solution', *The Guardian*, 8 August 2008, on www.guardian. co.uk/commentisfree/2008/ aug/08/nuclearpower.fossilfuels.

25 'Ask the Experts', op. cit., and 'Issues & Policy', *AmericasPower. org*, no date, on www. americaspower.org/Issues-Policy.

26 James F Roberts, quoted in 'US coal industry leader says technology is key to energy security, climate change', *International Mining*, 13 September 2008, on www. im-mining.com/2008/09/13/ us-coal-industry-leader-says-technology-is-key-to-energy-security-climate-change.

27 It's remarkably hard to find figures for the dollar size of coal mining in the US. Our rough estimates are derived from Energy Information Administration data for corporate market shares for 2007, 'Major US Coal Producers', September 2008, on www.eia.doe.gov/cneaf/coal/page/ acr/table10.html. This gives the Peabody Energy Corporation a

16.8 per cent share. Peabody itself declared $4.6bn revenues in 2007, on www.peabodyenergy.com. In scaling up the dollar figures from Peabody to the whole US coal mining industry, we've taken account of the fact that some of Peabody's revenues come from abroad.

28 James F Roberts, quoted in 'US coal industry leader says technology is key to energy security, climate change', op. cit.

29 See the Energy Improvement and Extension Act of 2008 (Engrossed Amendment as Agreed to by Senate), H.R.6049, 23 September 2008, and Energy Improvement and Extension Act of 2008 (Introduced in House), H.R.7201, 28 September 2008, available through http://thomas.loc.gov.

30 David Adam, 'Fuel made from coal ignites green row', *The Guardian,* 5 April 2008, on www.guardian.co.uk/environment/2008/apr/05/carbonemissions.climatechange.

31 Sasol, 'Sasol and Shenhua Ningxia Coal Industry contract engineering companies for CTL study', press release, 22 October 2008, on www.sasol.com/sasol_internet/frontend/navigation.jsp?articleId=23100002&navid=4&rootid=4.

32 Quoted in Zhang Qi, 'Is it the end of the line for coal-to-oil in China?', *China Daily*, 9 October 2008, on www.chinadaily.com.cn/bizchina/2008-10/09/content_7090441.htm.

33 MIT, *The Future of Coal*, op. cit.

p.ix. We have taken the millions of US tons mentioned in this report as equivalent to Megatonnes.

34 Rahul Banerjee and others, 'High-Throughput Synthesis of Zeolitic Imidazolate Frameworks and Application to CO_2 Capture', *Science*, Vol. 319, No. 5865, 15 February 2008, on http://yaghi.chem.ucla.edu/pdfPublications/2008hiThroughput.pdf.

35 Peter Styles, reported in Institute of Physics, *Carbon Capture and Storage*, report on a seminar held on 5 December 2007 to discuss methods of sequestering carbon dioxide produced from burning fossil fuels as a means of countering global warming, May 2008, p5, on www.iop.org/activity/policy/Events/Seminars/file_30843.pdf.

36 'Burning question', *The Engineer*, 25 February 2008, on www.theengineer.co.uk/Articles/304737/Burning+question.htm.

37 Royal Society, 'CO_2 capture & storage: international progress & future prospects', workshop co-hosted by the Royal Society, the Royal Academy of Engineering and the IEA Coal Industry Advisory Board, 7/8 November 2007, p3, on http://royalsociety.org/displaypagedoc.asp?id=29440.

38 See Clifford Krauss, 'US again becoming a major coal exporter', *International Herald Tribune*, 19 March 2008, on www.iht.com/articles/2008/03/19/business/coal.php.

39 BP, *Statistical Review of World Energy*, June 2008, p35, on www.bp.com/liveassets/bp_ internet/globalbp/globalbp_uk_ english/reports_and_publications/ statistical_energy_review_2008/ STAGING/local_assets/ downloads/pdf/statistical_review_ of_world_energy_full_ review_2008.pd, and historical data on www.bp.com/liveassets/ bp_internet/globalbp/globalbp_ uk_english/reports_and_ publications/statistical_energy_ review_2008/ STAGING/local_assets/ downloads/spreadsheets/ statistical_review_full_report_ workbook_2008.xls.

40 Ibid.

41 George Orwell, 'Down the mine', essay, 1937, available on www. orwell.ru/library/essays/mine/ english/e_dtm.

42 IEA, *Key World Energy Statistics 2008*, 2008, p7, on www.iea.org/ Textbase/publications/free_new_ Desc.asp?PUBS_ID=1199.

43 IEA and OECD, *Natural Gas Market Review 2008: Optimizing Investments and Ensuring Security in a High-priced Environment*, September 2008, p134.

44 IEA and OECD, *Natural Gas Market Review 2008*, op. cit., pp136-137.

45 Ibid., p138.

46 BERR, *Meeting the Energy Challenge*, 2007, paragraphs 4.68, 4.70, p123, on www.berr. gov.uk/files/file39568.pdf.

47 BERR, *Energy Markets Outlook*, October 2007, Chapter 5, paragraph 5.9.1, p52, on www. berr.gov.uk/files/file41998.pdf.

48 BERR, *Meeting the Energy Challenge*, op. cit., p116.

49 Jill Kirby, 'Dithering ministers saddle us with an energy crunch', *The Sunday Times*, 3 August 2008, on www.timesonline.co.uk/ tol/comment/columnists/guest_ contributors/article4449120.ece.

50 Jim Prevost, 'On Track at Kenai's LNG Plant', *Alaska Business Monthly*, 1 February 2000.

51 Anna Bergek and others, 'Technological capabilities and late shakeouts: industrial dynamics in the advanced gas turbine industry, 1987–2002', *Industrial and Corporate Change*, Vol. 17, Issue 2, April 2008.

52 Ross Tieman, 'Dash for gas has run out of steam', *Financial Times* report on the gas industry, 10 March 2008, p3.

53 Julian Darley, *High Noon for Natural Gas*, Chelsea Green Publishing company, 2004, p93.

54 Kenneth J Bird and others, *Circum-Arctic Resource Appraisal: Estimates of Undiscovered Oil and Gas North of the Arctic Circle*, US Geological Survey report No. 3049, 23 July 2008, Table 1, Summary of Results, p4, on http://pubs.usgs. gov/fs/2008/3049/fs2008-3049. pdf.

55 Statoil, 'Snøhvit: the world's northernmost LNG project', no date, on www.statoil.com/ statoilcom/snohvit/svg02699. nsf?OpenDatabase&lang=en.

56 Bob MacKnight, 'Promise in United States Gas', *The PFC Energy Quarterly*, Third Quarter 2008, on www.pfcenergy.com/quarterly/articles.aspx?issueID=107&id=50&title=Promise%20in%20United%20States%20; Wood Mackenzie, 'Wood Mackenzie: Significant Shale Gas Potential in the Rockies Numerous Plays', press release, 9 July 2008, on www.woodmacresearch.com/cgi-bin/corp/portal/corp/corpPressDetail.jsp?oid=1141400.

57 Greenpeace spokesman Juha Aromaa, quoted in Tristana Moore, 'Gas pipeline stirs up Baltic fears', *BBC News online*, 31 December 2007, on http://news.bbc.co.uk/1/hi/business/7153924.stm.

58 For the British case, see Martin Shipton, 'EU probe into LNG pipeline', *Western Mail*, 21 December 2007, on http://icwales.icnetwork.co.uk/news/politics-news/2007/12/21/eu-probe-into-lng-pipeline-91466-20275883/.

59 See the remarks of US politicians quoted in David Ivanovich, 'Study doubts ports' security', *Houston Chronicle*, 9 January 2008, on www.chron.com/CDA/archives/archive.mpl?id=2008_4492190.

60 'Ukraine and Russia reach gas deal', *BBC News online*, 4 January 2006, on http://news.bbc.co.uk/1/hi/world/europe/4579648.stm.

61 'Gas shortage sends prices soaring', *BBC News online*, 13 March 2006, on http://news.bbc.co.uk/1/hi/business/4802786.stm.

62 BERR, *Meeting the Energy Challenge*, op. cit., paragraph 4.58, pp120-121, on www.berr.gov.uk/files/file39568.pdf.

63 IEA, *Natural Gas Market Review 2008*, op. cit., pp 17, 22, 23.

64 Ibid., p26.

65 Ed Crooks, 'Uneasy reliance on Russia likely to persist', *Financial Times*, 5 September 2008, p7.

66 Petroleum Economist, *Gas in the CIS & Europe*, 2008, on www.petroleum-economist.com/default.asp?page=19&searchtype=17&productid=5631.

67 Ariel Cohen, senior research fellow at the Heritage Foundation, quoted in Bill Powell, 'Just how scary is Russia?', *Fortune*, 15 September 2008, p46.

68 Anonymous official quoted in ibid., p46.

69 IEA, *Natural Gas Market Review 2007: Security in a Globalizing Market to 2015*, 2007, p84, on www.iea.org/textbase/nppdf/free/2007/Gasmarket2007.pdf.

70 IEA, *Natural Gas Market Review 2008*, op. cit., pp28, 86.

71 Iain Dey, 'Price war threat to UK gas supplies', *The Sunday Times Business News*, 9 March 2008, on http://business.timesonline.co.uk/tol/business/industry_sectors/natural_resources/article3510672.ece.

72 Carl Mortishead, 'Cartel of key gas exporters could have a significant impact on Europe', *The Times*, 22 October 2008, on http://business.timesonline.co.uk/tol/business/columnists/article4988242.ece.

73 Zbigniew Brzezinski, *The Grand Chessboard: American Primacy and its Geostrategic Imperatives*, Basic Books, 1997, pp52, 53.

74 Three of the best-known books are Peter Odell, *Oil and World Power* (1970), Penguin Books, eighth edition, 1986; Anthony Sampson, *The Seven Sisters, the Great Oil Companies and the World They Made*, Viking Press, 1975; and Daniel Yergin, *The Prize: the Epic Quest for Oil, Money, and Power* (1991), Simon & Schuster, 1993. On North Sea oil, see Christopher Harvie, *Fool's Gold: The Story of North Sea Oil* (1994), Penguin, 1995. The poet Al Alvarez recounted his experiences on a North Sea oil platform in *Offshore: a North Sea Journey*, Hodder & Stoughton, 1986.

75 Two online trade websites with plenty of data are *Oil & Gas Journal*, on www.ogj.com/index.cfm, and *Oil and Gas International*, on www.oilandgasinternational.com.

76 See for example Matthew R Simmons, *Twilight in the Desert: the Coming Saudi Oil Shock and the World Economy*, Wiley, 2005, and Jeremy Leggett, *Half Gone: Oil, Gas, Hot Air and the Global Energy Crisis*, Portobello Books, 2005.

77 Matthew Yeomans, *Oil: Anatomy of an Industry*, The New Press, 2004; Sonia Shah, *Crude: the Story of Oil*, Seven Stories Press, 2004; Paul Roberts, *The End of Oil: the Decline of the Petroleum Economy and the Rise of a New Energy Order*, 2004.

78 Kenneth S Deffeyes, *Hubbert's Peak: The Impending World Oil Shortage*, Princeton University Press, 2001, and *Beyond Oil: the View from Hubbert's Peak*, Hill and Wang, 2005; David Goodstein, *Out of Gas: The End of the Age of Oil*, W W Norton, 2004. At Princeton for more than 40 years, Deffeyes has a background with Shell, and continues to consult for the petroleum industry.

79 The movie's website is www.oilcrashmovie.com/film.html.

80 For engineer and analyst Gregson Vaux, the peaking of oil in 2009 heralds a more intensive use of coal, so that at an annual rate of growth of 3.1 per cent – for world energy demand, he writes on one page, or for the general economy, he writes on the next – the 'year of total coal depletion' will fall no later than 2062. See Vaux, 'A projection of future coal demand given diminishing oil supplies', in Andrew McKillop with Sheila Newman, *The Final Energy Crisis*, Pluto Press, 2005, pp275, 276.

81 Richard Heinberg, *Peak Everything: Waking Up to the Century of Decline in Earth's Resources*, Clairview Books, 2007.

82 Daniel Yergin, *The Prize*, op. cit., p45.

83 Richard Hofstadter, *The Age of Reform*, Vintage Books, 1960, pp183, 193.

84 Tony Freyer, *Regulating Big*

Business: Antitrust in Great
Britain and America 1880-1990,
Cambridge University Press, p86.

85   Ibid., pp57, 174.

86   For a left-leaning recapitulation of
grievances against Rockefeller
and his ilk, influential in its time,
see Matthew Josephson, *The
Robber Barons: the Great
American Capitalists, 1861-1901*,
Harcourt, Brace and Company,
1934, especially Chapter XVI,
'Concentration: the great trusts.'

87   Richard Hofstadter, *The Age of
Reform*, op. cit., pp203-205.

88   Craig Unger, *House of Bush,
House of Saud: the Hidden
Relationship Between the World's
Two Most Powerful Dynasties*,
Gibson Square Books, 2004.

89   Ibid., pp273, 281.

90   Ibid., pp4, 272.

91   Jerry Useem, 'The Devil's
Excrement: Pérez Alfonzo's
different name for oil', *Fortune*,
21 January 2003, cited in Shah,
Crude, op. cit., p3.

92   Richard Lugar, Foreword to David
Sandalow, *Freedom from Oil*,
McGraw-Hill 2008, ppxi-xii.

93   Matthew Simmons, *Twilight in the
Desert*, op. cit., p19.

94   Ibid., p179.

95   Ibid., p334.

96   Ibid., p339.

97   Paul Roberts, *The End of Oil*, op.
cit., and pp6-9.

98   Toby Shelley, *Oil: Politics, Poverty
and the Planet*, Zed Books,
2005.

99   See for example Terry Lynn Karl,
*The Paradox of Plenty: Oil Booms
and Petro-States*, University of
California Press, 1997; Macartan
Humphreys, Jeffrey D Sachs and
Joseph E Stiglitz, editors,
Escaping the Resource Curse,
Columbia University Press, 2007.
For a different appraisal, see
Daniel Lederman and William F
Maloney, editors, *Natural
Resources, Neither Curse Nor
Destiny*, World Bank Publications,
2006.

100   Mary Kaldor, Terry Lynn Karl and
Yahia Said, editors, *Oil Wars*,
Pluto Press, 2007.

101   See for example Yergin, *The
Prize*, op. cit., p337.

102   Noam Chomsky and Gilbert
Achar, *Perilous Power: The
Middle East and US Foreign
Policy*, Penguin, 2007, p53.

103   Philip Hammond, *Framing
Post-Cold War Conflicts: The
Media and International
Intervention*, Manchester
University Press, 2007, p220.

104   Greg Jaffe and Thomas E Ricks,
'Military Exploits: Of Men and
Money And How the Pentagon
Often Wastes Both', *Wall Street
Journal*, 22 September 1999, on
www.pulitzer.org/archives/6359.

105   See for example Philip Hammond,
Media, War and Postmodernity,
Routledge, 2007.

106   David Chandler, *Empire in Denial:
the Politics of State-Building*,
Pluto Press, 2006.

107   Michael T Klare, *Resource Wars:*

The New Landscape of Global Conflict (2001), Henry Holt and Company, 2002.

108   See Lutz Kleveman, *The New Great Game: Oil and International Politics in Central Asia* (2002), Atlantic Books, 2004, p32.

109   Paul Sperry, *Crude Politics: How Bush's Oil Cronies Hijacked the War on Terrorism*, Thomas Nelson, 2003; Michael T Klare, *Blood and Oil: How America's Thirst for Petrol is Killing Us*, Hamish Hamilton, 2004; Dilip Hiro, *Blood of the Earth: The Global Battle for Vanishing Oil Resources*, Politico's, 2007.

110   Michael Klare, *Resource Wars*, op. cit., pp5-10.

111   See Lutz Kleveman, *The New Great Game*, op. cit., pp3, 5, and Dick Cheney and others, *National Energy Policy*, Report of the National Energy Policy Development Group, May 2001, Chapter 8, p7, on www. whitehouse.gov/energy/ National-Energy-Policy.pdf.

112   Michael Klare, *Resource Wars*, op. cit., pp5-10.

113   Michael Klare, *Blood and Oil*, op. cit., pp147-148.

114   Ibid., pp150-151.

115   Halford John Mackinder, 'The geographical pivot of history', *The Geographical Journal*, 25 January 1904, available for $5.95 from www.amazon.com/geographical-pivot-history-Geographical-Journal/dp/B00096T222. See also Brian W Blouet, editor, *Global Geostrategy: Mackinder and the Defence of the West*,

Frank Cass, 2005.

116   Andy Stern, *Who Won The Oil Wars?*, Collins & Brown, 2005, p7.

117   Ibid., pp75-83.

118   Ibid., p8.

119   Ibid., p189.

120   John Ghazvinian, *Untapped: The Scramble for Africa's Oil* (2007), Harvest Books, 2008, pp288-289.

121   Andy Stern, op. cit., p189.

122   Carl Mortishead, 'Shell deepwater platform attacked as Nigerian separatists step up protests', *The Times*, 20 June 2008.

123   Edward L Morse and Amy Myers Jaffe, *Strategic Energy Policy: Challenges for the 21st Century*, Report of an Independent Task Force Cosponsored by the James A Baker III Institute for Public Policy of Rice University and the Council on Foreign Relations, 2001, on www.cfr.org/content/ publications/attachments/ Energy%20TaskForce.pdf.

124   Colin J Campbell and Jean H Laherrère, 'The End of Cheap Oil', *Scientific American*, March 1998, p83.

125   Robert L Hirsch and others, *Peaking of World Oil Production: Impacts, Mitigation, & Risk Management*, February 2005, on www.netl.doe.gov/publications/ others/pdf/Oil_Peaking_NETL.pdf.

126   See www.chevron.com/ Documents/Pdf/ RealIssuesAdTrillionBarrels.pdf.

127   'First R&D Conference to address

challenges for the next trillion', *Journal of Petroleum Technology Online*, 12 March 2007, on http://updates.spe.org/index.php/2007/03/12/first-rd-conference-to-address-challenges-for-the-next-trillion.

128 Matthew R Simmons, *Twilight in the Desert*, op. cit., p53.

129 Sonia Shah, *Crude*, op. cit., p53.

130 Ibid., p150.

131 Darrell Stonehouse, 'The world gets heavier', *Oilsands Review*, March 2008, on www.oilsandsreview.com/articles.asp?ID=535.

132 World Energy Council, *2004 Survey of Energy Resources*, 2004, p99, on www.worldenergy.org/documents/ser2004.pdf.

133 M King Hubbert, *Nuclear Energy and the Fossil Fuels*, Shell, 1956, on www.hubbertpeak.com/hubbert/1956/1956.pdf.

134 'Am I promising war, famine, pestilence, and death? If we can keep the petrochemicals industry going we might avoid the pestilence part. The other three are serious possibilities.' Kenneth S Deffeyes, *Beyond Oil*, op. cit., p8.

135 For summaries of Pike's views, see Steve Connor, 'Oil shortage a myth, says industry insider', *The Independent*, 9 June 2008, on www.independent.co.uk/environment/climate-change/oil-shortage-a-myth-says-industry-insider-842778.html, and Andrew Orlowski, 'Peak oil: postponed', *The Register*, 17 September 2008, on www.theregister.co.uk/2008/09/17/richard_pike_rcs_interview.

136 Rafael Kandiyoti, 'Pipeline politics caused the war in Georgia', *The Daily Telegraph*, 20 August 2008, on www.telegraph.co.uk/opinion/main.jhtml?xml=/opinion/2008/08/20/do2005.xml.

137 Rafael Kandiyoti, *Pipelines: Oil Flows and Crude Politics*, IB Tauris, 2008.

138 Wharton Business School and the Boston Consulting Group, *The New Competition for Global Resources*, 2008, on http://knowledge.wharton.upenn.edu/papers/download/BCGReport_Competition_for_Global_Resources.pdf.

139 Ibid., p2.

140 Ibid., pp2, 3.

141 Ibid., p3.

142 Rick Peters, in ibid., pp2,3.

143 US Energy Information Administration, *International Energy Outlook 2008*, September 2008, p24, on www.eia.doe.gov/oiaf/ieo/pdf/0484(2008).pdf.

144 Ian Rutledge, *Addicted to Oil: America's Relentless Drive for Energy Security*, IB Tauris, 2005, p10.

145 Amory Lovins, *Winning the Oil Endgame* (2004), Earthscan, 2005, p8.

146 Bureau of Transportation Statistics, *National Transportation Statistics*, Table 1-37: US Passenger-Miles, December 2007, on www.bts.

gov/publications/national_ transportation _statistics/html/ table_01_37.html, and Table 1-7: 'Number of Stations Served by Amtrak and Rail Transit, Fiscal Year', July 2008, on www.bts.gov/ publications/national_ transportation_statistics/html/ table_01_07.html.

147 Department for Transport, *Transport Trends*, 2007 edition, p26, on www.dft.gov.uk/ 162259/162469/221412/ 190425/220778/trends2007a. pdf. and www.dft.gov.uk/ datatablespublications/trends/ current/section2pbtm.xls.

148 George Monbiot, 'They call themselves libertarians; I think they're antisocial bastards', *The Guardian*, 20 December 2005, on www.guardian.co.uk/uk/2005/ dec/20/politics.publicservices.

149 The key text advocating such a vision is Jeremy Rifkin, *The Hydrogen Economy*, Polity Press, 2002. A growing literature, both for and against hydrogen, has followed this book.

150 Indrajit Gupta and R Sriram, 'Interview with Ratan Tata: Making of Nano', *The Economic Times*, 11 January 2008, on http://economictimes.indiatimes. com/articleshow/msid-2690794,prtpage-1.cms.

151 Quoted in Andrew Buncombe, 'Can the world afford the Tata Nano?', *The Independent*, 12 January 2008, on www. independent.co.uk/environment/ climate-change/can-the-world-afford-the-tata-nano-769421. html.

152 Andrew McKillop, 'The Chinese car bomb', in Andrew McKillop with Sheila Newman, op. cit., pp228, 232.

153 'China needs to add 1mln km of highway by 2020', *China Daily*, 17 November 2007, on www. chinadaily.com.cn/ bizchina/2007-11/17/ content_6261374.htm.

154 'Performance indicators', *ICAO Journal*, Vol. 62, No. 2, 2007, p5. Between 1981 and 2006 the number of passengers passing through UK airports virtually quadrupled, from 58 million to 235 million: see Civil Aviation Authority, 'Main Outputs of UK Airports 1981-2006', Table 2.1, on www.caa.co.uk/docs/80/ airport_data/2006Annual/ Table_02_1_Main_Outputs_of_ UK_Airports_2006.pdf.

155 George Monbiot, 'Apocalypse now', *The Guardian*, 29 July 1999, on www.guardian.co.uk/ Columnists/ Column/0,,279973,00.html.

156 Kurt Kleiner, 'Civil aviation faces green challenge', *Nature*, Vol. 448, No. 7150, 12 July 2007, p120.

157 Guy Norris, 'Future eCore Foundation Plan Revealed', *Aviation Week and Space Technology*, 13 July 2008, on www.aviationweek.com/aw/ generic/story_generic. jsp?channel=awst&id=news/ aw071408p2. xml&headline=Future%20 eCore%20Foundation%20 Plan%20Revealed.

158 Gregory Polek, 'After two decades

of study, Pratt's PW1000G takes flight', *Aviation International News online*, 30 July 2008, on www.ainonline.com/news/single-news-page/article/after-two-decades-of-study-pratts-pw1000g-takes-flight.

159   Ben Webster, 'BA reprimanded over claim that new runway will reduce emissions', *The Times*, 5 January 2008, on www.timesonline.co.uk/tol/news/environment/article3134484.ece.

160   Friends of the Earth, 'Branson climate initiative – reaction', press release, 21 September 2006, on www.foe.co.uk/resource/press_releases/branson_climate_initiative_21092006.html.

161   Jamie, 'Virgin's biofuel flight is all spin and greenwash', Greenpeace UK, 25 February 2008, on www.greenpeace.org.uk/blog/climate/virgins-biofuel-flight-is-all-spin-and-greenwash-20080225.

162   Continental Airlines, 'Continental Airlines, Boeing and GE Aviation Announce Plans for Sustainable Biofuels Flight Demonstration', 13 March 2008, on www.continental.com/web/en-US/apps/vendors/default.aspx?i=PRNEWS

163   'Alternative fuel partnership teams Airbus with Honeywell, IAE and JetBlue', 15 May 2008, on www.airbus.com/en/presscentre/pressreleases/pressreleases_items/08_05_15_alternative_fuel_partnership.html.

164   Defense Advanced Research Projects Agency, 'Agency Seeks to Develop Military Aviation Biofuel', press release, 18 July 2006, on www.defenselink.mil/transformation/articles/2006-07/ta071806c.html.

165   David Biello, 'Jet fuel from algae passes first test', *Scientific American*, 9 September 2008, on www.sciam.com/blog/60-second-science/post.cfm?id=jet-fuel-from-algae-passes-first-te-2008-09-09.

166   WHO, *Indoor Air Pollution and Health*, June 2005, on www.who.int/mediacentre/factsheets/fs292/en/index.html.

167   Those standards are the 1993 Winter Oxyfuel Program and the 1995 Year-round Reformulated Gasoline Program (1995) under the 1990 Clean Air Act.

168   Shota Atsumi and others, 'Non-fermentative pathways for synthesis of branched-chain higher alcohols as biofuels', *Nature*, Vol. 451, No. 7174, 3 January 2008.

169   'Mad scientist who wants to put a microbe in your tank', *The Times*, June 1, 2007, on www.timesonline.co.uk/tol/news/uk/science/article2010049.ece.

170   ETC Group, 'Patenting Pandora's Bug: Goodbye, Dolly... Hello, Synthia! J Craig Venter Institute Seeks Monopoly Patents on the World's First-Ever Human-Made Life Form', news release, 7 June 2007, on www.etcgroup.org/en/materials/publications.html?pub_id=631.

171   Quoted in Ed Pilkington, 'I am creating artificial life, declares US gene pioneer', *The Guardian*, 6 October 2007, on www.guardian.

co.uk/science/2007/oct/06/
genetics.climatechange.

172   David Pimentel and Tad W
Patzek, 'Ethanol Production Using
Corn, Switchgrass, and Wood;
Biodiesel Production Using
Soybean and Sunflower', *Natural
Resources Research*, Vol. 14, No.
1, March 2005, on http://
petroleum.berkeley.edu/papers/
Biofuels/NRRethanol.2005.pdf.

173   Alexander E Farrell and others,
'Ethanol Can Contribute to Energy
and Environmental Goals',
Science, Vol. 311, Issue 5760,
27 January 2006, and M R
Schmer and others, 'Net energy
of cellulosic ethanol from
switchgrass', *Proceedings of the
National Academy of Sciences of
the United States*, Vol. 105, No.
2, 7 January 2008, on www.
pnas.org/content/105/2/464.full.
pdf+html?sid=662a2069-0bd3-
4cfc-b286-64b65759f693.

174   '"Slave" labourers freed in Brazil',
BBC News online, 3 July 2007,
on http://news.bbc.co.uk/1/hi/
world/americas/6266712.stm.

175   Simon Romero, 'Spoonfuls of
Hope, Tons of Pain; In Brazil's
Sugar Empire, Workers Struggle
With Mechanization', *The New
York Times*, 21 May 2000, on
http://query.nytimes.com/gst/
fullpage.html?res=
9F0CE3D8123AF93
2A15756C0A9669C8B63.

176   Robin Pagnamenta, 'Bioethanol
demand forces Brazil sugar cane
industry upheaval', *The Times*, 7
June 2008, on http://business.
timesonline.co.uk/tol/business/
industry_sectors/natural_
resources/article4083137.ece.

177   Terry Macalister, 'Sun sets on
Brazil's sugar cane cutters', *The
Guardian*, 5 June 2008, on www.
guardian.co.uk/
environment/2008/jun/05/
biofuels.carbonemissions.

178   Jack Chang, 'Brazil's sugar cane
mills race to keep up with ethanol
boom', *McClatchy Newspapers*,
19 May 2008, on www.
mcclatchydc.com/117/
story/37660.html.

179   See for example www.
precisionag.com.

180   Food and Agriculture Organization
of the United Nations (FAO), *The
State of Food and Agriculture
2007: Paying farmers for
environmental services*, 2007,
pp124-125, on ftp://ftp.fao.org/
docrep/fao/010/a1200e/
a1200e00.pdf.T

181   Human beings have also become
less dependent on food aid.
While absolute amounts of food
aid have remained roughly
constant for the past decade, it
has food aid has declined as a
proportion of the world food
economy. Food aid stands at
about 10 million tonnes of grain
equivalent per year, which
represents just 0.5 per cent of
total production. See FAO, *The
State of Food and Agriculture
2006: Food aid for food
security?*, 2006, pp4-6, on ftp://
ftp.fao.org/docrep/fao/009/
a0800e/a0800e.pdf.

182   FAO, *State of Food and
Agriculture 2007*, op cit,
pp131-132.

183   USDA National Agricultural
Statistics Service - Quick Stats,

on www.nass.usda.gov/
QuickStats/Create_Federal_All.
jsp.

184 World Bank, *World Development
Report 2008: Agriculture for
Development,* October 2007,
p15, $26 from http://publications.
worldbank.org/ecommerce/
catalog/product-detail?product_
id=6966252&.

185 Ibid., p52.

186 European Union, 'Agriculture:
Member States agree in principle
to abolish obligatory set-aside', 1
July 2008, on http://europa.eu/
rapid/pressReleasesAction.do?ref
erence=IP/08/1069&type=HTML
&aged=0&language=EN&guiLan
guage=en.

187 Charles Clover, 'Wildlife disaster
as uncropped land is ploughed',
The Daily Telegraph, 30 January
2008, on www.telegraph.co.uk/
earth/main.jhtml?xml=/
earth/2008/01/30/eafarm130.
xml.

188 Environmental Defense Fund, 'Ag
Secretary Urged to Reject Early
Release of Land in Conservation
Reserve Program', Press release,
9 July 2008, on www.edf.org/
pressrelease.
cfm?contentID=8048.

189 National Wildlife Federation,
'Judge orders USDA to Halt
Expanded Haying and Grazing on
Conservation Reserve Program
Lands', press release, 9 July
2008, on http://nwfaffiliates.org/
ht/display/ArticleDetails/i/11584.

190 Nick Kurczewski, 'BYD F3DM
Hybrid Concept – Auto Shows',
Car and Driver, March 2008, on
www.caranddriver.com/news/

auto_shows/2008_geneva_auto_
show_auto_shows/concept_
debuts/byd_f3dm_hybrid_
concept_auto_shows.

191 US Environmental Protection
Agency, *Renewable Fuel
Standard Implementation:
Frequently Asked Consumer
Questions*, November 2007,
pp.1-2, on www.epa.gov/otaq/
renewablefuels/420f07062.pdf.

192 Ian Traynor, 'EU set to scrap
biofuels target amid fears of food
crisis', *The Guardian,* 19 April
2008, on www.guardian.co.uk/
environment/2008/apr/19/
biofuels.food.

193 'India Sets Target of 20% Biofuels
by 2017', *Green Car Congress,*
12 September 2008, on www.
greencarcongress.com/2008/09/
india-sets-targ.html.

194 Giles Clark, 'Belarus biofuels
sector looking for foreign
investment', *Biofuel Review,* 12
June 2008, on www.
biofuelreview. om/content/
view/1613/1 and Russian
Biofuels Association, 'Russia to
Supply Biofuels', 8 June 2008,
on www.biofuels.ru/bioethanol/
news/russia_to_supply_biofuels.

195 G Charles Dismukes and others,
'Aquatic phototrophs: efficient
alternatives to land-based crops
for biofuels', *Current Opinion in
Biotechnology*, Vol. 19, 2008.

196 'BP and DuPont Announce
Partnership to Develop Advanced
Biofuels', 20 June 2006, on
www.bp.com/genericarticle.do?ca
tegoryId=2012968&contentId=7
018942. Cargill produces
polylactic acid plastics through its

subsidiary Natureworks LLC, now in partnership with Japan's Teijin. See www.natureworksllc.com.

197   George M Whitesides and George W Crabtree, 'Don't forget long-term fundamental research in energy', *Science*, Vol. 315, No. 5813, 9 February 2007.

198   Carol Turley, 'Carbon capture and storage by oceans', in Institute of Physics, 'Carbon Capture and Storage', op. cit..

199   D J Cooper and others, 'Large decrease in ocean-surface CO_2 fugacity in response to in situ iron fertilization', *Nature*, Vol. 383, No. 6600, 10 October 1996. Subsequently at least a dozen further experiments have been carried out.

200   See www.climos.com.

201   ETC Group, 'London Convention Puts Brakes on Ocean Geoengineering', news release, 9 November 2007, on www. etcgroup.org/upload/publication/pdf_file/661.

202   Barack Obama and Joe Biden, *New Energy for America*, fact sheet, op. cit., p7.

203   Barack Obama, Speech by videolink to Governors' Global Climate Summit in California, transcript in Nick Juliano, 'Obama promises return to global climate change negotiations', *The Raw Story*, 18 November 2008.

204   Barack Obama and Joe Biden, op. cit., p8.

205   Barack Obama, interview on energy with the *San Francisco Chronicle*, 17 January 2008, YouTube video, on http://uk.

youtube.com/watch?v=SMwBbl6 RoIs&feature=related.

206   Joe Biden, on the stump in Ohio, 2008 campaign, YouTube video, on http://uk.youtube.com/watch?v=iJ55UzAsp6M&feature =related.

207   Barack Obama, *New Energy for America*, speech in Lansing, Michigan, 4 August 2008, transcript on www.pbs.org/newshour/bb/politics/july-dec08/obamaenergy_08-04.html.

208   Ibid.

209   Barack Obama and Joe Biden, *New Energy*, op. cit., p3.

210   Ibid., p4.

211   Ibid., p4.

CHAPTER 6

1   World Wind Energy Association (WWEA), 'Wind turbines generate more than 1 per cent of the global electricity', 21 February 2008, p1, on www.wwindea.org/home/images/stories/pr_statistics2007_210208_red.pdf.

2   Vaclav Smil, *Energy at the Crossroads: Global Perspectives and Uncertainties*, MIT Press, 2003, p274-5.

3   Adapted from ibid., pp240-242.

4   Ibid, p241.

5   Michael Brower, *Cool Energy*, MIT Press, 1992, p76.

6   Vaclav Smil, op. cit., p272.

7   GE Energy, *3.6MW Offshore Series Wind Turbine*, 2005, on www.gepower.com/prod_serv/

products/wind_turbines/en/ downloads/ge_36_brochure_new. pdf.

8   'New Record: World's Largest Wind Turbine (7+ Megawatts)', *Metaefficient*, 3 February 2008, on www.metaefficient.com/news/ new-record-worlds-largest-wind-turbine-7-megawatts.html.

9   Colin Fernandez, 'Up on the roof ... Cameron's wind turbine arrives', *The Daily Mail*, 21 March 2007, on www.dailymail.co.uk/ pages/live/articles/news/news. html?in_article_id=443686&in_ page_id=1770.

10   Energy & Environment Ltd, *The D400 Wind Generator*, on www. energyenv.co.uk/ D400WindTurbine.asp.

11   See BERR, *Windspeed Database*, on www.berr.gov.uk/energy/ sources/renewables/explained/ wind/windspeed-database/ page27326.html.

12   Eclectic Energy Limited, *The D400 Wind Generator*, on www. d400.co.uk/.

13   Daniel Bellamy, 'How biking Cameron takes the green vote for a ride', *Western Mail*, 29 April 2006.

14   'Cameron forced to remove turbine', *BBC News online*, 29 March 2007, on http://news.bbc. co.uk/1/hi/uk_politics/6505807. stm.

15   The World Commission on Sustainable Development, *Our Common Future*, Oxford University Press, 1987, pp194-195.

16   The £2bn London Array is set to have 341 turbines in the Thames estuary, and is designed to supply one per cent of Britain's electricity. See London Array, 'The project', on www. londonarray.com/about/.

17   Lars Paulsson and Paul Dobson, 'Shell, E.ON Stall Offshore Wind Projects EU Needs (Update 1)', *Bloomberg.com*, 14 May 2008, on www.bloomberg.com/apps/ne ws?pid=20601109&refer=&sid= alnc6d8ZAOoE.

18   Amory Lovins, 'Energy Strategy: The Road Not Taken?', *Foreign Affairs*, October 1976, p92. Available on www.rmi.org/images/ PDFs/Energy/E77-01_ TheRoadNotTaken.pdf.

19   Amory Lovins and L Hunter Lovins, *Brittle Power: Energy Strategy for National Security* (1982), RMI, new edition, 2001, on www.rmi.org/images/PDFs/ EnergySecurity/S82-03_ BrPwrParts123.pdf.

20   John Arquilla and David Ronfeldt, eds, *Networks and Netwars: The Future of Terror, Crime, and Militancy,* RAND Corporation, November 2001, on www.rand. org/pubs/monograph_reports/ MR1382/.

21   Robert L Park, 'End of the world?', review of Martin Rees, *Our Final Hour: How Terror, Error, and Environmental Disaster Threaten Humankind's Future in this Century, On Earth and Beyond*, Basic Books, 2003, in *Issues in Science and Technology*, Fall 2003, on www. issues.org/20.1/br_park.html.

22 Commission of the European Communities, *On a European Programme for Critical Infrastructure protection*, Green Paper, 17 November 2005, on http://eur-lex.europa.eu/LexUriServ/site/en/com/2005/com2005_0576en01.pdf.

23 Ted G Lewis, *Critical infrastructure Protection in Homeland Security: defending a networked nation*, Wiley-Interscience, 2006, p29.

24 See Richard G Lugar and R James Woolsey, 'The New Petroleum', *Foreign Affairs*, January/February 199, and, Woolsey, interviewed in Laura Rozen, 'James Woolsey, Hybrid Hawk', *Mother Jones*, May/June 2008, on www.motherjones.com/news/outfront/2008/05/balance-of-power-the-hybrid.html. 'A renewable-energy economy', one campaigning Green journalist adds, 'would have far more independent sources of power – home-based fuel cells, stand-alone solar systems, regional wind farms – which would make the nation's electricity grid a far less strategic target for future guerrilla attacks'. See Ross Gelbspan, 'Climate change as Number One?', *The Globalist*, 1 December 2005, on www.theglobalist.com/dbweb/StoryId.aspx?StoryId=4947, adapted from Gelbspan, *Boiling Point: How Politicians, Big Oil and Coal, Journalists and Activists Have Fueled the Climate Crisis – and What We Can Do to Avert Disaster*, Basic Books, 2004.

25 James Woolsey, interviewed in Patrick Tucker, 'James Woolsey on ending the oil era', *Energy Bulletin*, 13 June 2007, on http://www.energybulletin.net/node/31004.

26 'T Boone Pickens' Energy Independence Pledge', *Pickens Plan*, on www.pickensplan.com/thepledge/.

27 'Texas to Spend Billions on Wind Power Transmission Lines, *Environment News Service*, 18 July 2008, on www.ens-newswire.com/ens/jul2008/2008-07-18-094.asp.

28 Hermann Scheer, *Energy Autonomy: The Economic, Social and Technological Case for Renewable Energy*, Earthscan, 2007, pp231-233.

29 Ibid., p234.

30 J B S Haldane, *DAEDALUS, or Science and the Future: a paper read to the Heretics, Cambridge, on February 4th, 1923*, on http://cscs.umich.edu/~crshalizi/Daedalus.html. Since their purpose is not grinding, Haldane should not, strictly, have referred to his rows of machines as wind *mills* – but no matter.

31 The operator at Delabole, Good Energy, has recently proposed halving the number of turbines, doubling their height, and so more than doubling their output. See 'Wind turbines could double height', *BBC News online*, on http://news.bbc.co.uk/1/hi/england/cornwall/7246250.stm.

32 Lewis Smith, 'Wind farm 'marks step towards cleaner energy'', *The Times*, February 10, 2007 www.timesonline.co.uk/tol/news/uk/article1362761.ece.

33   According to the British Wind Energy Association, in September 2008 there were 2546.78MW of wind in operation and 1673.10MW under construction in the UK. See BWEA, 'Statistics', on www.bwea.com/statistics/. In 2007 a total of 453.85MW were built, giving a rate of 1GW every 26.4 months. See http://www.bwea.com/statistics/2007.asp.

34   WWEA, op. cit. p1.

35   Ibid., p2.

36   American Wind Energy Association, 'US Wind energy installations surpass 20,000 megawatts', 3 September 2008, on www.awea.org/newsroom/releases/Wind_Installations_Surpass_20K_MW_03Sept08.html.

37   The Crown Estate, 'The Crown Estate to purchase the world's largest offshore wind turbine from Clipper Windpower for deployment in UK waters', press release, 17 April 2008, on www.thecrownestate.co.uk/newscontent/92-clipper-wind-turbine.htm.

38   Catherine Elsworth, 'World's largest offshore wind turbine to be in UK', *The Daily Telegraph*, 5 October 2007, on www.telegraph.co.uk/Earth/main.jhtml?xml=/Earth/2007/10/05/eawind105.xml.

39   National Renewable Energy Laboratory (NREL), 'Large Wind Turbine Blade Test Facilities to be in Mass., Texas,' news release NR-1607, 25 June 2007, on www.nrel.gov/news/press/2007/519.html.

40   Gustave P Corten and Herman F Veldkamp, 'Aerodynamics: Insects can halve wind-turbine power', *Nature*, Vol. 412, No. 6842, 5 July 2001, pp41-42.

41   N Dalili and others, 'A review of surface engineering issues critical to wind turbine performance', *Renewable and Sustainable Energy Reviews*, 5 December 2007.

42   Technological progress is reviewed in US Department of Energy, *20 per cent Wind Energy by 2030: Increasing Wind Energy's Contribution to U.S. Electricity Supply*, July 2008, on www1.eere.energy.gov/windandhydro/pdfs/41869.pdf.

43   WWEA, op. cit.; WWEA, 'New World Record in Wind Power Capacity: 14,9 GW added in 2006 – Worldwide Capacity at 73,9 GW', 29 January 2007, on www.wwindea.org/home/images/stories/pdfs/pr_statistics2006_29010.pdf; and WWEA, 'Worldwide wind energy boom in 2005: 58.982 MW capacity installed', 6 March 2006, on www.wwindea.org/home/index.php?option=com_content&task=view&id=88&Itemid=43.

44   T Boone Pickens, quoted in Ed Pilkington, 'Big oil to Big wind: Texas veteran sets up $10bn clean energy project', *The Guardian*, 14 April 2008, on www.guardian.co.uk/environment/2008/apr/14/windpower.energy.

45   Clifford Krauss, 'Move Over, Oil, There's Money in Texas Wind', *The New York Times*, 23 February

2008, on www.nytimes.com/2008/02/23/business/23wind.html?pagewanted=all.

46 Edison International, 'Southern California Edison Starts Construction on the Nation's Largest Wind Transmission Project', 7 March 2008, on www.edison.com/pressroom/pr.asp?bu=&year=0&id=6992.

47 Steve Smith, managing director of networks at Ofgem, cited in Jonathan Leake, 'Host of new pylons to carry wind farm power', *The Sunday Times*, 3 August 2008, p12.

48 See BERR, *UK Renewable Energy Strategy: Consultation Document*, June 2008, p11, on www.opinionsuite.com/berr/download?filename=uk-renewable-energy-strategy-consultation-document.

49 Ditlev Engel, statement to investors, May 2007, quoted in Keith Johnson, 'Alternative Energy Hurt By a Windmill Shortage', *The Wall Street Journal*, 9 July 2007.

50 See Joseph Ogando, 'Wind Energy's Manufacturing Crunch', *Design News*, 17 September 2008, on www.designnews.com/article/48283-Wind_Energy_s_Manufacturing_Crunch.php. This article is an excellent summary of the state of the art in wind turbine manufacture.

51 BTM Consult, 'International Wind Energy Development World Market Update 2007 Forecast 2008-2012', press release, 27 March 2008, on www.btm.dk/Documents/Pressrelease.pdf. For a commentary, see Edward Milford, 'Record Growth for Wind: What Comes Next', *Renewable Energy World Magazine*, Vol. 11, Issue 4, July/August 2008, on www.renewableenergyworld.com/rea/news/story?id=53436.

52 Naazneen Karmali, 'India's 40 Richest', *Forbes.com*, 14 November 2007, on www.forbes.com/2007/11/13/india-billionaires-richest-biz-07india-cx_nk_1114india_land.html.

53 WWEA, 'Acquisition of REpower by Suzlon is important step in international cooperation', on www.wwindea.org/home/index.php?option=com_content&task=view&id=175&Itemid=40.

54 Suzlon, 'Milestones of Suzlon Group', on www.suzlon.com/Milestone.aspx?cp=1_5#.

55 Suzlon, 'Manufacturing facilities', on www.suzlon.com/ManufacturingFacilities.html?cp=1_7.

56 'Turbines "no risk to farm birds"', *BBC News*, 1 October 2008, on http://news.bbc.co.uk/1/hi/sci/tech/7646142.stm.

57 The Scottish Government, *Decision on Lewis Wind Farm*, 21 April 2008, on www.scotland.gov.uk/News/Releases/2008/04/21102611.

58 Jonny Hughes, 'Not just for peat's sake', *The Guardian*, 20 April 2008, on http://commentisfree.guardian.co.uk/jonny_hughes/2008/04/not_just_for_peats_sake.html.

59 Ibid.

60 S Butterfield and others, 'Engineering Challenges for Floating Offshore Wind Turbines', paper to the 2005 Copenhagen Offshore Wind Conference, Copenhagen, Denmark, October 2005 NREL conference paper NREL/CP-500-38776, September 2007, on www.nrel.gov/wind/pdfs/38776.pdf.

61 Paul D Sclavounos and others, 'Floating Offshore Wind Turbines: Responses in a Seastate Pareto Optimal Designs and Economic Assessment', October 2007 MIT working paper, on http://oe.mit.edu/flowlab/pdf/Floating_Offshore_Wind_Turbines.pdf.

62 See for example Danny Fortson, 'Wind farm firms and MoD agree on radar costs deal', *The Independent*, 6 June 2008 www.independent.co.uk/news/business/news/wind-farm-firms-and-mod-agree-on-radar-costs-deal-841462.html.

63 Massimo Canale and others, 'Power Kites for Wind Energy Generation', *IEEE Control Systems Magazine*, December 2007, p25.

64 Dan Frosch, 'Citing Need for Assessments, US Freezes Solar Energy Projects', *The New York Times*, 27 June 2008, on www.nytimes.com/2008/06/27/us/27solar.html?ref=business.

65 Quoted in Dan Frosch, 'US Lifts Moratorium on New Solar Projects', *The New York Times*, 3 July 2008, on www.nytimes.com/2008/07/03/us/03solar.html?scp=1&sq=US per cent20Lifts per cent20Moratorium per cent20on

per cent20New per cent20Solar per cent20Projects&st=cse.

66 Travis Bradford, *Solar Revolution: The Economic Transformation of the Global Energy Industry*, The MIT Press, 2006, pp178-179.

67 BERR, 'What is the Renewables Obligation?', no date, on www.berr.gov.uk/whatwedo/energy/sources/renewables/policy/renewables-obligation/what-is-renewables-obligation/page15633.html.

68 See Union of Concerned Scientists, 'Production Tax Credit for Renewable Energy', no date, on www.ucsusa.org/clean_energy/solutions/big_picture_solutions/production-tax-credit-for.html.

69 Angel Gurría, 'Energy, Environment, Climate Change: Unlocking the Potential for Innovation', Keynote speech during the World Energy Council Energy Leaders Summit, London, 16 September 2008, on www.oecd.org/document/18/0,3343,en_2649_33717_41329298_1_1_1_1,00.html.

70 Progress Energy, 'Progress Energy Carolinas, SunEdison Plan Solar Project at Plant Site', 22 August 2008, on www.progress-energy.com/aboutus/news/article.asp?id=19282.

71 SunEdison, 'SunEdison Simplifies Solar by Pushing Adoption of Utility-Industry Standards', 18 January 2008, on www.sunedison.com/images/press/011808-IHS.pdf.

72 General Motors, 'General Motors To Add 1.2 Megawatt Rooftop

Solar Installation At Baltimore Powertrain Plant', news release, 21 August 2008, on http://media.gm.com/servlet/GatewayServlet?target=http://image.emerald.gm.com/gmnews/viewmonthlyreleasedetail.do?domain=74&docid=48012.

73 Communities and Local Government, 'Energy Performance of Buildings: Overview', no date, on www.communities.gov.uk/planningandbuilding/theenvironment/energyperformance/overview.

74 Jonathan Johns, partner at E&Y, quoted in Fiona Harvey and Andrew Bolger, 'BT wind farms to supply green energy', *Financial Times*, 19 October 2007.

75 Hanif Lalani, quoted in ibid.

76 See Communities and Local Government, *Code for Sustainable Homes: Technical Guide*, 17 April 2008, on www.planningportal.gov.uk/uploads/code_for_sustainable_homes_techguide.pdf.

77 Communities and Local Government, *Cost Analysis of The Code for Sustainable Homes*, London, CLG, 21 July 2008, on www.communities.gov.uk/documents/planningandbuilding/pdf/codecostanalysis.pdf.

78 Xuemei Bai, 'China's Solar-Powered City', *Renewable Energy World.com*, 22 May 2007, on www.renewableenergyworld.com/rea/news/story?id=48605.

79 A useful source on CSP is *CSP Today*, on http://social.csptoday.com/index.php.

80 'First EU Commercial Concentrating Solar Power Tower Opens in Spain', *Environment News Service*, 20 March 2007, on www.ens-newswire.com/ens/mar2007/2007-03-30-02.asp.

81 Figures derived from NREL, 'US Parabolic Trough Power Plant Data', no date, on www.nrel.gov/csp/troughnet/power_plant_data.html.

82 Masdar, 'Sener and Masdar Announce Joint Venture to Develop Concentrating Solar Power Plants in the "Sunbelt"', 22 March 2008, on www.masdaruae.com/text/news-d.aspx?_id=55; 'Sener and Masdar form solar JV to develop CSP plants', Technology For Life, 15 March 2008, on http://technology4life.wordpress.com/2008/03/15/sener-and-masdar-form-jv-to-develop-csp-plants/.

83 Stirling Energy Systems (SES), 'Application Filed for World's Largest Solar Energy Generating Plant', 20 June 2008, on www.stirlingenergy.com/downloads/30-June-2008-Application-Filed-for-Worlds-Largest-Solar-Energy-Generating-Plant.pdf. With SES, it's not steam that drives a turbine, but heated hydrogen sealed in a Stirling engine.

84 European Solar Thermal Electricity Association president José Alfonso Nebrera, quoted in 'Industry seeks uniformity and certainty in policies pertaining to CSP and CPV', *CSP today*, 22 September 2008, on http://social.csptoday.com/content/industry-seeks-uniformity-and-certainty-policies-pertaining-csp-

and-cpv.

85 TREC, *Summary of the Concept & the Studies,* 5 August 2008, p6, on www.desertec.org/downloads/ summary_en.pdf. For more on CSP, see Sorin Grama and others, *Concentrating Solar Power – Technology, Cost, and Markets*, Prometheus Institute and Greentech Media, 31 March 2008, executive summary on www.greentechmedia.com/ assets/pdfs/executivesummaries/ Concentrating-Solar-Power-Final-Executive-Summary.pdf.

86 The European Photovoltaic Industry Association claims to be the world's largest industry association devoted to solar electricity, and is a useful source on PVs. See www.epia.org/index. php?id=3.

87 Paula Mints, 'Is Booming Growth Sustainable? The Global Photovoltaic Industry', *Renewable Energy World Magazine*, Vol. 11, Issue 4, July/August 2008, on www.renewableenergyworld.com/ rea/news/story?id=53437.

88 Federal Ministry for the Environment, *Nature Conservation and Nuclear Safety, EEG – The Renewable Energy Sources Act: The success story of sustainable policies for Germany*, July 2007, on www.invest-in-germany.com/uploads/media/ EEG_Brochure_01.pdf.

89 Paula Mints, op. cit.

90 IEEE Virtual Museum, 'Russell Ohl', on www.ieee-virtual-museum.org/collection/people. php?taid=&id=1234770&lid=1.

91 John Gartner, 'Silicon Shortage Stalls Solar', *Wired*, 28 March 2005. www.wired.com/science/ planetEarth/ news/2005/03/67013.

92 Michael R Splinter, quoted in 'Solar power looks into a brighter future', *China Daily*, 3 June 2008, on www.chinadaily.com.cn/ bizchina/2008-06/03/ content_6732354.htm.

93 Martin LaMonica, 'Intel spins off solar cell maker SpectraWatt', *cnet news*, 16 June 2008, on http://news.cnet com/8301-11128_3-9969631-54.html, and LaMonica, 'Intel Capital spreads its solar bets with Sulfurcell', *cnet news*, 9 July 2008, on http:// news.cnet.com/8301-11128_ 3-9986577-54.html.

94 National Semiconductor, 'National Semiconductor Enters Photovoltaic Market with Technology That Maximizes Solar Energy Production', news release, 30 June 2008, on www. national.com/news/item/ 0,1735,1344,00.html.

95 Ucilia Wang, 'New Research Predicts End to Silicon Shortage', *greentechmedia*, 26 June 2008, on www.greentechmedia. com/articles/new-research-predicts-end-to-silicon-shortage-1055.html; Travis Bradford, *Polysilicon: Supply, Demand & Implications for the PV industry*, Greentech Media and the Prometheus Institute, June 2008, executive summary on www. greentechmedia.com/assets/ pdfs/executivesummaries/ Polysilicon-Executive-Summary. pdf.

96   General Motors, 'GM Adding World's Largest Rooftop Solar Power Installation to Zaragoza Plant', News release, 8 July 2008, on www.gm.com:80/corporate/responsibility/environment/news/2008/solar_070808.jsp.

97   First Solar, 'Company overview', 2008, on www.firstsolar.com/company_overview.php.

98   Martin Roscheisen, 'Nanosolar Ups Funding to $1/2B; Partners Strategically for Solar Utility Power', Nanosolar, 27 August 2008, on www.nanosolar.com/blog3/?p=138.

99   Martin LaMonica, 'IBM muscles into CIGS solar-cell market', *cnet news*, 16 June 2008, on http://news.cnet.com/8301-11128_3-9966992-54.html?tag=mncol;txt; 'IBM and TOK team on CIGS thin-film materials and processes; target 15 per cent efficiencies', *PV-tech.org*, 16 June 2008, on www.pv-tech.org/power_generation/article/ibm_and_tok_team_on_cigs_thin_film_materials_and_processes_target_15_effici.

100   Solar Systems, '154MW Victorian Project', on www.solarsystems.com.au/154MWVictorianProject.html.

101   NREL, 'NREL Solar Cell Sets World Efficiency Record at 40.8 Percent', 13 August 2008, on www.nrel.gov/news/press/2008/625.html.

102   'Emcore wins orders worth $40m for CPV solar cells and receivers', *Semiconductor Today*, 6 August 2008, on www.semiconductor-today.com/news_items/2008/AUGUST/EMCORE_060808.htm.

103   IBM, 'IBM research unveils breakthrough in solar farm technology', 15 May 2008, on www-03.ibm.com/press/us/en/pressrelease/24203.wss.

104   'Thin-film solar market to reach 9GW in 2012', *Semiconductor Today*, 9 June 2008, on www.semiconductor-today.com/news_items/2008/JUNE/PROM_090608.htm.

105   IEA, *Energy Technology Perspectives 2008*, June 2008, p387. Executive Summary on www.iea.org/textbase/npsum//ETP2008SUM.pdf.

106   'AfDB approves funds for Inga studies', *International Water Power and Dam Construction*, 14 May 2008, on www.waterpowermagazine.com/story.asp?sectioncode=130&storyCode=2049645.

107   'Grand Inga Dam, DR Congo', *International Rivers*, no date, on www.internationalrivers.org/en/africa/grand-inga-dam-dr-congo.

108   Dr Latsoucabé Fall, *Report on the high level WEC workshop on financing the Inga hydropower projects, held in London 21–22 April 2008*, World Energy Council, 8 May 2008, p4, on www.worldenergy.org/documents/report_on_inga_financing_workshop_final_8_may_2008.pdf.

109   IEA, ibid., and World Energy Council, *Survey of World Energy Resources 2007*, p272, on www.worldenergy.org/documents/ser2007_final_online_version_1.pdf.

110 Dr Latsoucabé Fall, *Inga Social & Environmental Issues: Introduction to the Discussions*, 2007, pp3, 4, on http://cesenet.org/documents/latsoucabe_fall_social__environment_isses_20_04_2008.pdf.

111 International Rivers, '10 Questions on Grand Inga', 21 April 2008, on www.internationalrivers.org/en/node/2734, and Terri Hathaway, 'Report from WEC's Inga Financing Workshop in London, 21-22 April 2008', *International Rivers*, on www.internationalrivers.org/en/africa/grand-inga-dam-dr-congo/report-wecs-inga-financing-workshop-london-21-22-april-2008.

112 International Rivers, 'Mission', no date, on www.internationalrivers.org/en/mission.

113 Sarah J Wachter, 'Giant dam projects aim to transform African power supplies', *International Herald Tribune*, 19 June 2007, on www.iht.com/bin/print.php?id=6204822.

114 L S Blunden and A S Bahaj, 'Tidal energy resource assessment for tidal stream generators', *Proceedings of the Institution of Mechanical Engineers, Part A: Journal of Power and Energy*, Vol. 221, No. 2, 2007, abstract on http://journals.pepublishing.com/content/b2t04810304q8l50/?p=7c7b6a342e8e40578169b605373d87fa&pi=2.

115 Ibid., and 'First sea-bed mounted tidal turbine deployed at EMEC', *New Energy Focus*, 12 September 2008, on http://newenergyfocus.com/do/ecco.py/view_item?listid=1&listcatid=32&listitemid=1687&section=Hydro%20%26%20Marine, and 'SeaGen completed: world's first megawatt-scale tidal turbine installed', *Marine Current Turbines*, 20 May 2008, on www.marineturbines.com/3/news/article/9/seagen_completed__world_s_first_megawatt_scale_tidal_turbine_installed/.

116 Metoc plc, *Tidal Power in the UK – Research Report 1: UK tidal resource assessment*, Sustainable Development Commission, October 2007, pp9-11, 15, 38, on www.sd-commission.org.uk/publications.php?id=608.

117 S H Salter, 'Wave power', *Nature*, Vol. 249, No. 5459, 21 June 1974.

118 'The Untimely Death of Salter's Duck', *Green Left Weekly*, 29 July 1992, on www.greenleft.org.au/1992/64/2832.

119 See www.pelamiswave.com.

120 Michael Pollitt, 'New wave power generation', *The Guardian*, 7 August 2008, on www.guardian.co.uk/technology/2008/aug/07/research.waveandtidalpower.

121 See 'Volcanoes and geothermal energy', *Encyclopaedia Britannica online*, on www.britannica.com/EBchecked/topic/632130/volcano/253608/Volcanoes-and-geothermal-energy.

122 John Berger, *Charging Ahead*, University of California Press, 1998, p224.

123 International Geothermal

Association, 'Installed Generating Capacity', updated 5 September 2008, on http://iga.igg.cnr.it/geoworld/geoworld.php?sub=elgen.

124 Based on John W Lund and others, 'World-Wide Direct Uses of Geothermal Energy 2005', *Proceedings of the World Geothermal Congress 2005, Antalya, Turkey, 24-29 April 2005*, extracted on 'Welcome to our page with data for the United States', International Geothermal Association, updated 5 September 2008, on http://iga.igg.cnr.it/geoworld/geoworld.php?sub=elgen&country=usa.

125 A Satman, U Serpen, ED Korkmaz Basel, *Proceedings*, Thirty-Second Workshop on Geothermal Reservoir Engineering, 22-24 January 2007, on http://pangea.stanford.edu/ERE/pdf/IGAstandard/SGW/2007/satman1.pdf.

126 An assessment by an MIT-led interdisciplinary panel, *The Future of Geothermal Energy: Impact of Enhanced Geothermal Systems (EGS) on the United States in the 21st Century*, MIT, 2006, on www1.eere.energy.gov/geothermal/pdfs/future_geo_energy.pdf. An eight-month study by the US Department of Energy confirmed the potential identified by the MIT report, although it suggested that the report had underestimated the investment in new technology required. See US Department of Energy, *An Evaluation of Enhanced Geothermal Systems Technology*, 2008, on www1.eere.energy.gov/geothermal/pdfs/evaluation_egs_tech_2008.pdf.

127 US Department of Energy, 'R&D Successes', Energy Efficiency and Renewable Energy Geothermal Technologies Program, updated 2 March 2006, on www1.eere.energy.gov/geothermal/awards.html.

128 Alyssa Kagel, *The State of Geothermal Technology Part II: Surface Technology*, Geothermal Energy Association, January 2008, p49, on www.geo-energy.org/publications/reports/Geothermalper cent20Technologyper cent20-per cent20Partper cent20IIper cent20(Surface).pdf.

129 Potter Drilling, 'Benefits for Developers', no date, on www.potterdrilling.com/technology/benefits/.

130 Australian Geothermal Energy Association, 'Geothermal energy: an important part of Australia's energy future', media release, 20 August 2008, on www.agea.org.au/dyn/media/news/attachment/3.

131 See www.energystorageandpower.com/home.html.

132 J Lynn Lunsford, 'Solar Venture Will Draw on Molten Salt', *Wall Street Journal*, 2 January 2008, on http://online.wsj.com/article/SB119924708042261755.html?mod=hpp_us_whats_news.

133 José C Martín. 'Solar Tres – First Commercial Molten-Salt Central Receiver Plant', NREL CSP Technology Workshop, Denver, 7 March 2007, on www.nrel.gov/csp/troughnet/pdfs/2007/martin_solar_tres.pdf.

134 'Understanding the potential of ammonia-based thermochemical energy storage', *CSP today*, 22 September 2008, on http://social.csptoday.com/content/understanding-potential-ammonia-based-thermochemical-energy-storage.

135 Jigar Shah, quoted in Elisa Wood, 'Price parity for US solar: Is the goal within sight?', *Renewable Energy World*, Vol. 10, Issue 6, November/December 2007, on www.renewableenergyworld.com/rea/news/story?id=51441.

136 Adam Smith, *The Wealth of Nations* (1776), Book 1, Chapter 2, on www.readprint.com/chapter-8608/Adam-Smith.

137 See Paul Seabright, *The Company of Strangers: A Natural History of Economic Life*, Princeton University Press, 2004.

138 Amory Lovins, 'Energy Strategy', op. cit., p92.

139 More than 8GW of wind power projects in the UK is in the planning system, generally languishing – even if a recent decision in favour of a 315MW offshore wind farm, Sherringham Shoal off the Norfolk coast, makes a refreshing change. Certainly the UK Government's Planning White Paper (see Chapter 4) has little relevance in terms of accelerating the development of wind. See BWEA, 'Statistics', on www.bwea.com/statistics; BWEA, 'Wind industry welcomes major approvals for new wind farms off and on shore', 8 August 2009, on www.bwea.com/media/news/articles/wind_industry_welcomes_

major_a.html, and BWEA, 'UK no nearer to 2010 renewables target with Planning White Paper', 21 May 2007, on www.bwea.com/media/news/070521.html.

140 Thomas L Friedman, 'Eight strikes and you're out', *The New York Times*, 12 August 2008, on www.nytimes.com/2008/08/13/opinion/13friedman.html?_r=1&scp=1&sq=Eight per cent20strikes per cent20and per cent20you per cent92re per cent20out&st=cse&oref=slogin.

141 BTM Consult, op. cit.

142 Mike Scott, 'Cloud over clean energy', *FTFM*, 17 September 2007.

143 Masdar, 'Abu Dhabi's Masdar Initiative to Invest Up to $2 Billions in Photovoltaic Solar Energy', 28 May 2008, on www.masdaruae.com/text/news-d.aspx?_id=57; General Electric and Mubadala, 'GE and Mubadala launch multi-billion dollar global business partnership', press release, 22 July 2008, on www.ge.com/pdf/news/GE_Mubadala_press_release.pdf.

144 Mike Scott, op. cit.

145 New Energy Finance, 'Renewables dive in the wake of Lehman Brothers, as an ebbing tide lowers all boats', *Week in Review*, Vol. IV Issue 67, 9-15 September 2008, on www.newenergymatters.com/download.php?n=NEF_Week_in_Review_2008-09-16.pdf&f=WiR_pdffile&t=weeklybriefing.

146 Claire Cain Miller, 'Thin Film Solar Companies Raise Hundreds of Millions in Financing', *The New*

York Times, 11 September 2008, on http://bits.blogs.nytimes. com/2008/09/11/another-thin-film-solar-company-rakes-in-venture-capital.

147   Barack Obama, *New Energy for America*, Speech in Lansing, MIchigan, 4 August 2008, Transcript on www.pbs.org/newshour/bb/politics/july-dec08/obamaenergy_08-04.html.

148   James Dehlsen, cited in Fiona Harvey, 'Solar Power sees light at end of tunnel', and Jamie Webster, senior consultant in the Gas & Upstream Group of PFC Energy, a consultancy, cited in Sheila McNulty, 'Pickens Plan blows a break of fresh air into the debate', both in *Financial Times Report Modern Energy 2008, 16 September 2008*, p4. Similarly the winds, like coal-fired plants, are strong electricity generators at or near off-peak hours – at the beginning and end of each day. Without large-scale storage, again, this means that a big expansion of wind in the US is likely to displace coal, not gas, from the energy supply mix. Webster, in McNulty, ibid.

CHAPTER 7

1   Bill Joy, 'Why the future doesn't need us', *Wired*, April 2000, on www.wired.com/wired/archive/8.04/joy.html.

2   Lewis Smith, 'World-leading telescopes face being shut down to save £2.5m', *The Times*, 6 March 2008, on www.timesonline.co.uk/tol/news/uk/science/article3492504.ece.

3   Engineering and Physical Sciences Research Council, '£48M for Fusion Research – EPSRC's largest ever grant allocation', press release, 16 February 2004, on www.epsrc. ac.uk/PressReleases/FusionResearch48Million.htm.

4   See King's remarks to reporters, quoted in Jonathan Amos, '"Climate crisis" needs brain gain', *BBC News online*, 8 September 2008, on http://news.bbc.co. uk/1/hi/sci/tech/7603257.stm.

5   Fermilab News, 'Robert Rathbun Willson, founding director of Fermilab, dies at age 85', press release, 17 January 2000, on www.fnal.gov/pub/presspass/press_releases/wilson.html.

6   Charts from H-Holger Rogner, Dadi Zhou and others, 'Introduction', in B Metz and others, editors, *Climate Change 2007: Mitigation. Contribution of Working Group III to the Fourth Assessment Report of the Intergovernmental Panel on Climate Change*, IPCC, p112, on www.ipcc.ch/pdf/assessment-report/ar4/wg3/ar4-wg3-chapter1.pdf.

7   The comparisons are rough because there are three different economic ratios at issue here. Nevertheless, the figures are still suggestive. See US Department of Commerce, *Survey of Current Business*, January 2008, pD-53, on www.bea.gov/scb/pdf/2008/01%20January/D-Pages/0108dpg_d.pdf; Paola Caselli and others, *Investment and Growth in Europe and in the United States in the Nineties*, Bank of Italy, no date, Figure 1, on http://digilander.libero.it/fschivardi/images/CaselliPaganoSchivardi.pdf; National Bureau of Statistics of China, *The Figure of GDP by Expenditure Approach in the Year of Economic Census, Announcement of the National Bureau of Statistics of China No. 2*, 2006, on www.stats.gov.cn/was40/gjtjj_en_detail.jsp?searchword=gross+fixed+capital+formation&channelid=9528&record=1.

8   US Department of Commerce, ibid.

9   This section is based on Daniel Ben-Ami, *Cowardly Capitalism: the Myth of the Global Financial Casino*, John Wiley & Sons, 2001.

10   Ben Hunt, *The Timid Corporation: Why Business is Terrified of Taking Risk*, John Wiley & Sons, 2003.

11   Vijay V Vaitheeswaran, *Power to the People*, Farrar, Straus and Giroux, 2003, p54.

12   Jason Makansi, *Lights Out: The Electricity Crisis, the Global Economy,* and *What It Means To You*, John Wiley, 2007.

13   In America in 2007, for example, a private-equity consortium consisting of KKR, TPG Capital and Goldman Sachs performed a leveraged buyout of TXU, Dallas, for $45bn, forming Energy Future Holdings Corporation; in June 2008 XTO Energy Inc paid a much smaller sum – $4.19bn – for Hunt Petroleum Corporation.

14   Over 1998-2002, Exxon merged with Mobil, BP acquired Amoco and Arco, Total purchased Petrofina and Elf, and Conoco merged with Phillips. In 2005 France's Électricité de France bought control of Italy's Edison, and Italy's Enel did the same to Spain's Endesa in 2007.

15   Robert Hughes, *The Culture of Complaint: The Fraying of America*, Oxford University Press, 1993.

16   See www.ewon.com.au/about_us/8_2.html.

17   See www.oeb.gov.on.ca/html/en/abouttheoeb/index.htm and www.oeb.gov.on.ca/html/en/consumers/complaint/alleged_forgery.htm.

18   In October 2008, Energywatch was merged into Consumer Focus, 'the new consumer champion working in England, Wales, Scotland and, for post, Northern Ireland'. See www.consumerfocus.org.uk.

19   'Consumer Watchdogs Slated by MPs', *BBC News online*, 29 November 2005, on http://news.bbc.co.uk/1/hi/business/4478426.stm.

20   Gabriel Kahn, 'Energy Rookies Take On Titans As Rules Ease', *The Wall Street Journal*, 10 March 2008.

21   See for example www.uswitch.com and www.confused.com.

22   Department of Trade and Industry (DTI), *Our Energy Future – Creating a Low Carbon Economy*, February 2003, p39, on www.berr.gov.uk/files/file10719.pdf.

23   Amory Lovins, *The Negawatt*

Revolution: Solving the CO₂ Problem, Keynote Address at the Green Energy Conference, Montreal 1989, on www.ccnr.org/amory.html.

24 Harry Wallop, 'Household energy bills could top £1,300 by the end of year, experts warn', *The Daily Telegraph*, 10 June 2008, on www.telegraph.co.uk/news/2106294/Household-energy-bills-could-top-and1631%2C300-by-the-end-of-year%2C-experts-warn.html.

25 Press Association, 'PM vows to ease energy cost burden', *The Guardian*, 10 September 2008, on www.guardian.co.uk/uk/feedarticle/7788279.

26 See 'Home Energy Saving Programme', *Number 10*, 11 September 2008, on www.number10.gov.uk/Page16807.

27 See T Yates, *Sustainable Refurbishment of Victorian Housing: Guidance, Assessment Method and Case Studies*, Building Research Establishment Press, 2006.

28 LD Shorrock and JI Utley, *Domestic Energy Fact File 2003*, Building Research Establishment Press, 2003, pp17-18.

29 See www.passiv.de.

30 Communities and Local Government, *Code for Sustainable Homes; Technical Guide*, CLG, 2007, developed by the Building Research Establishment, on www.communities.gov.uk.

31 Communities and Local Government, *Building Regulations; Energy Efficiency Requirements for New Dwellings*, CLG, 2007, on www.communities.gov.uk.

32 On top of that, the government worries that, as microgenerators multiply over many years, the overall performance of the National Grid might be adversely affected. It adds that clusters of microgenerators might make electricity distribution networks need reinforcement. See DTI, *Our Energy Challenge: Power From the People – microgeneration strategy, 2006*, pp18, 30, 31, on www.berr.gov.uk/files/file27575.pdf.

33 UK Green Building Council, *Zero Carbon Task Group Report: The Definition of Zero Carbon*, May 2008, pp4-5, on www.ukgbc.org/site/document/download/?document_id=180.

34 DEFRA, *Climate Change: The UK Programme 2006*, DEFRA 2006, p75, paragraph 3, on www.defra.gov.uk/environment/climatechange/uk/ukccp/index.htm.

35 DTI, *Our Energy Future*, op. cit., p39. British functionaries are not alone in favouring energy services. After French and Dutch voters rejected the EU constitution in 2005, the Brussels Commission found a new grand project with which to try to legitimise itself: making Europe 'the most energy-efficient region in the world'. See the EU's Directive 2006/32/EC of the European Parliament and of the Council of 5 April 2006 on energy end-use efficiency and energy

services and repealing Council Directive 93/76/EEC, published in the *Official Journal of the European Union*, 27 April 2006, on www.cen.eu/cenorm/workarea/sectorfora/energy+management/directive200632energyenduseefficiency.pdf.

36   DTI, *Meeting the Energy Challenge: A White Paper on Energy*, May 2007, p60, on www.berr.gov.uk/files/file39387.pdf.

37   Gary Hamel, *Leading the Revolution*, Harvard Business School Press, 2000, p19.

38   Ibid., pp211-223.

39   Joseph Schumpeter, *Capitalism, Socialism and Democracy* (1942), Unwin Paperbacks, 1987.

40   Ibid., p88.

41   Ibid., pp83-85.

42   Gary Hamel, op. cit., pp18, 66, 69.

43   Ibid., p7.

44   Ibid., p213.

45   Ibid., p17.

46   See Nicholas Carr, 'Why IT doesn't matter', *Harvard Business Review*, May 2003; *Does IT Matter? Information Technology and the Corrosion of Competitive Advantage*, Harvard Business School Press, 2003, and *The Big Switch: Rewiring the World, from Edison to Google*, Harvard Business School Press, 2008.

47   Carl Schram and others, *Innovation Measurement: Tracking the State of Innovation in the American Economy, Report to the Secretary of Commerce by The Advisory Committee on Measuring Innovation in the 21st Century Economy*, January 2008, on www.innovationmetrics.gov/Innovation%20Measurement%2001-08.pdf.

48   Ibid., p1.

49   On the history of prepayment meters, see 'A brief history of meter companies and meter evolution', on http://watthourmeters.com/history.html.

50   Booz Allen Hamilton Vice President Barry Jaruzelski, interviewed in Amy Bernstein, 'Making Innovation Strategy Succeed', *Strategy and Business*, 8 January 2008, on www.strategy-business.com/li/leadingideas/li00057?pg=0.

51   See Donald Norman, *The Psychology of Everyday Things*, Basic Books, 1988; John Seely Brown and others, *Storytelling in Organizations: Why Storytelling Is Transforming 21st Century Organizations and Management*, Butterworth-Heinemann, paperback edition, 2004; Eric Von Hippel, *Democratizing Innovation*, MIT Press, 2005; Jakob Nielsen and Hoa Loranger, *Prioritizing Web Usability*, New Riders, 2006.

52   HM Treasury, 'Speech by the Rt Hon Gordon Brown MP, Chancellor of the Exchequer, to United Nations Ambassadors, New York, 20 April 2006', on www.hm-treasury.gov.uk/newsroom_and_speeches/press/2006/press_31_06.cfm.

53   Henry Chesbrough, *Open

Innovation: the New Imperative for Creating and Profiting From Technology, Harvard Business School Press, 2003, and *Open Business Models: How to Thrive in the New Innovation Landscape*, Harvard Business School Press, 2006.

54   Department for Innovation, Universities & Skills and BERR, *The 2007 R&D Scoreboard: The Top 850 UK and 1250 Global Companies by R&D Investment*, November 2007, pp 116, 124, 138, on www.innovation.gov.uk/rd_scoreboard/downloads/2007_rd_scoreboard_data.pdf. The energy firms are among the 1250 general companies most active in R&D worldwide.

55   Meagan C Dietz and Jeffrey J Elton, 'Getting more from intellectual property', *McKinsey Quarterly*, November 2004.

56   Ibid.

57   See James Woudhuysen, *It Makes Sense to Share*, Amadeus, October 2005, on www.woudhuysen.com/documents/JW_It_Makes_Sense_To_Share.pdf.

58   Pete Engardio and Bruce Einhorn 'Outsourcing Innovation', *Business Week*, 21 March 2005, on www.businessweek.com/magazine/content/05_12/b3925601.htm?chan, and chart, www.businessweek.com/magazine/content/05_12/b3925603.htm.

59   Tony Davila, Marc J. Epstein and Robert Shelton, *Making Innovation Work: How to Manage It, Measure It, and Profit from It*, Wharton School Publishing, 2005.

60   See for example 'The love-in', *The Economist*, 11 October 2007.

61   See for example W Chan Kim and Renée Mauborgne, *Blue Ocean Strategy: How to Create Uncontested Market Space and Make the Competition Irrelevant*, Harvard Business School Press, 2005.

62   See OECD, *Presentations for OECD Business Symposium on Open Innovation in Global Networks*, Copenhagen, 25-26 February 2008, on www.oecd.org/document/11/0,3343,en_2649_37417_40199243_1_1_1_3 7417,00.html; 'Innovation Networks: Looking for Ideas Outside the Company', *Knowledge @ Wharton*, 14 November 2007, on http://knowledge.wharton.upenn.edu/article.cfm?articleid=1837, and Charles Leadbeater and James Wilson, *The Atlas of Ideas: China, India and the New Geography of Science*, Project Summary, Demos, 12 October 2005, on www.demos.co.uk/projects/currentprojects/atlasproject/.

63   Mario Cervantes, 'Policy Issues emerging from Globalisation and Open Innovation', slide 8, and Els Van de Velde, 'Insights from the Company Case Studies', slides 11 and 12, both in OECD, ibid.

64   Scott Berkun, *The Myths of Innovation,* O'Reilly UK, 2007; Frans Johansson, *The Medici Effect: What Elephants and Epidemics Can Teach Us About Innovation*, Harvard Business

School Press, 2004.

65 See www.lifeaftertheoilcrash.net.

66 We might add that the world has also not entered either the attention economy or the support economy. See Thomas H Davenport and John C Beck, *The Attention Economy: Understanding the New Currency of Business*, Harvard Business School Press, 2001, and Shoshana Zuboff and James Maxmin, *The Support Economy: Why Corporations are Failing Individuals and The Next Epsiode of Capitalism* (2002), Allen Lane, 2003.

67 Joseph Schumpeter, op. cit., p83.

68 Ibid.

69 IEA, *World Energy Outlook 2008*, 12 November 2008, Executive Summary, p12, on www.iea.org/Textbase/npsum/WEO2008SUM.pdf.

70 International Telecommunications Union, 'ITU and climate change'. on www.itu.int/themes/climate.

71 See for example www.sustainit.org.

72 Matt Naumann, 'Google CEO Eric Schmidt Offers Energy Plans', *Mercury News*, 9 September 2008, on www.mercurynews.com/olympics/ci_10419245.

73 John Reed, 'Electric dreams: Plug-in cars are picking up speed and credibility', *Financial Times*, 7 January 2008, on http://search.ft.comftArticle?queryText=Silicon+Valley+electric+cars&y=0&aje=true&x=0&id=080107000362&ct=0.

74 BP, 'Group Income statement', Annual Review, 2007, p96, on www.bp.com/liveassets/bp_internet/annual_review/annual_review_2007/STAGING/local_assets/downloads_pdfs/a/ara_2007_annual_review.pdf and 'Group Income statement', *Annual Review, 2006, p32, on* www.bp.com/liveassets/bp_internet/annual_review/annual_review_2006/STAGING/local_assets/downloads_pdfs/a/ara_2006_annual_review.pdf.

75 American Wind Energy Association, *Wind Energy Fast Facts*, on www.awea.org/newsroom/pdf/Fast_Facts.pdf. The annual increases from 2002 to 2007 are for the ends of those years. The AWEA's figure of 48 billion kWh for 2008 is based on an estimated capacity factor of 33 per cent.

76 Energy Information Administration, *Planned Capacity Additions from New Generators, by Energy Source*, 22 October 2007, on www.eia.doe.gov/cneaf/electricity/epa/epat2p4.html. Total includes negligible capacity additions from 'Other Gas' and 'Hydroelectric conventional'. Data reflect plans as of 1 January 2006.

77 EIA, *Annual Energy Outlook 2008 (early release)*, on www.eia.doe.gov/oiaf/aeo/pdf/table1.pdf. Renewables includes hydroelectricity, biomass and other.

78 Green Car Congress, 'Reported US Sales of Hybrids Up 27.3 per cent in January 2008', 7

February 2008, on www.
greencarcongress.com/2008/02/
reported-us-sal.html. The figure
does not include HEVs made by
GM.

79 Anja Hartmann, Jens Riese and
Thomas Vahlenkamp, 'Cutting
carbon, not economic growth:
Germany's path', *McKinsey
Quarterly*, April 2008.

80 Ibid.

81 Deutsche Bank, 'Deutsche Bank
Asset Management launches
climate change investment
initiative', press release, 17
October 2007, on www.db.com/
presse/en/content/press_
releases_2007_3670.
htm?month=3; Deutsche Bank,
'Deutsche Bank's Asset
Management division publishes
major climate change research',
press release, 22 October 2008,
on www.db.com/presse/en/
content/press_
releases_2008_3794.
htm?month=3.

82 Douglas G Cogan, *Corporate
Governance and Climate Change:
The Banking Sector*, Ceres
January 2008, pp14-15, on www.
ceres.org/NETCOMMUNITY/
Document.Doc?id=269.

83 Mark Fulton, 'Overview', of
*Investing in Climate Change
2009: Necessity and Opportunity
in Turbulent Times*, October
2008, on www.dbadvisors.com/
deam/dyn/globalResearch/1113_
index.jsp; Deutsche Bank,
*Economic stimulus: the case for
"Green infrastructure", Energy
Security and "Green" jobs*,
November 2008, on www.
dbadvisors.com/deam/stat/

globalResearch/1113_
GreenEconomicStimulus.pdf.

84 For a consumerist critique of
Greenwashing in the domain of
marketing, see www.
greenwashingindex.com.

85 E F Schumacher, *Small is
Beautiful: Economics as if People
Mattered* (1973),
HarperPerennial, 1989, p15.

86 Ibid., p15.

87 Ibid., pp17-19.

88 Ibid., p169.

89 Ibid., pp22, 92, 102.

90 See CBD, *Convention on
Biological Diversity (with
annexes). Concluded at Rio de
Janeiro on 5 June 1992*, on www.
cbd.int/doc/legal/cbd-un-en.pdf
and CBD, *The Convention on
Biological Diversity: From
Conception to Implementation,
2002*, on www.cbd.int/doc/
publications/CBD-10th-
anniversary.pdf.

91 Robert Costanza, and others,
'The value of the world's
ecosystem services and natural
capital', *Nature*, Vol. 387, No.
6630, 15 May 1997, pp253-60.

92 E F Schumacher, *Small is
Beautiful*, op. cit., pp37, 38.

93 Paul Hawken, Amory Lovins and
Hunter Lovins, *Natural
Capitalism: the Next Industrial
Revolution*, Earthscan, 1999, p8.

94 Ibid., p8.

95 Ibid., p251.

96 Ibid., p251.

97 Ibid., pp25-37, 41.

98 Ibid., p46.

99 Ibid., pp8, 10.

100 United Nations Human Settlements Programme, *Financing Urban Shelter – Global Report on Human Settlements 2005*, 2005, Tables 1.2 and 1.3, p5, on www.unhabitat.org/pmss/getElectronicVersion. asp?nr=1818&alt=1. The Programme's figures, it will be noted, exclude the homes needed to replace what the Programme describes as 'deteriorated and substandard housing stocks'. Ibid., pp4-5.

101 Ibid. The Programme observes that in its projection, which is the 'starting point' for its report, precision is 'not really very important. What is critical, however, is the order of magnitude. Close to 3 billion people, or about 40 per cent of the world's population by 2030, will need to have housing and basic infrastructure services. The Programme's figures, it will be noted, exclude the homes needed to replace what the Programme describes as 'deteriorated and substandard housing stocks'. Ibid., pp4-5.

102 Yuri Kageyama, 'Toyota banking on famed production ways in housing business', *The Seattle Times,* 15 June 2006, on http://seattletimes.nwsource.com/html/businesstechnology/2003062192_toyotahousing15.html.

103 See MA, 'Guide to the Millennium Assessment Reports', 2005, on www.millenniumassessment.org/en/Index.aspx.

104 MA, *Living Beyond Our Means: Natural Assets and Human Well-being, Statement from the Millennium Ecosystem Assessment Board*, 2005, p5, on www.millenniumassessment.org/documents/document.429.aspx.pdf.

105 Ibid., p7.

106 MA and World Health Organisation, *Ecosystems and Human Well-Being – Health Synthesis*, 2005, p1, on www.millenniumassessment.org/documents/document.357.aspx.pdf.

107 MA, *Current States and Trends Assessment*, Chapter 1, p27, on www.millenniumassessment.org/documents/document.765.aspx.pdf.

108 On top of the two examples of this tendency mentioned in reference 8 of Chapter 3, see also UN Secretary General Ban Ki Moon, 'A Climate Culprit In Darfur', *The Washington Post*, 16 June 2007, pA15, on www.washingtonpost.com/wp-dyn/content/article/2007/06/15/AR2007061501857.htm and French President Nicolas Sarkozy, quoted in AFP, 'Climate change driving Darfur crisis: Sarkozy', 18 April 2008, on http://afp.google.com/article/ALeqM5h7l_NjlMjZF-QWDOwxIbibX5AEuA.

109 Jean-Jacques Rousseau, *Discourse on the Origin of Inequality*, opening lines of Part 2, 1754, on www.constitution.org/jjr/ineq_04.htm.

110   Bill Becker, 'Time on geo-engineering: What are they thinking? Part 1', *Grist*, 20 March 2008, on http://gristmill.grist.org/story/2008/3/20/11858/7089.

111   Philip Stott, '"Sustainable development" is just dangerous nonsense', *The Daily Telegraph*, 16 August 2002, on www.telegraph.co.uk/opinion/main.jhtml?xml=/opinion/2002/08/16/do1601.xml.

112   James R Fleming, 'Fixing the Weather and Climate: Military and Civilian Schemes for Cloud Seeding and Climate Engineering', in Lisa Rosner (ed) *The Technological Fix: How People Use Technology To Create and Solve Problems*, Routledge, 2004, p190.

113   Royal Society, *Philosophical Transaction of the Royal Society A*, Vol. 366, No. 1882, 13 November 2008.

114   Interview, 'The Wrong way to a warmer world?', BBC Radio 4 Analysis, transcript on http://news.bbc.co.uk/nol/shared/spl/hi/programmes/analysis/transcripts/03_04_08.txt.

115   Roger Pelke Jr and others, 'Normalised Hurricane Damage in the United States: 1900-2005', *Natural Hazards Review*, Vol. 9, No. 1, 1 February 2008, p29.

116   Michael Williams, *Deforesting the Earth: From Prehistory to Global Crisis – An Abridgement*, University of Chicago Press, 2006, p25. In his remarkable study of the German landscape over the past 250 years, the Harvard historian David

Blackbourn gives the example of the moorland painter Otto Modersohn (1865-1943) who 'confided to his diary that "nature should be our teacher", only to go on to note that this not very original thought occurred to him "on the bridge that leads over the canal"'. See David Blackbourn, *The Conquest of Nature: Water, landscape and the making of modern Germany*, Jonathan Cape, 2006, p13.

117   Lord Nicholas Stern, *Stern Review*, op. cit., Part 1, 'Approach', Chapter 1, p2, paragraph 1.1, on www.hm-treasury.gov.uk./media/3/6/Chapter_1_The_Science_of_Climate_Change.pdf.

118   William Denevan, 'The Americas before and after 1492: Current Geographical Research', *Annals of the Association of American Geographers*, Vol. 82, No. 3, September 1992, pp 369-385, 373, on http://jan.ucc.nau.edu/~alcoze/for398/class/pristinemyth.html.

119   Michael Williams, op. cit., p34.

120   Steve Connor, 'Money is not the answer to this problem', *The Independent*, 18 April 2008, p2, on www.independent.co.uk/opinion/commentators/steve-connor-money-is-not-the-answer-to-this-problem-814738.html.

121   See for example the website Water Footprint, maintained by the University of Twente and the UNESCO-IHE Institute for Water Education, the Netherlands, on www.waterfootprint.org/?page=files/home.

122 Nassim Nicholas Taleb, *The Black Swan*, Penguin, 2007, p116.

123 Ibid.

124 See 'Barack Obama talks to Rachel Maddow 5 days before election', *MSNBC,* 30 October 2008. Transcript on www.msnbc.msn.com/id/27464980.

125 Alastair Darling, Budget speech, in *Financial Times*, 13 March 2008, and on http://search.ft.com/ftArticle?queryText=budget+speech&y=0&aje=true&x=0&id=080313000131&ct=0.

126 For the figures, see Department for Transport, 'Public road length: by road type: 1996-2006', TSGB *2007: Road Lengths – Data Tables*, Table 7.8, on www.dft.gov.uk/pgr/statistics/datatablespublications/roadstraffic/roadlengths/tsgbchapter7roadlengthsdatat1875.

127 Federal Highways Administration, 'Deficient Bridges by State and Highway System', December 2007, on www.fhwa.dot.gov/bridge/defbr07.cfm.

128 Andrew Bounds, 'EU plans air traffic control shake-up', *Financial Times*, 19 March 2008.

129 In Bangladesh, contamination of groundwater by arsenic from naturally occurring rocks has proved a major and lengthy disaster for public health. See Fred Pearce, 'Bangladesh's arsenic poisoning: who is to blame?' *The Courier*, January 2001, UNESCO, on www.unesco.org/courier/2001_01/uk/planet.htm.

130 Mark A Shannon and others, 'Science and technology for water purification in the coming decades', *Nature*, Vol. 452, 20 March 2008, p301.

131 IPCC, 'Summary for Policymakers', in M L Parry and others, editors, *Climate Change 2007: Impacts, Adaptation and Vulnerability. Contribution of Working Group II to the Fourth Assessment Report of the Intergovernmental Panel on Climate Change, 2007*, on www.ipcc.ch/pdf/assessment-report/ar4/wg2/ar4-wg2-spm.pdf, p19, and IPCC, 'Summary for Policymakers', in B Metz and others, editors, *Climate Change 2007: Mitigation. Contribution of Working Group III to the Fourth Assessment Report of the Intergovernmental Panel on Climate Change, 2007*, on www.ipcc.ch/pdf/assessment-report/ar4/wg3/ar4-wg3-spm.pdf, p12.

132 B Metz and others, op. cit., Table SPM3, 'Key mitigation technologies and practices by sector', p10; Parry and others, op. cit., p19, and M L Parry and others, *Technical Summary, A report accepted by Working Group II but not approved in detail*, Table TS6, 'Examples of current and potential options for adapting to climate change for vulnerable sectors', p70, on www.ipcc.ch/pdf/assessment-report/ar4/wg2/ar4-wg2-ts.pdf. The material in this report, we're told, 'has not been subject to line-by-line discussion and agreement, but nevertheless presents a comprehensive, objective and balanced view of the subject matter'.

133 M L Parry and others, *Technical Summary*, op. cit., p70.

134 Diana Farrell and others, 'Making the most of the world's energy resources', *McKinsey Quarterly*, 2007 Number 1, February 2007.

135 Ivo J H Bozon, Warren J Campbell and Mats Lindstrand, 'Global trends in energy', *McKinsey Quarterly*, 2007 Number 1, February 2007.

136 Ibid.

137 Jurriaan Ruijs, Olivier Scheele and Niels van Buuren, 'Building capabilities for success', both in *McKinsey Quarterly*, 2007 Number 1, February 2007.

138 Per-Anders Enkvist, Tomas Nauclér and Jerker Rosander, 'A cost curve for greenhouse gas reduction', *McKinsey Quarterly*, 2007 Number 1, February 2007.

139 Per-Anders Enkvist and others, op. cit.

140 Gert Jan Nabuurs and others, 'Forestry', Chapter 9 in B Metz and others, editors, *Climate Change 2007: Mitigation. Contribution of Working Group III to the Fourth Assessment Report of the Intergovernmental Panel on Climate Change*, p544, on www.ipcc.ch/pdf/assessment-report/ar4/wg3/ar4-wg3-chapter9.pdf.

141 Susan E Page and others, 'The amount of carbon released from peat and forest fires in Indonesia during 1997', *Nature*, Vol. 420, No. 6911, 7 November 2002, p61. Abstract on www.nature.com/nature/journal/v420/n6911/full/nature01131.html.

142 J E Michael Arnold and others, 'Woodfuels, livelihoods, and policy interventions: Changing Perspectives', *World Development*, Vol. 34, Issue 3, March 2006, pp596-611.

143 Per-Anders Enkvist and others, op. cit.

144 Ibid.

145 Per-Anders Enkvist, Tomas Nauclér, and Jens Riese, 'What countries can do about cutting carbon emissions', *McKinsey Quarterly*, April 2008.

146 Ibid.

147 See for example Stuart Pimm and Jeff Harvey, 'No need to worry about the future', *Nature*, Vol. 414, 8 November 2001, p149; John Rennie, Stephen Schneider, John P Holdren, John Bongaarts and Thomas Lovejoy, 'Misleading Math about the Earth', *Scientific American*, January 2002, and 'Arguments that don't hold water', *The Guardian*, 20 August 2001, on www.guardian.co.uk/environment/2001/aug/20/climatechange.physicalsciences.

148 Press release, '"Pies for damn lies and statistics"' as Danish anti-green author gets his just desserts', 5 September 2001, on http://risingtide.org.uk/pages/news/lomborg.htm.

149 Stuart Pimm and Jeff Harvey, op. cit., p149.

150 Bjørn Lomborg, *The Skeptical Environmentalist*, Cambridge University Press, 2001, p331.

151 Kathryn Schulz, 'Let us not praise infamous men: On Bjørn

Lomborg's hidden agenda', *Grist*, 12 December 2001, on www.grist.org/advice/books/2001/12/12/infamous/index.html.

152 Bjørn Lomborg, *Cool It*, Cyan, 2007, pp5-6.

153 Ibid., pp178-179, 206.

154 Ibid., p184.

155 Ibid., p203-204, 226.

156 Ibid., pp194, 213.

157 Ibid., p216.

158 Ibid., p211.

159 Ibid., pp61, 174-176.

160 Ibid., p136.

161 Ibid., p136.

162 Ibid., p167.

163 See Roll back Malaria, *The Global Malaria Action Plan: For a Malaria Free World*, September 2008, p73, on www.rollbackmalaria.org/gmap/gmap.pdf.

164 See 'Malaria's watershed', editorial, *Nature*, Vol. 455, No. 7214, 9 October 2008, p707.

165 Quoted in 'Hackers and Spending Sprees, *Newsweek Web Exclusive*, 5 November 2008, p2, on www.newsweek.comid/167581/page/2.

166 Barack Obama, *New Energy for America*, Speech in Lansing, Michigan, 4 August 2008, Transcript onwww.pbs.org/newshour/bb/politics/july-dec08/obamaenergy_08-04.htm.

167 National Intelligence Council, *Global trends 2025: a transformed world*, 20 November 2008, on www.dnigov/nic/PDF_2025/2025_Global_Trends_Final_Report.pdf.

168 Ibid., p51.

169 Al Gore, 'The Climate for Change', *The New York Times*, 9 November 2008, on www.nytimes.com/2008/11/09/opinion/09gore.html?_r=2&pagewanted=all.

170 IEA, *World Energy Outlook 2008*, op. cit., p5.

171 Ibid., pp3, 7.

172 Ibid., p5.

173 B Metz and others, editors, *Climate Change 2007: Mitigation. Contribution of Working Group III to the Fourth Assessment Report of the Intergovernmental Panel on Climate Change, 2007*, Table SPM3, 'Key mitigation technologies and practices by sector', p10, on www.ipcc.ch/pdf/assessment-reportar4/wg3/ar4-wg3-spm.pdf.

174 Ibid.

175 M L Parry and others, *Technical Summary, A report accepted by Working Group II but not approved in detail*, Table TS6, 'Examples of current and potential options for adapting to climate change for vulnerable sectors', p70, on www.ipcc.ch/pdf/assessment-report/ar4/wg2/ar4-wg2-ts.pdf.

176 Bjørn Lomborg, *Cool It*, op. cit., p176.

177 Per-Anders Enkvist, Tomas Nauclér and Jerker Rosander, 'A cost curve for greenhouse gas reduction', *McKinsey Quarterly*, February 2007. The figures for Gigatonnes saved are McKinsey's estimate of the annual feasible volumes of GHG emissions that each measure could eliminate if GHG emissions trading reaches prices up to 40 €/tonne in 2030. The article assumes GHG 'abatement measures' (1) beginning in 2008 or before (2) centring on a 450-parts-per-million scenario by 2030 (3) saving 26 Gt of GHG emissions a year by 2030, relative to IEA and US EPA 'business as usual' emissions projections.

178   Per-Anders Enkvist, Tomas Nauclér and Jens Riese, 'What countries can do about cutting carbon emissions', *McKinsey Quarterly*, April 2008. The remarks on biofuels assume 'the right combination of land use, feedstocks, and technology, as well as policies to limit potential indirect emissions resulting from land-use change'.

179   Bjørn Lomborg, 'The Green Inquisition', *The Guardian*, 14 July 2008, on www.guardian.co.uk/commentisfree/2008/jul/14/climatechange; Bjørn Lomborg, *Cool It*, op. cit., p224.

180   Diana Farrell and others, 'Making the most of the world's energy resources', *McKinsey Quarterly*, February 2007.

181   Per-Anders Enkvist, Tomas Nauclér and Jens Riese, 'What countries can do about cutting carbon emissions', *McKinsey Quarterly*, April 2008.

182   Bjørn Lomborg, *Cool It*, op. cit., p89.

183   Ibid., pp23-4.

184   Diana Farrell and Jaana Remes, 'How much capital is required for energy- and carbon-abatement investments?', *McKinsey Quarterly*, April 2008.

185   Diana Farrell and others, op. cit.

186   Per-Anders Enkvist, Tomas Nauclér and Jens Riese, 'What countries can do about cutting carbon emissions', *McKinsey Quarterly*, April 2008.

187   Bjørn Lomborg, *Cool It*, op. cit., p51.

188   Per-Anders Enkvist, Tomas Nauclér and Jerker Rosander, 'A cost curve for greenhouse gas reduction', *McKinsey Quarterly*, February 2007.

189   Bjørn Lomborg, *Cool It*, op. cit., pp145-6.

190   Ibid., p150.

191   Per-Anders Enkvist, Tomas Nauclér and Jerker Rosander, 'A cost curve for greenhouse gas reduction', *McKinsey Quarterly*, February 2007.

192   Bjørn Lomborg, *The Skeptical Environmentalist: Measuring the Real State of the World*, Cambridge University Press, 2001, p117.

193   Bjørn Lomborg, *Cool It*, op. cit., p52.

THE AUTHORS

James Woudhuysen

is visiting professor of forecasting and innovation at De Montfort University, Leicester, UK, and a contributor to *Computing* magazine. He read physics at the University of Sussex, and at the Science Policy Research Unit, Sussex, did postgraduate research in the political economy of nuclear energy. After a spell in journalism and consulting, he worked for the Henley Centre for Forecasting, London, and went on to head worldwide market intelligence at Philips Consumer Electronics, the Netherlands, before returning to the UK. His website is www.woudhuysen.com

Joe Kaplinsky

is pursuing postgraduate research in chemical biology at Imperial College London. He read theoretical physics at the University of Manchester, staying there to do experimental research in low temperature physics. He then took masters degrees in structural molecular biology (Birkbeck, University of London) and protein and membrane chemical biology (Imperial). On becoming a patent analyst, he wrote about a wide range of energy technologies, from the handling of nuclear waste, the liquefaction of coal, gas turbine generators and drilling for oil through to the management of power in consumer electronics.